# GCSE
# Physics

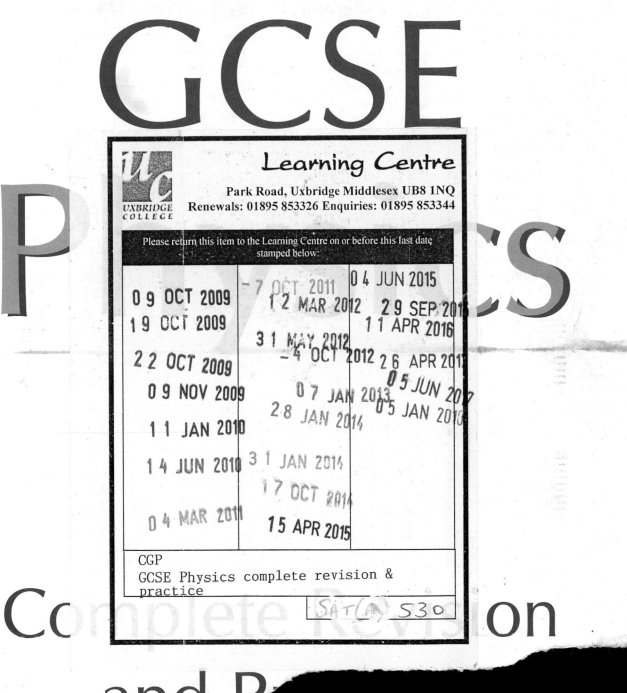
## Complete Revision
## and Practice

# Contents

# Contents

Published by Coordination Group Publications Ltd

*Editors:*
Richard Parsons
Claire Thompson

*Contributors:*
Tony Aldridge, Stuart Barker, Jenny Bilborough, Jonathon Grange, Dominic Hall, Jason Howell, Sharon Keeley, Diana Lazarewicz
Richard Man, Barbara Mascetti, John Maunder, Becky May, John Myers, Andy Park, Alan Rix, Emma Singleton, Luke Waller
James Paul Wallis, Sharon Watson, Suzanne Worthington

Illustrations by Sandy Gardner (e-mail: illustrations@sandygardner.co.uk) and Bowser, Colorado, USA

With thanks to Deborah Dobson, Steve Parkinson and Glenn Rogers for the proofreading.
With thanks to Dr. David Stockdale, Durham University, for help with page 136.
Photograph on page 136 courtesy of NASA http://nix.nasa.gov

AQA (NEAB)/AQA examination questions are reproduced by permission of the Assessment and Qualifications Alliance.
Edexcel (London Qualifications Ltd) examination questions are reproduced by permission of London Qualifications Ltd.
OCR examination questions are reproduced by permission of OCR.

Please note that AQA questions from pre-2003 papers are from legacy syllabuses and not from the current specification.
CGP has carefully selected the questions contained in the practice paper to cover subject areas which are still relevant to the current
specification. As such the exam provides a good test of your knowledge of the syllabus areas you
need to understand to get a good grade in the real thing.

ISBN 1 84146 379 5
Website: www.cgpbooks.co.uk
Printed by Elanders Hindson, Newcastle upon Tyne.
Clipart source: CorelDRAW and VECTOR

# Units and Formulas

This is all very basic stuff and you need to learn it all pretty thoroughly.

If you don't, and you then try and do Physics, it'll be like trying to write stories without learning the alphabet first.

So as long as this is all just a load of weird symbols and nonsense to you then you won't find Physics very easy at all.  This is the Physics alphabet — without it you'll be in trouble.

| | Quantity | Symbol | Standard Units | Formula |
|---|---|---|---|---|
| 1 | Potential Difference | V | volts, V | $V = I \times R$ |
| 2 | Current | I | amperes, A | $I = V / R$ |
| 3 | Resistance | R | ohms, $\Omega$ | $R = V / I$ |
| 4 | Charge | Q | coulombs, C | $Q = I \times t$ |
| 5 | Power | P | watts, W | $P = V \times I$ or $P = I^2 R$ |
| 6 | Energy | E | joules, J | $E = QV$ or $V = E/Q$ |
| 7 | Time | t | seconds, s | $E = P \times t$ or $E = IVt$ |
| 8 | Force | F | newtons, N | $F = ma$ |
| 9 | Mass | m | kilograms, kg | |
| 10 | Weight (a force) | W | newtons, N | $W = mg$ |
| 11 | Density | D | kg per m³, kg/m³ | $D = m/V$ |
| 12 | Moment | M | newton-metres, Nm | $M = F \times r$ |
| 13 | Velocity or Speed | v or s | metres/sec,  m/s | $s = d/t$ |
| 14 | Acceleration | a | metres/sec², m/s² | $a = \Delta v/t$ or $a = F/m$ |
| 15 | Pressure | P | pascals, Pa (N/m²) | $P = F/A$ |
| 16 | Area | A | metres², m² | |
| 17 | Volume | V | metres³, m³ | $P_1 V_1 = P_2 V_2$ |
| 18 | Frequency | f | hertz, Hz | $f = 1/T$ (T=time period) |
| 19 | Wavelength (a distance) | λ or d | metres, m | $V = f \times \lambda$  (wave formula) |
| 20 | Work done | Wd | joules, J | $Wd = F \times d$ |
| 21 | Power | P | watts, W | $P = Wd / t$ |
| 22 | Potential Energy | PE | joules, J | $PE = m \times g \times h$ |
| 23 | Kinetic Energy | KE | joules, J | $KE = \frac{1}{2}mv^2$ |

There's also <u>efficiency</u>, which has no units:

$$\text{Efficiency} = \frac{\text{Useful work output}}{\text{Total energy input}}$$

And also the <u>transformer equation</u>:

$$\frac{\text{Primary Coil Voltage}}{\text{Secondary Voltage}} = \frac{\text{No. of turns on Primary Coil}}{\text{No. of turns on Secondary Coil}}$$

## *Physics — you won't get anywhere without formulas*

Your task is <u>simplicity itself</u>.  Leave the "Quantity" column exposed and cover up the other three.  Then simply <u>fill in the three columns</u> for each quantity: "Symbol", "Units", "Formula".  And just keep practising and practising till you can <u>do it all</u>.   This really is so important.  So do it.

# Units and Formulas

## You Always Use the *Same Old Routine* with Formulas

The thing about using formulas in Physics is that it's <u>always the same old routine</u>. Once you've learnt how to do it for <u>one</u> formula, you can do it for <u>any other</u>.

And that makes the whole thing <u>really simple</u> — but there are still a lot of people who seem to make it much harder than it needs to be. So take it nice and slowly...

## *Formula Triangles* Make Sure You *Get it Right*

<u>All</u> the formulas on the previous page (except "$P_1V_1=P_2V_2$") can be put into <u>formula triangles</u>. It's <u>important</u> to learn how to put any formula into a triangle.

There are <u>two easy rules</u>:

> 1) If the formula is "<u>A = B×C</u>":
>    <u>A goes on the top</u> and <u>B×C goes on the bottom</u>.
>
> 2) If the formula is "<u>A = B/C</u>":
>    <u>B must go on the top</u>
>    (because that's the only way it'll give "B divided by something")
>    — so obviously <u>A and C must go on the bottom</u>.

### *Three Examples:*

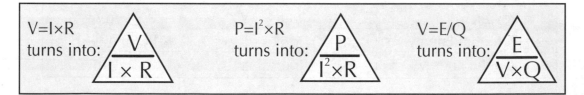

$V=I\times R$
turns into:

$P=I^2\times R$
turns into:

$V=E/Q$
turns into:

<u>How to use them</u>:
Cover up the thing you want to find and write down what's left showing.

<u>Example</u>:
To find Q from the last one, cover up Q and you get E/V left showing, so "$Q = E/V$".

## Always use a Formula Triangle for equations

You really really must learn this method of dealing with formulas. They're all the same — once you can do one, you can do anything they give you in the exam.

# Units and Formulas

## Using Formulas — The Three Rules

1) Find a formula which contains the thing you want to find together with the other things which you've got values for. Convert that formula into a formula triangle.
2) Stick the numbers in and work out the answer.
3) Think very carefully about all the units — and check that the answer is sensible.

Example: A hairdrier is rated at 700 W and draws a current of 3 A. Find its resistance.
Answer: The three quantities mentioned are power (700 W), current (3 A) and resistance.

1. The formula with these three in is "$P = I^2 R$", and the formula triangle version gives us $R = P / I^2$

2. Sticking the numbers in: $R = 700/3^2 = 700/9 = 77.78$, or 78 Ω (to 2 sig. fig.)

3. The power and current are already in their proper units of Watts and Amps, so that's OK. The answer for R must be given in its proper units too, namely Ω, which we've done. The value of 78 Ω is fine. If it was 1 000 000 Ω or 0.00034 Ω you'd worry and check it.

## Watch out for the Units

Once you've got the hang of formula triangles there's only one thing left to get wrong — units. There's two things about units that you have to really watch out for:

1) Make sure that the numbers you put into the formula are in standard (SI) units.
2) When you write the answer down, make sure your answer has its proper units.

Important examples:

500 g must be turned into 0.5 kg, 2 minutes into 120 seconds, 700 kJ into 700 000 J,

145 cm into 1.45 m, etc. before putting them into a formula.

If you don't put SI units in then the answer won't come out with SI units, which can get tricky unless you know what you're doing.

## Formulas — they're all the same

Physics formulas are amazingly repetitive. You really must get it into your head that they're basically all the same. This page has the simple rules that would allow anyone to work out the answers without really knowing anything about Physics at all.

# Warm-Up and Worked Exam Questions

These warm-up questions check you've got the basics — then you can go on to the exam questions.

## Warm-up Questions

1) Write down the letter that is most commonly used as a symbol for the following quantities, paying attention to whether it is a capital letter or a small letter:
   (a) voltage  (b) time  (c) charge  (d) mass  (e) current  (f) force

2) Write down the full names and abbreviations for the standard (SI) units used to measure the following quantities.  a) current  b) velocity  c) energy  d) time  e) charge  f) pressure

3) Write down the following formulas using the usual symbols.  The first one has been done.
   a) force in terms of mass and acceleration:      $F = ma$
   b) pressure in terms of force and area:
   c) power in terms of voltage and current:

4) a) Rearrange the formula $E = Pt$ to make P the subject.
   b) Rearrange the formula $D = m/V$ to make m the subject.
   c) Rearrange the formula $P = F/A$ to make A the subject.

   *(i.e. "get P on its own" so that the formula reads P = something.)*

## Worked Exam Questions

1)     In this question, assume acceleration due to gravity = 10 m/s²
   a)  A brick of mass 2 kilograms is held 4 metres above the ground.
       What is its potential energy?

   *Use the formula P.E. = mgh:  h = 2 × 10 × 4 = 80 J*

   *[2 marks]*

   b)  How high does a weightlifter have to lift a mass of 150 kg in order to increase its potential energy by 2625 J?

   *Use P.E. = mgh in the form h = P.E./mg: h = 2625/(150 × 10) = 1.75 m*

   *[2 marks]*

   c)  A car of mass 1100kg is travelling at 25 m/s.  What is its kinetic energy?

   *Use K.E. = ½ mv² = ½ × 1100 × 25² = ½ × 1100 × 625 = 343 750 J = 344 kJ*

   *[2 marks]*

2)  a)  A current of 3A flows through a television.
       How long would it take 60 000C of charge to flow through it?

   *Use formula Q = It in the form t = Q/I:      t = 60 000 / 3 = 20 000 s*

   *[2 marks]*

   b)  A 2.3 kW kettle is connected to the mains supply at 230V.  What current does it take?

   *Take care — the power is given in kW. Convert it into W to get answer in SI units.*

   *Use formula P = IV in the form I = P/V:  I = 2300/230 = 10 A*

   *(This answer looks sensible.)*   *[2 marks]*

   c)  A low-energy lightbulb has an efficiency of 0.85.
       How much light energy will it give out if 500J of electrical energy is supplied to it?

   *efficiency = energy output / energy input,*

   *so energy output = energy input × efficiency = 500 × 0.85 = 425 J*

   *[2 marks]*

# Exam Questions

1)  a)  What force is needed to accelerate a shopping trolley of mass 50kg at 0.9 m/s²?

    ........................................................................................................................................
    *[2 marks]*

    b)  Another trolley, pushed with the same force, accelerates at 0.5m/s². What is its mass?

    ........................................................................................................................................
    *[2 marks]*

    c)  Nathan weighs 660 N. The soles of his shoes have a total area of 0.03m². What pressure is he exerting on the ground?

    ........................................................................................................................................
    *[2 marks]*

2)  a)  What is the frequency of a sound wave which completes one cycle in 0.00357 seconds?

    ........................................................................................................................................
    *[1 mark]*

    b)  What is the wavelength of that same sound wave on a day when the velocity of sound in air is 330 m/s?

    ........................................................................................................................................
    *[2 marks]*

3)  A car of mass 1500kg is travelling at 97 kilometres per hour.    *(Hint: Be careful with the units* What is its kinetic energy in kJ?                                   *of speed.)*

    ........................................................................................................................................

    ........................................................................................................................................
    *[3 marks]*

4)  a)  A fixed mass of gas has a volume of 120cm³ at a pressure of one atmosphere. What is its volume at 3.1 atmospheres, given that the temperature remains constant?

    ........................................................................................................................................
    *Hint: stick to the units the question gives you.*  *[2 marks]*

    b)  Air is trapped at a pressure of 50 kPa in a cylinder of volume 0.0005m³ with a piston at one end. The piston is pushed inwards, compressing the volume of the gas to 0.0003m³. What is the pressure of the air now, given that the temperature remains constant?

    ........................................................................................................................................
    *[2 marks]*

# Electrical Circuits

Electricity's great — but it's not much fun if the words don't mean anything to you.

1) Current is the flow of electrons round the circuit.

2) Voltage is the driving force that pushes the current round.
It's a bit like "electrical pressure".

3) Resistance is anything in the circuit which slows the flow down.

4) There's a balance:  the voltage is trying to push the current round the circuit,
and the resistance is opposing it — the relative sizes of the voltage and
resistance decide how big the current will be:

> If you increase the VOLTAGE — then MORE CURRENT will flow.
> If you increase the RESISTANCE — then LESS CURRENT will flow.

## It's Just Like the **Flow of Water** Around a Set of **Pipes**

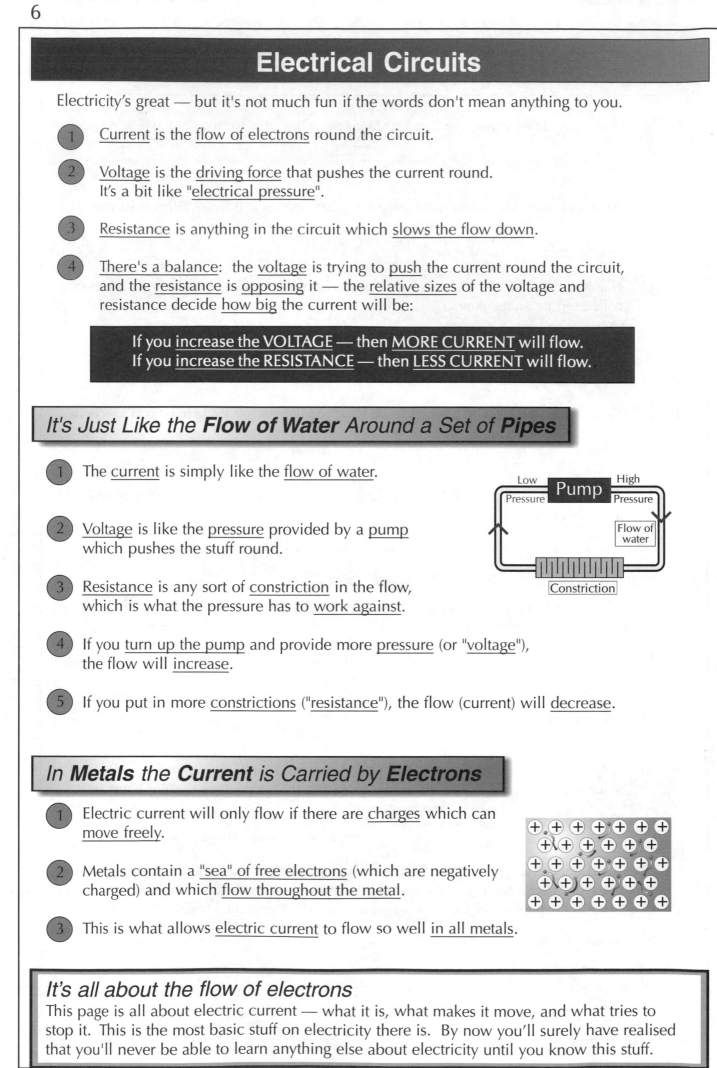

1) The current is simply like the flow of water.

2) Voltage is like the pressure provided by a pump
which pushes the stuff round.

3) Resistance is any sort of constriction in the flow,
which is what the pressure has to work against.

4) If you turn up the pump and provide more pressure (or "voltage"),
the flow will increase.

5) If you put in more constrictions ("resistance"), the flow (current) will decrease.

## In **Metals** the **Current** is Carried by **Electrons**

1) Electric current will only flow if there are charges which can
move freely.

2) Metals contain a "sea" of free electrons (which are negatively
charged) and which flow throughout the metal.

3) This is what allows electric current to flow so well in all metals.

## It's all about the flow of electrons

This page is all about electric current — what it is, what makes it move, and what tries to
stop it.  This is the most basic stuff on electricity there is.  By now you'll surely have realised
that you'll never be able to learn anything else about electricity until you know this stuff.

# Electrical Circuits

## Electrons Flow the **Opposite Way** to **Conventional Current**

We <u>normally</u> say that current in a circuit flows from <u>positive to</u> <u>negative</u>. Alas, electrons were discovered long after that was decided and they turned out to be <u>negatively charged</u> — <u>unlucky</u>. This means they <u>actually flow</u> from –ve to +ve, <u>opposite</u> to the flow of "<u>conventional current</u>".

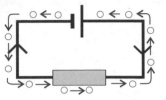

## In **Electrolytes**, Current is Carried by Both **+ve and –ve Charges**

1. <u>Electrolytes</u> are <u>liquids</u> which contain charges which can <u>move freely</u>.

2. They are either <u>ions dissolved in water</u>, like salt solution, or <u>molten ionic liquids</u>, like molten sodium chloride.

3. When a voltage is applied the <u>positive</u> charges move towards the <u>–ve</u>, and the <u>negative</u> charges move towards the <u>+ve</u>. This is an <u>electric current</u>.

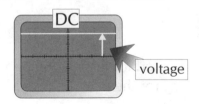

## **AC** Changes Direction but **DC** Doesn't

Direct current <u>keeps flowing</u> in the <u>same direction</u> all the time. The <u>Cathode Ray Oscilloscope (CRO) trace</u> is a <u>horizontal line</u>.

Alternating current keeps <u>reversing its direction</u> back and forth. The <u>CRO trace</u> is <u>always a wave</u>.

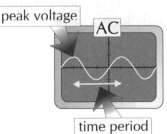

You need to <u>learn</u> these CRO traces.

## Electrons one way, current the other

This stuff about electrons moving the opposite way to current is pretty confusing, for sure. But, as with most of Physics, learning the facts (properly) is the only way you'll be able to answer questions on it. Once you've learnt it, actually *understanding* will be far far easier.

# Electrical Circuits

This is without doubt the most standard circuit the world has ever known.  So know it.

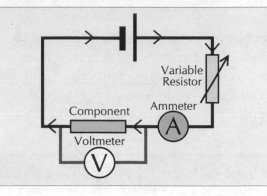

## The **Ammeter**

1) Measures the <u>current</u> (in <u>amps</u>) flowing through the component.
2) Must be placed <u>in series</u>.
3) Can be put <u>anywhere</u> in series in the <u>main circuit</u>,
   but never <u>in parallel</u> like the voltmeter.

## The **Voltmeter**

1) Measures the <u>voltage</u> (in <u>volts</u>) across the component.
2) Must be placed <u>in parallel</u> around the <u>component under test</u> —
   <u>NOT</u> around the variable resistor or the battery.
3) The <u>proper</u> name for "voltage" is "<u>potential difference</u>" or "<u>p.d.</u>"

## Five Important Points

1   This <u>very basic circuit</u> is used for <u>testing components</u>,
    and for getting <u>I-V graphs</u> for them.

2   The <u>component</u>, the <u>ammeter</u> and the <u>variable resistor</u> are all <u>in series</u>, which
    means they can be put <u>in any order</u> in the main circuit.  The <u>voltmeter</u>, on the
    other hand, can only be placed <u>in parallel</u> around the <u>component under test</u>,
    as shown.  Anywhere else is definitely <u>wrong</u>.

3   As you <u>vary</u> the <u>variable resistor</u> it alters the <u>current</u> flowing through the circuit.

4   This allows you to take several <u>pairs of readings</u> from the <u>ammeter</u> and <u>voltmeter</u>.

5   You can then <u>plot</u> these values for <u>current</u> and <u>voltage</u> on a <u>I-V graph</u>.

## Remember — Ammeter in series, Voltmeter in parallel
You absolutely <u>must learn</u> this basic circuit.  Once you've got this firmly in your head,
all the other more complicated circuits will be a whole lot easier.

# Electrical Circuits

## The **Four** Most Important **Current-Voltage Graphs**

I-V graphs show how the current varies as you change the voltage.
Learn these four very well:

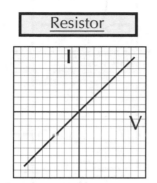

Resistor

The current through a <u>resistor</u> (at constant temperature) is <u>proportional to voltage</u>.

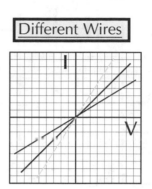

Different Wires

<u>Different wires</u> have different <u>resistances</u>, hence the different <u>slopes</u>.

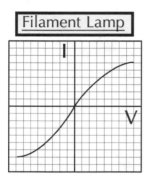

Filament Lamp

As the <u>temperature</u> of the filament <u>increases</u>, the <u>resistance increases</u>, hence the <u>curve</u>.

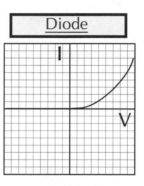

Diode

Current will only flow through a diode <u>in one direction</u>, as shown.

## Calculating Resistance: *R =V/I, (or R ="1/gradient")*

For the <u>straight-line graphs</u> the resistance of the component is <u>steady</u> and is equal to the <u>inverse</u> of the <u>gradient</u> of the line, or "<u>1/gradient</u>". In other words, the <u>steeper</u> the graph, the <u>lower</u> the resistance.
If the graph <u>curves</u>, it means the resistance is <u>changing</u>. In that case R can be found for any point by taking the <u>pair of values</u> (V,I) from the graph and sticking them in the formula <u>R =V/I</u> (see page 1). Easy.

## *In the end, you'll have to learn this — resistance is futile...*
There's a lot of important stuff here and you need to <u>learn all of it</u>. The only way to be really sure you know it is to <u>cover up the page</u> and see how much of it you can <u>write down</u> by heart.

# Electrical Circuits

## Circuit Symbols *You Should* **Know:**

You have to know <u>all</u> these circuit symbols for the Exam.

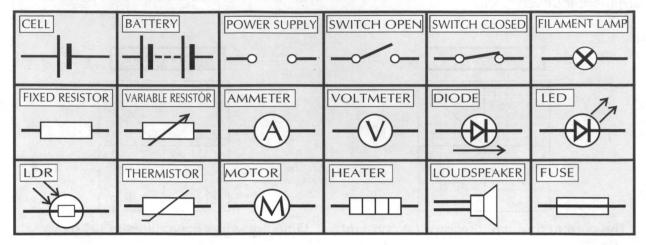

| CELL | BATTERY | POWER SUPPLY | SWITCH OPEN | SWITCH CLOSED | FILAMENT LAMP |
| FIXED RESISTOR | VARIABLE RESISTOR | AMMETER | VOLTMETER | DIODE | LED |
| LDR | THERMISTOR | MOTOR | HEATER | LOUDSPEAKER | FUSE |

## 1) Variable Resistor

1) A <u>resistor</u> whose resistance can be <u>changed</u> by turning a knob, for example.
2) The old-fashioned ones are <u>huge coils of wire</u> with a <u>slider</u> on them.
3) They're great for <u>altering the current</u> flowing through a circuit.
   Turn the resistance <u>up</u>, the current <u>drops</u>. Turn the resistance <u>down</u>, the current goes <u>up</u>.

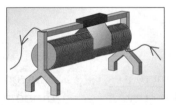

## 2) "Semiconductor Diode" or just "Diode"

A special device made from <u>semiconductor</u> material such as <u>silicon</u>. It lets current flow freely through it <u>in one direction</u>, but <u>not</u> in the other (i.e. there's a very high resistance in the <u>reverse</u> direction). This turns out to be really useful in various <u>electronic circuits</u>.

## *Turn the book over and write it all down*

Another page of basic, important details about circuits. You need to know all those circuit symbols as well as the extra details for the five special devices on this and the next page.

# Electrical Circuits

## 3) Light Emitting Diode or "LED"

1) A diode which gives out light. It only lets current go through in one direction.

2) When it does pass current, it gives out a pretty coloured light, e.g. red, green or yellow.

3) Stereos usually have loads of small LEDs which light up as the music's playing.

## 4) Light-Dependent Resistor or "LDR"

1) In bright light, the resistance falls.

2) In darkness, the resistance is highest.

3) This makes it a useful device for various electronic circuits, e.g. automatic night lights; burglar detectors.

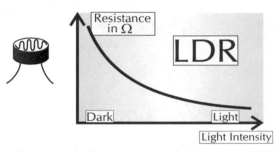

## 5) Thermistor (Temperature-dependent Resistor)

1) In hot conditions, the resistance drops.

2) In cool conditions, the resistance goes up.

3) Thermistors make useful temperature detectors, e.g. car engine temperature sensors and electronic thermostats for central heating.

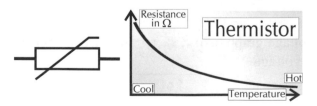

---

*Learn it all — graphs, symbols, uses and everything*
When you think you know it all try covering the page and writing it all down, and drawing all the symbols. See how you did, and then try again.

# Warm-Up and Worked Exam Question

You should have learnt all this basic stuff by now, but just to check you're up to speed with it, make sure you can do these warm-up questions. Then there's a worked example, so you can see what kind of exam questions you're likely to get. Once you're happy with that, have a go at the questions on pages 13-14.

## Warm-up Questions

1) Complete the blanks in this passage: An electric current is a flow of _electrons_ round a circuit. These particles carry a _____ electric charge. They are pushed away from the _____ terminal of the cell and round to the _____ terminal, the opposite direction to _____ current. No current will flow through a circuit unless the circuit is _____.

2) Draw the following circuit symbols: a) fuse b) filament lamp c) motor d) thermistor

3) When you increase the resistance of a variable resistor, what happens to the current flowing through it?

4) When you increase the temperature of a filament lamp in a circuit, what happens to its resistance? _increases_

5) Give one practical use for each of the following: LED, light dependent resistor, thermistor.

## Worked Exam Question

1) Draw a circuit diagram showing a series circuit containing:

a) a cell

*[1 mark]*

b) a variable resistor

*[1 mark]*

c) an ammeter

*[1 mark]*

d) a diode connected in such a way that current will flow

*[2 marks]*

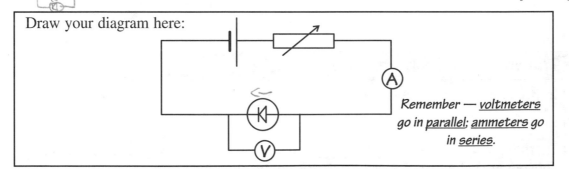

Draw your diagram here:

Remember — _voltmeters_ go in _parallel_; _ammeters go_ in _series_.

e) Add to the diagram a voltmeter placed so as to measure the voltage across the diode.
*[1 mark]*

f) On the Current-Voltage graph below, sketch the way the current through the diode will vary with the voltage.

*[1 mark]*

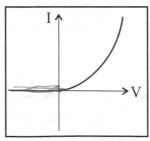

*Make sure your line only goes through the quadrant where both I and V are positive. Remember it's a diode, so the current can only flow in one direction.*

# Exam Questions

1) a) (i) In a metal wire, what type of particles carry the electric current?

...................... *electrons* ..............................................................................

*[1 mark]*

(ii) Describe in terms of moving charged particles what is happening in a metal wire when the ends are connected to the positive and negative poles of an electric cell.

*the freely moving electrons move around an electric cell, to the which flows through the metal wire.*

*[2 marks]*

(iii) If the voltage across the cell is increased, what happens to the current?

*the current increases*

*[1 mark]*

b) Describe how the electric current flows in an electrolyte.

*The positive particles in a liquid more to the negatively charged electrolyte -ve and negative more to the positively changed +ve.*

*[2 marks]*

c) A rotating bicycle dynamo produces alternating current in a circuit. A cathode ray oscilloscope (CRO) is connected to the dynamo to display its voltage output.

On the screen below, draw the CRO trace that will be produced.

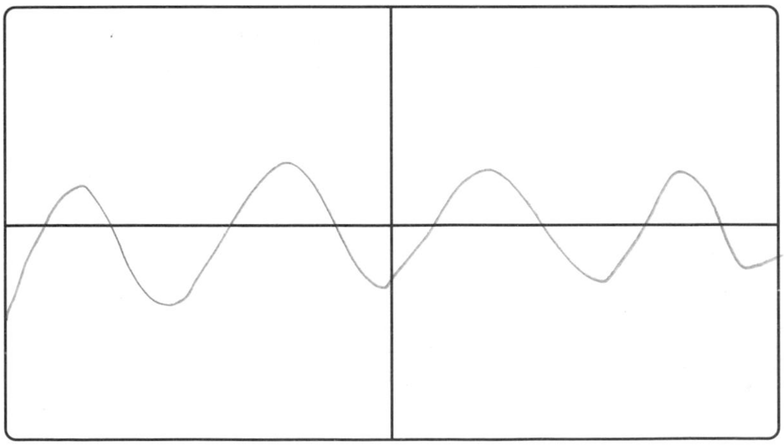

*[1 mark]*

# Exam Questions

2)

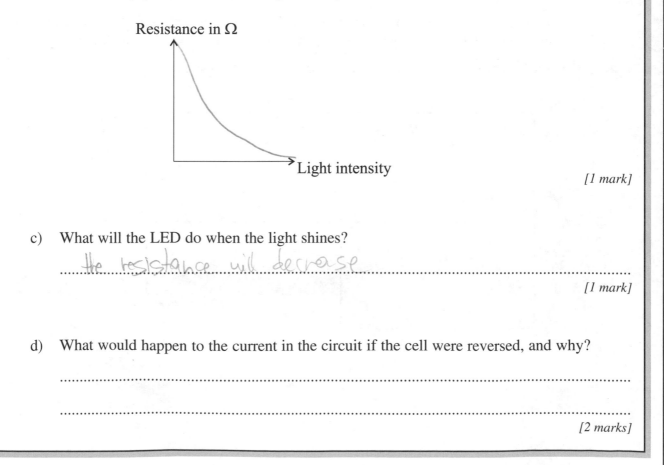

a) In the circuit diagram shown above,

    (i) which component (A, B or C) is a Light Dependent Resistor (LDR)?

         ......................C......................................................................................................

    (ii) which component is the cell?

         ......................A......................................................................................................

    (iii) which component is the Light Emitting Diode (LED)?

         ......................B......................................................................................................

*[3 marks]*

b) A bright light is shone on the circuit. On the axes below sketch how the resistance of the LDR will vary as the amount of light falling on it increases.

Resistance in $\Omega$

Light intensity

*[1 mark]*

c) What will the LED do when the light shines?

     ............the resistance will decrease......................................................................

*[1 mark]*

d) What would happen to the current in the circuit if the cell were reversed, and why?

     ..........................................................................................................................

     ..........................................................................................................................

*[2 marks]*

# Types of Circuits

You need to be able to tell the difference between series and parallel circuits <u>just by looking at them</u>. You also need to know the <u>rules</u> about what happens with both types. Read on.

## *Series Circuits — all or nothing*

1) In <u>series circuits</u>, the different components are connected <u>in a line</u>, <u>end to end</u>, between the +ve and –ve terminals of the power supply (except for <u>voltmeters</u>, which are always connected <u>in parallel</u>, but they don't count as part of the circuit).

2) If you remove or disconnect <u>one</u> component, the circuit is <u>broken</u> and current <u>stops</u> flowing.

3) This is generally <u>not very handy</u>, and in practice, <u>very few things</u> are connected in series.

## *In **Series Circuits**:*

1) The <u>total resistance</u> is just the <u>sum</u> of all the resistances.

2) The <u>same current</u> flows through <u>all parts</u> of the circuit.

3) The <u>size of the current</u> is determined by the <u>total p.d. of the cells</u> and the <u>total resistance</u> of the circuit:  i.e.  $I = V/R$

4) The <u>total p.d.</u> of the <u>supply</u> is <u>shared</u> between the various <u>components</u>, so the <u>voltages</u> round a series circuit <u>always add up</u> to equal the <u>total voltage</u> of the supply.

5) The <u>bigger</u> the <u>resistance</u> of a component, the bigger its <u>share</u> of the <u>total p.d.</u>

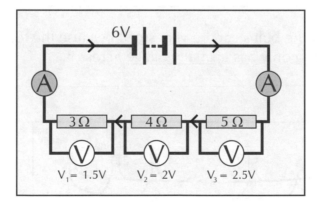

Voltages add to equal the <u>supply</u>: 1.5 + 2 + 2.5 = 6V
<u>Total resistance</u> = 3 + 4 + 5 = 12 Ω
<u>Current</u> =V/R = 6 / 12 = 0.5 A

## *Series circuits — it's just one thing after another*
They really do want you to know the difference between series and parallel circuits. It's not that hard but you do have to make a real effort to <u>learn all the details</u>. That's what this page is for. Learn all those details, then <u>cover the page</u> and <u>write them all down</u>. Then try again...

# Types of Circuits

## Total p.d., Voltmeters and Ammeters

1) The <u>total p.d.</u> provided by cells in <u>series</u> is the <u>sum</u> of the individual p.ds.

2) <u>Voltmeters</u> are always connected <u>in parallel</u> around components.
   In a <u>series circuit</u>, you can put voltmeters <u>around each component</u>.

3) The readings from all the components will <u>add up</u> to equal the reading from the <u>voltage source</u> (the cells).

4) <u>Ammeters</u> can be placed <u>anywhere</u> in a <u>series circuit</u> and will <u>all give the same reading</u>.

## Christmas Fairy Lights are Wired in Series

<u>Christmas fairy lights</u> are about the <u>only real-life example</u> of things connected in <u>series</u>, and everyone's seen the <u>whole lot go out</u> just because <u>one</u> of the bulbs is faulty.

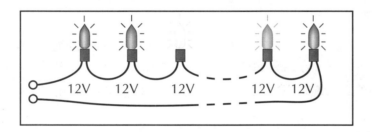

The only <u>advantage</u> is that the bulbs can be <u>very small</u> because the total 230 V is <u>shared out between them</u>, so <u>each bulb</u> only has a <u>small voltage</u> across it.

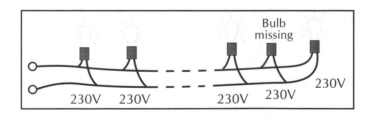

<u>By contrast</u>, a string of lights as used on a <u>building site</u> are connected in <u>parallel</u> so that each bulb receives the <u>full 230 V</u>. If <u>one</u> is removed, <u>the rest stay lit</u>. Which is most <u>convenient</u>.

Make sure you know the <u>difference</u> between the two wiring diagrams above.

---

*Series circuits — great in theory but annoying at Christmas*
What you really must get from this page are the advantages and the one big disadvantage of using series circuits... and hence why they're only really ever used for Christmas lights.

# Types of Circuits

Parallel circuits are much more <u>sensible</u> than series circuits and so they're <u>much more common</u> in <u>real life</u>.

## Parallel Circuits — *Independence* and *Isolation*

1) In <u>parallel circuits</u>, each component is <u>separately connected</u> to the +ve and –ve terminals of the <u>supply</u>.

2) If you remove or disconnect <u>one</u> of them, it will <u>hardly affect the others at all</u>.

3) This is <u>obviously</u> how <u>most things</u> must be connected, for example in <u>cars</u> and in <u>household electrics</u>. You have to be able to switch everything on and off <u>separately</u>.

## In *Parallel Circuits*:

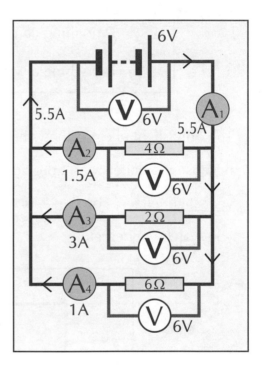

1) <u>All components</u> get the <u>full source p.d.</u>, so the voltage is the <u>same</u> across all components.

2) The <u>current</u> through each component <u>depends on its resistance</u>.
The <u>lower</u> the resistance, the <u>bigger</u> the current that'll flow through it.

3) The <u>total current</u> flowing around the circuit is equal to the <u>total</u> of all the currents in the <u>separate branches</u>.

4) In a parallel circuit, there are <u>junctions</u> where the current either <u>splits</u> or <u>rejoins</u>. The total current going <u>into</u> a junction <u>always equals</u> the total currents <u>leaving</u> — fairly obviously.

5) The <u>total resistance</u> of the circuit is <u>tricky to work out</u>, but it's <u>always less</u> than the branch with the smallest <u>resistance</u>.

---

Voltages all equal to <u>supply voltage</u>: = 6V
Total R is <u>less than</u> the <u>smallest</u>, i.e. <u>less than</u> 2 Ω
Total Current ($A_1$) = <u>sum</u> of all branches = $A_2+A_3+A_4$

---

## *Parallel circuits are much more useful*
Learn the big advantages of parallel circuits. Cover the book and write down everything you can remember about them. Once you've got it all, learn the circuit diagram *and* the numbers.

# Types of Circuits

## Connection of **Voltmeters** and **Ammeters**

1) Once again the voltmeters are always connected in parallel around components.

2) Ammeters can be placed in each branch to measure the different currents flowing through each branch, as well as one near the supply to measure the total current flowing through it.

## Everything **Electrical** in a **Car** is Connected in **Parallel**

Parallel connection is essential in a car to give these two features:

> 1 Everything can be turned on and off separately.
>
> 2 Everything always gets the full voltage from the battery.

The only slight effect is that when you turn lots of things on the lights may go dimmer

because the battery can't provide full voltage under heavy load. This is normally a

very slight effect. You can spot the same thing at home when you turn a kettle on, if

you watch very carefully.

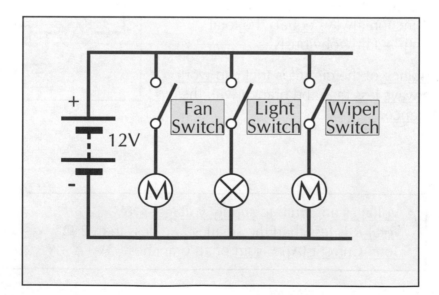

## Electric circuits — unparalleled joy...

Learn all the details for connecting ammeters and voltmeters, and also what two features
make parallel connection essential in a car. Then cover the page and write it all down.

# Warm-Up and Worked Exam Question

## Warm-up Questions

1) A circuit consists of a cell and two resistors connected in series. The potential difference (p.d.) across the cell is 6V. The p.d. across one of the resistors is 4V. What is the p.d. across the other resistor?

2) State whether each of the three circuits shown is series or parallel.

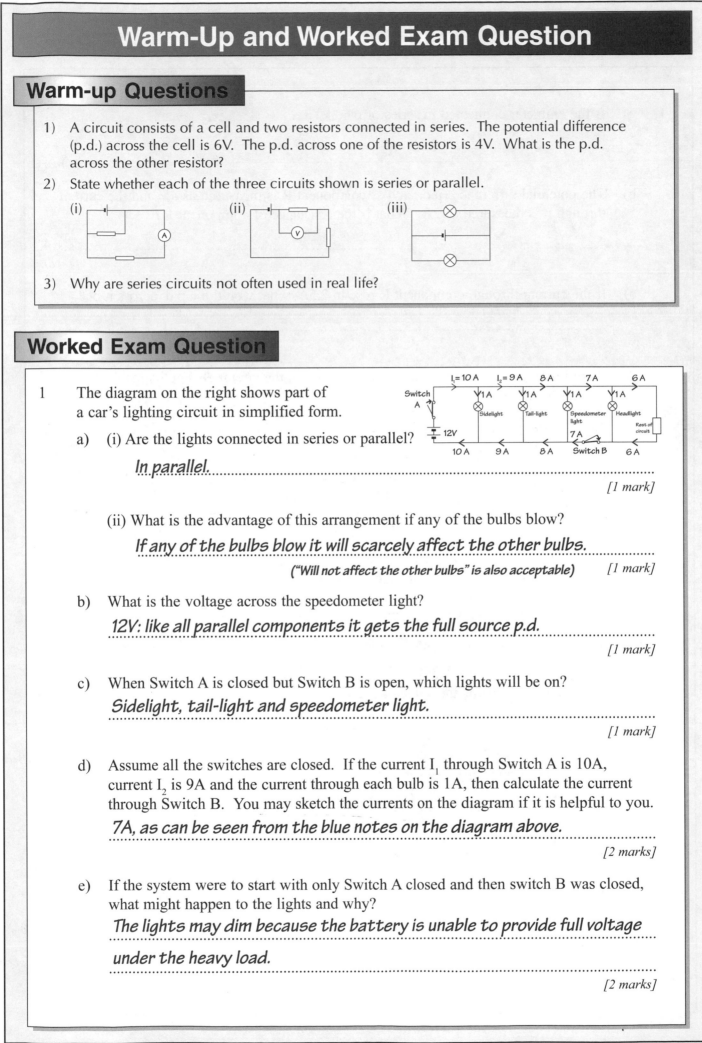

(i)   (ii)   (iii)

3) Why are series circuits not often used in real life?

## Worked Exam Question

1   The diagram on the right shows part of a car's lighting circuit in simplified form.

a)   (i) Are the lights connected in series or parallel?

*In parallel.*

*[1 mark]*

(ii) What is the advantage of this arrangement if any of the bulbs blow?

*If any of the bulbs blow it will scarcely affect the other bulbs.*

("Will not affect the other bulbs" is also acceptable)   *[1 mark]*

b)   What is the voltage across the speedometer light?

*12V: like all parallel components it gets the full source p.d.*

*[1 mark]*

c)   When Switch A is closed but Switch B is open, which lights will be on?

*Sidelight, tail-light and speedometer light.*

*[1 mark]*

d)   Assume all the switches are closed. If the current $I_1$ through Switch A is 10A, current $I_2$ is 9A and the current through each bulb is 1A, then calculate the current through Switch B. You may sketch the currents on the diagram if it is helpful to you.

*7A, as can be seen from the blue notes on the diagram above.*

*[2 marks]*

e)   If the system were to start with only Switch A closed and then switch B was closed, what might happen to the lights and why?

*The lights may dim because the battery is unable to provide full voltage under the heavy load.*

*[2 marks]*

*SECTION ONE — ELECTRICITY AND MAGNETISM*

# Exam Questions

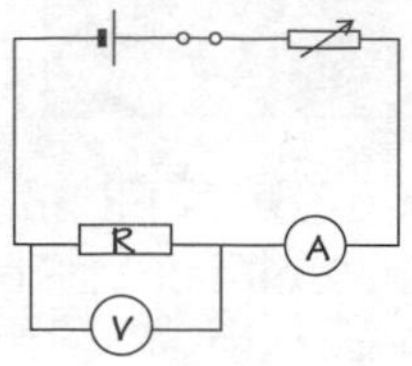

1   a)   Is the ammeter connected in series or parallel?

...............series.................................................

*[1 mark]*

b)   The potential difference (p.d.) across component R is measured as 4V and the current through it is measured as 2A.  What is the resistance of component R?

.................2.........................................................

*[1 mark]*

c)   If the current through component R rose to 2.5A, what would the p.d. across it be?

.................5.........................................................

*[1 mark]*

d)   Would moving the ammeter to another position in the circuit change its reading? Explain your answer.

.......................................................................

*[1 mark]*

e)   If the switch were open, how much current would flow through the circuit?

.......................................................................

*[1 mark]*

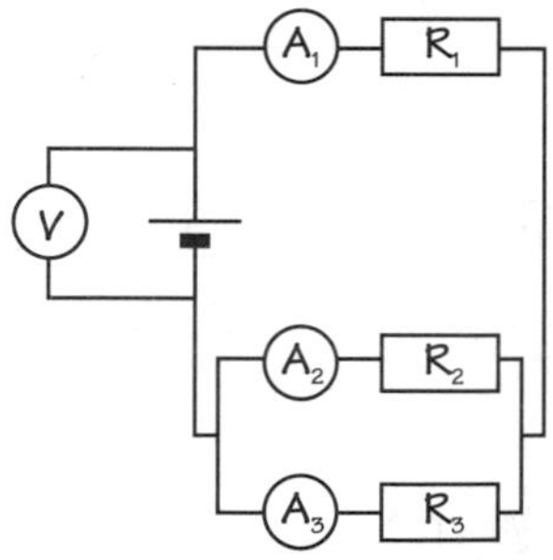

2   a)   In the circuit shown, if ammeter $A_2$ reads 2A and resistor $R_2$ is $6\Omega$, what is the p.d. across $R_2$?

.................................................................

.................................................................

*[1 mark]*

b)   If $A_3$ reads 3A, what is the value of $R_3$?

.......................................................................

*[1 mark]*

c)   What is the current measured at $A_1$?

.......................................................................

*[1 mark]*

d)   If $R_1 = 0.5\Omega$, what is the voltage across it?

.......................................................................

*[1 mark]*

e)   What is the voltage across the cell?

.......................................................................

*[2 marks]*

# Static Electricity

Static electricity is all about charges which are <u>not</u> free to move. This causes them to build up in one place and it often ends with a <u>spark</u> or a <u>shock</u> when they do finally move.

## 1) Build up of **Static** is Caused by **Friction**

1) When two <u>insulating</u> materials are <u>rubbed together</u>, electrons will be <u>scraped off one</u> and <u>dumped on the other</u>.
2) This'll leave a <u>positive static charge</u> on one and a <u>negative</u> static charge on the other.
3) <u>Which way</u> the electrons are transferred <u>depends</u> on the <u>two materials</u> involved.
4) The classic examples are <u>polythene</u> and <u>acetate</u> rods being rubbed with a <u>cloth duster</u>, as shown in the diagrams on the right:

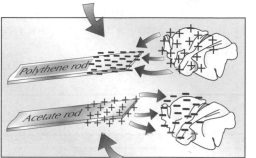

*With the <u>polythene rod</u>, electrons move <u>from the duster</u> to the rod.*

*With the <u>acetate rod</u>, electrons move <u>from the rod</u> to the duster.*

## 2) **Only Electrons Move** — Never the Positive Charges

<u>Watch out for this in Exams</u>. Both +ve and –ve electrostatic charges are only ever produced by the <u>movement of electrons</u>. The positive charges <u>definitely do not move</u>. A positive static charge is always caused by electrons <u>moving away elsewhere</u>, as shown above. Don't forget.

## 3) **Like** Charges Repel, **Opposite** Charges Attract

This is <u>easy</u> and <u>kind of obvious</u>.
Two things with <u>opposite electric charges</u> are <u>attracted</u> to each other.
Two things with the <u>same electric charge</u> will <u>repel</u> each other.
These <u>forces get weaker</u> the <u>further apart</u> the two things are — pretty obviously.

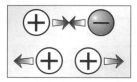

## 4) Charging by **Induction** is quite **Difficult**

When something which is <u>charged</u> comes near something <u>which isn't</u>, it tends to <u>induce charge</u>, because electrons in the <u>uncharged</u> object <u>move towards or away</u> from the charged object.

The <u>result</u> is always the same — the new arrangement of charge always makes the two objects <u>pull together</u> because the <u>repelling charges</u> are now <u>further apart</u> than the <u>attracting charges</u>. It's <u>tricky</u>, but you <u>can</u> understand it — and you can <u>learn</u> it.

attraction

---

## Remember — opposites attract

The way to tackle this page is to first <u>learn the four headings</u> till you can <u>write them all down</u>. Then learn the details for each one, and keep practising by <u>covering the page</u> and writing down each heading with as many details as you can remember for each one. Just <u>keep trying</u>...

# Static Electricity

## 5) As **Charge** Builds Up, So Does the **Voltage** — Causing **Sparks**

The greater the charge on an isolated object, the greater the voltage between it and the Earth.  If the voltage gets big enough there's a spark which jumps across the gap.  High voltage cables can be dangerous for this reason.   Big sparks have been known to leap from overhead cables to earth. But not often.

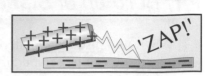

A charged conductor can be discharged safely by connecting it to earth with a metal strap.

They like asking you to give quite detailed examples in Exams. Make sure you learn all these details.

## Static Electricity can be **Helpful**:

### 1) **Inkjet Printer**:

1) Tiny droplets of ink are forced out of a fine nozzle, making them electrically charged.
2) The droplets are deflected as they pass between two metal plates. A voltage is applied to the plates — one is negative and the other is positive.
3) The droplets are attracted to the plate of the opposite charge and repelled from the plate with the same charge.
4) The size and direction of the voltage across each plate changes so each droplet is deflected to hit a different place on the paper.
5) Loads of tiny dots make up your printout.  Clever.

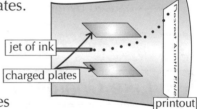

### 2) **Photocopier**:

1) The metal plate is electrically charged. An image of what you're copying is projected onto it.
2) Whiter bits of the thing you're copying make light fall on the plate and the charge leaks away.
3) The charged bits attract black powder, which is transferred onto paper.
4) The paper is heated so the powder sticks.
5) Voilà, a photocopy of your piece of paper (or whatever else you've shoved in there).

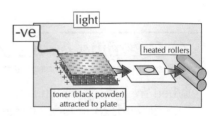

### 3) **Spray Painting** and **Dust Removal** in Chimneys...

These are other uses but photocopiers and inkjet printers are what they really want you to know.

## Learn the main uses of Static Electricity
You really need to learn those two big examples.  All the exam boards mention photocopiers and inkjet printers so I bet there'll be a question on them.  It's almost relevant to real-life too.

# Static Electricity

## Static Electricity can Cause **Crackles** and **Shocks**

### 1) *Car Shocks*

Air rushing past your car can give it a +ve charge. When you get out and touch the door it gives you a real buzz — in the Exam make sure you say "electrons flow from earth, through you, to neutralise the +ve charge on the car." Some cars have conducting strips which hang down behind the car. This gives a safe discharge to earth, but spoils all the fun.

### 2) *Clothing Crackles*

When synthetic clothes are dragged over each other (like in a tumble drier) or over your head, electrons get scraped off, leaving static charges on both parts, and that leads to the inevitable — attraction (they stick together) and little sparks / shocks as the charges rearrange themselves.

## Static Electricity can be **Really Dangerous**

### 1) *Lightning*

Rain droplets fall to Earth with positive charge. This creates a huge voltage and a big spark.

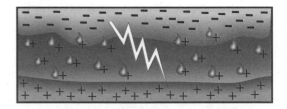

### 2) *Grain Shoots*, *Paper Rollers* and the Potential *Fuel Filling Disaster*:

1) As fuel flows out of a filler pipe, or paper drags over rollers, or grain shoots out of pipes, then static can build up.
2) This can easily lead to a spark — which in dusty or fumey places could cause an explosion.
3) The solution: make the nozzles or rollers out of metal so that the charge is conducted away, instead of building up.
4) It's also good to have earthing straps between the fuel tank and the fuel pipe.

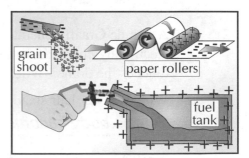

---

## *And learn all the problems too*

More and more examples of static electricity — learn all the numbered points and keep writing them down to make absolutely sure that you know it all.

# Warm-Up and Worked Exam Questions

## Warm-up Questions

1  What usually causes a build-up of static electricity?

2  There are three types of particles found inside an atom — neutrons, electrons and protons.
   a) From the three types of particle above, name the one that has a positive electric charge.
   b) Which type of particle has to move in order to create a build up of static electricity?

3  Name two devices used in an office which rely on static electricity to work.

4  Describe two real-life situations in which a build-up of static electricity can be dangerous.

## Worked Exam Questions

1  a)  A positively charged acetate rod is placed near a stream of water, and the stream veers towards the rod. Explain why.

   *The stream veers towards the rod because the positive charge on the rod*

   *attracts the electrons nearest to it in the water. This means the water*

   *nearest the rod becomes negatively charged so that it is attracted towards*

   *the positive rod. When the approach of a charged object separates the*

   *charges on another nearby object in this way it is called charging by induction.*

   *[2 marks]*

   b)  A child's balloon is floating just below a cloud that has become negatively charged in stormy weather. On the diagram below, sketch in the induced pattern of charges on the balloon.

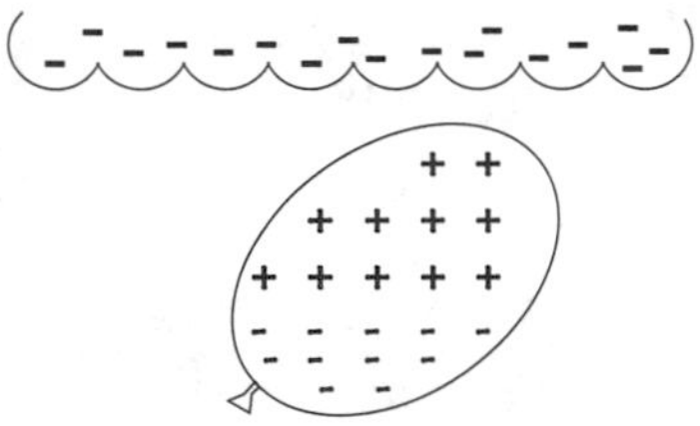

   *[2 marks]*

2)  A Van de Graaff generator is a machine that allows a metal dome to become electrically charged. In a class experiment, Anna stands on a sheet of polythene and touches the dome. When she touches it her hair rises up to stand on end.

   a)  Why does her hair stand on end?

   *Anna's body becomes charged. All the hairs on her head have the*

   *same type of charge. Hence they repel each other.*

   *[2 marks]*

   b)  What happens if she steps off the sheet of polythene? (Explain your answer.)

   *The charge would flow away to earth through her body.*

   *[2 marks]*

# Exam Questions

1    Paul holds a polythene rod in his hand and rubs it with a duster.
It becomes negatively charged.

    a)    What has happened in terms of the movement of subatomic particles to make the rod negatively charged?

    ....................................................................................................................................................

*[2 marks]*

    b)    State whether the electric charge on the duster is now positive, negative or zero.
Explain your answer.

    ....................................................................................................................................................

*[2 marks]*

    c)    Janine now holds a metal rod in her hand and rubs it with the duster. The rod does not become charged. Why not?

    ....................................................................................................................................................

*[3 marks]*

2    The passage below describes how a photocopier works. Read it and answer the questions.

    "A cylindrical drum is given a negative electrical charge. The picture is projected onto the charged drum. Charge leaks away where light strikes it, corresponding to the white parts of the image. Only the dark parts of the image stay charged. These charged areas attract black toner powder. When the drum is rolled against a piece of paper, toner is transferred to the paper to make a copy of the picture. The paper is heated to make the toner powder stick to it permanently."

    a)    Will there be more or fewer electrons than protons on the surface of the drum?

    ....................................................................................................................................................

*[1 mark]*

    b)    Explain why the grains of toner powder are attracted to the charged drum even though they are not electrically charged themselves.

    ....................................................................................................................................................

*[2 marks]*

    c)    Explain why it might improve the design to make the piece of paper electrically charged as well.

    ....................................................................................................................................................

*[2 marks]*

    d)    Sanjay is given a party balloon, a woollen glove, a piece of paper and some black toner powder on a saucer. Explain how he could use these things to demonstrate to his class in a simple way how a photocopier transfers ink to the paper.

    ....................................................................................................................................................

*[2 marks]*

# Electrical Energy

You can look at <u>electrical circuits</u> in <u>two ways</u>. The first is in terms of <u>a voltage pushing the current round</u> and the resistances opposing the flow, as on page 6.

The <u>other way</u> of looking at circuits is in terms of <u>energy transfer</u>. Learn them <u>both</u> and be ready to tackle questions about <u>either</u>.

## Energy is Transferred from Cells and Other Sources

1) Anything which <u>supplies electricity</u> is also supplying <u>energy</u>.
   There are <u>four sources</u> you need to <u>learn</u>: <u>cells</u>, <u>batteries</u>, <u>generators</u> and <u>solar cells</u>.

2) The energy is <u>transferred</u> by the <u>electric circuit</u> to <u>components</u> such as lamps, resistors, bells, motors, LEDs, buzzers, etc.

3) These components perform their own <u>energy transfer</u> and <u>convert</u> the <u>electrical energy</u> in the circuit into <u>other</u> forms of energy: e.g. <u>heat</u>, <u>light</u>, <u>sound</u> or <u>movement</u>.

4) Don't forget that a <u>complete circuit</u> is needed for the current to flow.
   If the circuit is <u>broken</u> there will be <u>no current flow</u> and <u>no transfer of energy</u>.

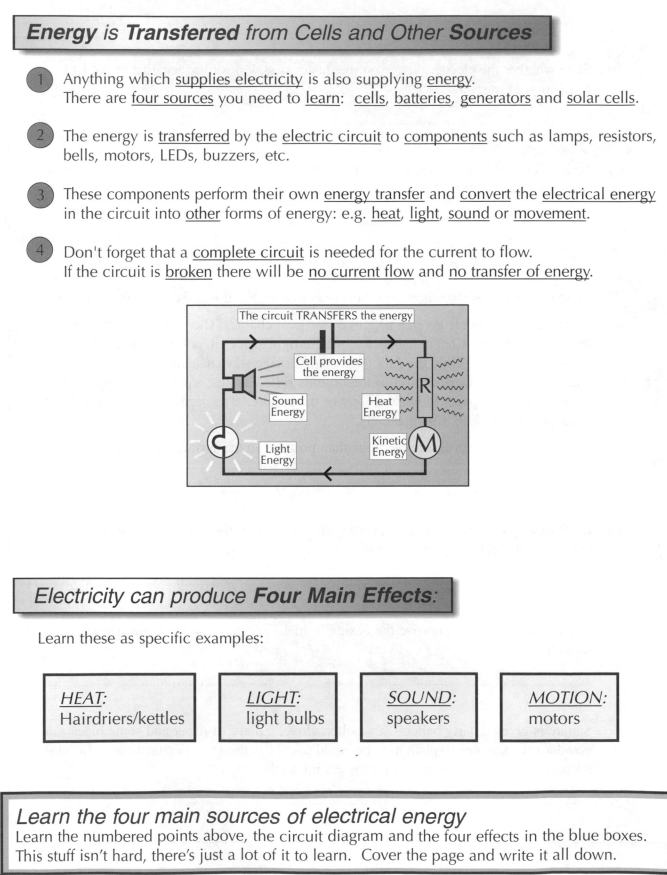

## Electricity can produce Four Main Effects:

Learn these as specific examples:

| HEAT:<br>Hairdriers/kettles | LIGHT:<br>light bulbs | SOUND:<br>speakers | MOTION:<br>motors |
|---|---|---|---|

---

### Learn the four main sources of electrical energy

Learn the numbered points above, the circuit diagram and the four effects in the blue boxes. This stuff isn't hard, there's just a lot of it to learn. Cover the page and write it all down.

# Electrical Energy

## All **Resistors** produce **Heat** when a **Current** flows through them

(1) This is important. Whenever a <u>current</u> flows through anything with <u>electrical resistance</u> (which is more or less <u>everything</u>) then <u>electrical energy</u> is converted into <u>heat energy</u>.

(2) The <u>more current</u> that flows, the <u>more heat</u> is produced.

(3) Also, a <u>bigger voltage</u> means <u>more heating</u>, because it pushes <u>more current</u> through.

(4) However, the <u>higher</u> you make the <u>resistance</u>, the <u>less heat</u> is produced. This is because a higher resistance means <u>less current</u> will flow, and that <u>reduces</u> the heating.

(5) The <u>amount of heat</u> produced can be <u>measured</u> by putting a resistor in a known amount of water or inside a solid block and measuring the <u>increase in temperature</u>.

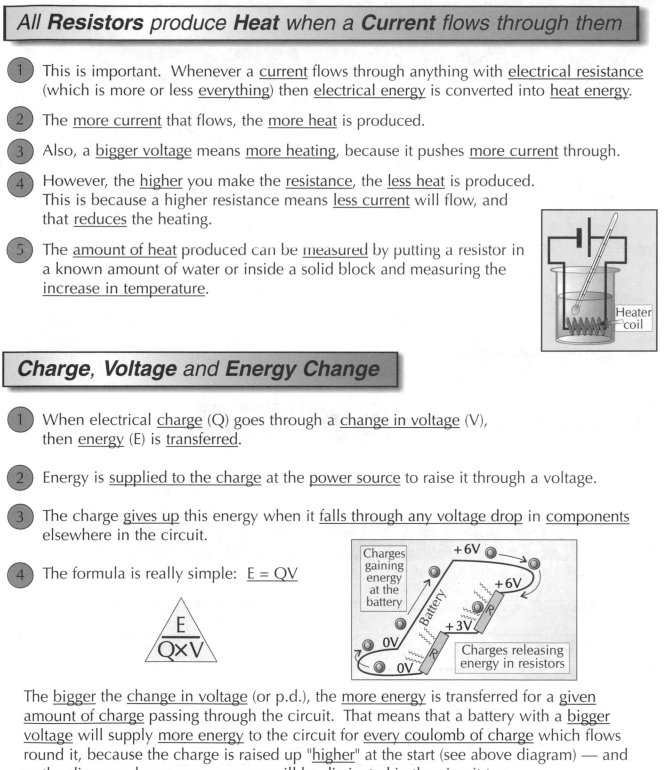

## **Charge**, **Voltage** and **Energy Change**

(1) When electrical <u>charge</u> (Q) goes through a <u>change in voltage</u> (V), then <u>energy</u> (E) is <u>transferred</u>.

(2) Energy is <u>supplied to the charge</u> at the <u>power source</u> to raise it through a voltage.

(3) The charge <u>gives up</u> this energy when it <u>falls through any voltage drop</u> in <u>components</u> elsewhere in the circuit.

(4) The formula is really simple:  <u>E = QV</u>

The <u>bigger</u> the <u>change in voltage</u> (or p.d.), the <u>more energy</u> is transferred for a <u>given amount of charge</u> passing through the circuit. That means that a battery with a <u>bigger voltage</u> will supply <u>more energy</u> to the circuit for <u>every coulomb of charge</u> which flows round it, because the charge is raised up "<u>higher</u>" at the start (see above diagram) — and as the diagram shows, <u>more energy will be dissipated</u> in the circuit too.
This gives rise to <u>two definitions</u> which you should learn:

> 1) <u>ONE VOLT</u> is <u>ONE JOULE PER COULOMB</u>
> 2) <u>VOLTAGE</u> is the <u>ENERGY TRANSFERRED PER UNIT CHARGE</u> passed

## ENERGY = CHARGE × VOLTAGE

Learn the first chunk of this page, then cover the book up and write a mini-essay about current flowing through resistors. Then do the same for charge and voltage — including that equation.

# Electrical Energy

Electricity is by far the <u>most useful</u> form of energy. Compared to gas or oil or coal it's <u>much easier</u> to turn it into the <u>four main types</u> of useful energy: <u>Heat</u>, <u>light</u>, <u>sound</u> and <u>motion</u>.

## Reading Your **Electricity Meter** and Working out the **Bill**

This is in the specifications — though you never actually need to do it in real life.

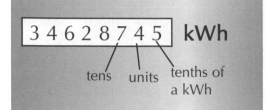

tens     units     tenths of a kWh

### Electricity Bill

| | |
|---|---|
| Previous meter reading............. | 345412.3 |
| This meter reading.................... | 346287.5 |
| Number of units used...................... | 875.2 |
| | |
| Cost per unit.............................. | 6.3p |
| Cost of electricity used................... | £55.14 |
| (875.2 units × 6.3p) | |
| Fixed Quarterly charge..................... | £7.50 |
| Total Bill................................. | £62.64 |
| VAT @ 8%................................... | £5.01 |
| Final total............................... | **£67.65** |

The reading on your meter shows the <u>total number of units</u> (kWh) used since the meter was fitted. Each bill is worked out from the <u>increase</u> in the meter reading since it was <u>last read</u> for the previous bill.

You need to <u>study</u> this bill until you know what all the different bits <u>are for</u>, and how it all works out. They could give you one <u>very similar</u> in the Exam.

## Kilowatt-hours (kWh) are "UNITS" of Energy

1) Your electricity meter counts the number of "<u>UNITS</u>" used.
2) A "<u>UNIT</u>" is otherwise known as a <u>kilowatt-hour</u>, or <u>kWh</u>.
3) A "<u>kWh</u>" might sound like a unit of power, but it's not — it's an <u>amount of energy</u>.

> A <u>KILOWATT-HOUR</u> is the amount of electrical energy
> used by a <u>1 kW appliance</u> left on for <u>1 HOUR</u>.

4) Make sure you can turn <u>1 kWh</u> into <u>3 600 000 Joules</u> like this:
   "E=P×t" = 1kW × 1 hour = 1000W × 3600 s = <u>3 600 000 J</u>   (=3.6 MJ)
(The formula is "Energy = Power×Time", and the units must be converted to SI first. See pages 1-3)

## The **Two Easy Formulas** for Calculating The **Cost of Electricity**

These are the two very easy formulas you need to learn:

| No. of <u>units</u> (kWh) used = <u>Power</u> (in kW) × <u>Time</u> (in hours) | Units = kW × hours |
|---|---|

| <u>Cost</u> = No. of <u>UNITS</u> × <u>price</u> per UNIT | Cost = Units × Price |
|---|---|

<u>Example</u>: *Find the cost of leaving a 60 W light bulb on for  a) 30 minutes   b) one year.*
<u>Answer:</u>a)   <u>No. of Units = kW × hours</u>  = 0.06kW × ½hr = 0.03 units.
           <u>Cost = Units × price per unit</u>(6.3p)  = 0.03 × 6.3p = <u>0.189p</u> for 30 mins.
        b)   <u>No. of Units = kW × hours</u>  = 0.06kW × (24×365)hr = 525.6 units.
           <u>Cost = Units × price per unit</u>(6.3p)  = 525.6 × 6.3p = <u>£33.11</u> for one year.

*N.B. Always turn the <u>power</u> into <u>kW</u> (not watts) and the <u>time</u> into <u>hours</u> (not minutes)*

## Remember — kWh measures energy, NOT power

This page has three sections and you need to learn the stuff in all of them. Start by memorising the headings, then learn the details under each heading. Then <u>cover the page</u> and <u>write down</u> what you know. Check back and see what you missed, and then <u>try again</u>. And keep trying.

# Warm-Up and Worked Exam Questions

## Warm-up Questions

1   Name a device that causes the following energy transfers:
    a) Electrical energy to light energy
    b) Chemical energy to electrical energy

2   Joshua does the vacuuming for half an hour using a 1300W vacuum cleaner.  How much energy has he used in a) kilowatt-hours and b) joules?

3   This picture shows an electricity meter on May 1st.    $\boxed{0}\,\boxed{3}\,\boxed{7}\,\boxed{8}\,\boxed{8} \cdot \boxed{1}$ kWh

    This picture shows the same meter on June 1st.    $\boxed{0}\,\boxed{4}\,\boxed{8}\,\boxed{1}\,\boxed{4} \cdot \boxed{5}$ kWh

    a) How many units of electricity have been used?
    b) If the cost of electricity is 5p per unit, how much did the electricity used in this period cost?

4   Which is the correct statement about the voltage of a circuit?
    a) Voltage = resistance per unit charge.          c) Voltage = energy transferred per second.
    b) Voltage = resistance per second.               d) Voltage = energy transferred per unit charge.

## Worked Exam Questions

1   A current of 7A runs for 5 minutes through a resistor.

a)   What charge is transferred in these 5 minutes?

    $Q = It$:          $7 \times 5 \times 60 = 2100 \ coulombs$

    *Hint: remember that the time must be expressed in seconds.*

    *[2 marks]*

b)   In these 5 minutes, the resistor transfers 12.6kJ of electrical energy to heat energy.  What is the voltage across the resistor?

    $E = QV \ rearranged \ gives \ V = E/Q$:    $V = 12,600 \ / \ 2100 = 6 \ V$

    *[2 marks]*

c)   When an electric fire is turned on the wire in its heating element glows red hot.  However the wire in the cable connecting the fire to the mains scarcely changes temperature.  What does this tell you about the resistances of the two wires?

    *It tells you that the resistance of the wire in the element is higher than the resistance of the wire in the cable.*

    *[1 mark]*

2   Your present electricity company, Company A, charge 5.96 pence per unit plus a standing charge of £6.37 per quarter.  Their rivals, Company B, are offering a rate of 5.78 pence per unit plus a standing charge of £7.50 per quarter.  If your family uses 1250 units each quarter is it worth your while to change suppliers to Company B?  Show your workings.

    *Company A: basic cost is £0.0596 x 1250 = £74.50*

    *plus standing charge of £6.37 — Total = £80.87*

    *Company B: basic cost is £0.0578 x 1250 = £72.25*

    *plus standing charge of £7.50 — Total = £79.75 (cheaper)  Yes it's worth swapping.*

    *[4 marks]*

# Exam Questions

1  a)  The following electrical devices can all convert electrical energy into other forms of
       energy.  Write down which type of energy is the main output of each of the devices
       below.
       (i) Buzzer          (ii) Kettle          (iii) Resistor          (iv) Motor          (v) Lamp

       .............................................................................................................................................
                                                                                              *[5 marks]*

   b)  (i)   Write down the formula that shows how much energy E is transferred when a
             charge Q undergoes a change of voltage V.

             .......................................................................................................................................
                                                                                              *[1 mark]*

       (ii)  Does the charge gain or lose electrical energy when it passes through a resistor?

             .......................................................................................................................................
                                                                                              *[1 mark]*

2  The Meadowes family's electricity bill for the last quarter looks like this:

| Present Reading | Previous Reading | Units Used | p per Unit | Amount |
|---|---|---|---|---|
| 16969 | 14303 | A | 4.680 | B |

| Service charge | £8.64 |
|---|---|
| Total charges excluding VAT | C |
| Plus VAT at 17.5% | D |
| Total payment due | E |

Calculate A, B, C, D and E.

*(Don't be put off by the number of dotted lines.
We've sometimes included less than you get in the
real exam so we could squeeze more questions in.)*

.............................................................................................................................................

.............................................................................................................................................

.............................................................................................................................................
                                                                                              *[5 marks]*

3  The headlamp of a car runs from a battery.
   The potential difference across the lamp is 12V and the current through it is 3.3A.

   a)  How much charge in coulombs passes through the bulb in one hour?

       .............................................................................................................................................
                                                                                              *[2 marks]*

   b)  How much energy in joules does it take to run the bulb for one hour?

       .............................................................................................................................................
                                                                                              *[2 marks]*

# Mains Electricity

Electricity is dangerous. It can kill you. Well just watch out for it, that's all.

## Hazards in The Home — Eliminate Them before They Eliminate You

A likely Exam question will show you a picture of a room inside a house, but with various electrical hazards in the picture — and they'll ask you to list all the hazards. This should be mostly common sense, but it will definitely help if you've already learnt this list:

 Long cables or frayed cables.

2 Cables in contact with something hot or wet.

3 Pet rabbits or children (always hazardous).

4 Water near sockets, or shoving things into sockets.

5 Damaged plugs, or too many plugs into one socket.

6 Lighting sockets without bulbs in.

7 Appliances without their covers on.

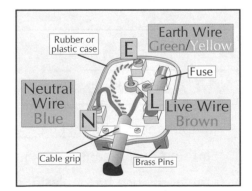

## Plugs and Cables — Learn the Safety Features

### Get the Wiring Right:

1 The right coloured wire is connected to each pin, and firmly screwed in.

2 No bare wires showing inside the plug.

3 Cable grip tightly fastened over the cable outer layer.

### Plug Features:

1 The metal parts are made of copper or brass because these are very good conductors.

2 The case, cable grip and cable insulation are all made of plastic because this is a really good insulator and is flexible too.

3 This all keeps the electricity flowing where it should.

---

## Hazards and plugs — useful stuff for exams and life

Make sure you can list all those hazards in the home. Make sure you know all the details for wiring a plug. Once you've learnt it all, cover the page and write it all down again.

# Mains Electricity

## Earthing and Fuses Prevent Fires and Shocks

The live wire alternates between a high +ve and –ve voltage, with an average of about 230V. The neutral wire is always at 0 V. Electricity normally flows in and out through the live and neutral wires only. The earth wire and fuse (or circuit breaker) are just for safety and work together like this:

1) If a fault develops in which the live somehow touches the metal case, then because the case is earthed, a big current flows in through the live, through the case and out down the earth wire.

2) This surge in current blows the fuse (or trips the circuit breaker), which cuts off the live supply.

3) This isolates the whole appliance making it impossible to get an electric shock from the case. It also prevents the risk of fire caused by the heating effect of a large current.

4) Fuses should be rated as near as possible but just higher than the normal operating current (see page 34).

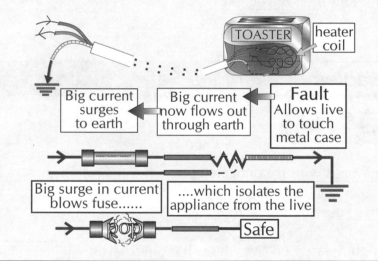

## All Appliances must be Earthed

All appliances with metal cases must be "earthed" to avoid the danger of electric shock.

"Earthing" just means the metal case must be attached to the earth wire in the cable.

If the appliance has a plastic casing and no metal parts showing then it's said to be double insulated.

Anything with double insulation like that doesn't need an earth wire, just a live and neutral.

## We're talking 230 volts — worth being careful
This stuff's a bit complicated, it's true. But you have to learn it all the same.
Learn all four steps and check you know them by drawing the diagram and annotating it.

# Mains Electricity

## The National Grid supplies the Whole Country with Electricity

1. The <u>National Grid</u> is the <u>network</u> of pylons and cables which covers <u>the whole country</u>.

2. It takes electricity from the <u>power stations</u>, to just where it's needed in <u>homes</u> and <u>industry</u>.

3. It enables power to be <u>generated</u> anywhere on the grid, and to then be <u>supplied</u> anywhere else on the grid.

The National Grid

## All Power Stations are Pretty Much the Same

They all have a <u>boiler</u> of some sort, which makes <u>steam</u>, which drives a <u>turbine</u>, which drives a <u>generator</u>. The generator produces <u>electricity</u> (by <u>induction</u>) by <u>rotating</u> an <u>electromagnet</u> within coils of wire (see pages 44-45).

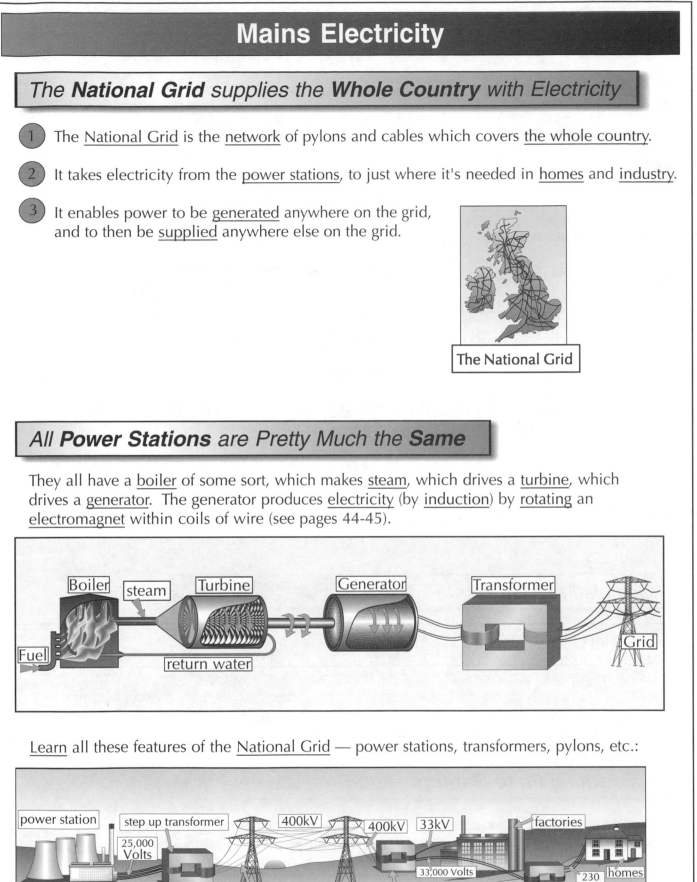

<u>Learn</u> all these features of the <u>National Grid</u> — power stations, transformers, pylons, etc.:

## Fuel, boiler, steam, turbine, generator, transformer, grid, toaster

It's easy to forget how much actually happens before the electricity appears out of your plug point. Make sure **you** don't forget. Learn the list of stages, then go back and fill in the details.

# Mains Electricity

## Pylon Cables are at **400 000 V** to keep the **Current Low**

You need to understand why the voltage is so high and why it's AC. Learn these points:

1. The formula for power supplied is: Power = Voltage × Current or: P = V×I

2. So to transmit a lot of power, you either need high voltage or high current.

3. The problem with high current is the energy loss (as heat) due to the resistance of the cables.

4. The formula for power loss due to resistance in the cables is: $P = I^2R$ (see below).

5. Because of the $I^2$ bit, if the current is 10 times bigger, energy losses will be 100 times bigger.

6. It's much cheaper to boost the voltage up to 400 000 V and keep the current very low.

7. This requires transformers as well as big pylons with huge insulators, but it's still cheaper.

8. The transformers have to step the voltage up at one end, for efficient transmission, and then bring it back down to safe useable levels at the other end.

9. This is why it has to be AC on the National Grid — so that the transformers will work (see page 46).

10. Mains electricity in your house is AC 50 Hz — the voltage changes direction 100 times a second.

## Calculating **Electrical Power** and **Fuse Ratings**

1) The standard formula for electrical power is: P=VI
2) If you combine it with V=I×R, and replace the "V" with "I×R", you get: $P=I^2R$
3) If instead you use V=I×R and replace the "I" with "V/R", you get: $P=V^2/R$
4) You choose which one of these formulas to use, purely and simply by seeing which one contains the three quantities which are involved in the problem you're looking at.

## Calculating **Fuse Ratings** — Always Use the Formula: **"P=VI"**

Most electrical goods indicate their power rating and voltage rating. To work out the fuse needed, you need to work out the current that the item will normally use. That means using "P=VI", or rather, "I=P/V".

$$\frac{P}{V \times I}$$

Example: A hairdrier is rated at 240V, 1.1 kW. Find the fuse needed.
Answer: I = P/V = 1100/240 = 4.6 A. Normally, the fuse should be rated just a little higher than the normal current, so a 5 amp fuse is ideal for this one.

## Remember — pylon cables have high voltage and low current

Fully explaining why pylon cables are at 400 000 V is a bit tricky — but you do need to learn it. The same goes for the power formulas and working out fuse ratings. Learn it, then write it out.

# Warm-Up and Worked Exam Questions

## Warm-up Questions

1  Describe three possible electrical hazards in the home that involve electrical cables.
2  State the correct colours for the following wires in a plug:
   a) neutral      b) earth        c) live
3  What do we mean when we describe an appliance as "double insulated"?
4  a) Write down the formula for electrical power, P, that also involves current I and voltage V.
   b) Write down the formula for electrical power that also involves current I & resistance R.
5  Does the National Grid use alternating or direct current?  Is the electrical transmission at a voltage of 400V or 400kV?

## Worked Exam Questions

1  Describe briefly how a power station works.  The following key words may be useful: turbine, electromagnet, steam, boiler, coils, electricity, fuel, generator.

   *A boiler turns water into steam. The steam drives a turbine, which in turn drives a generator. The generator works by rotating an electromagnet within coils of wire, which produces electricity.*  Or any reasonable phrasing that uses the key words correctly. 4 marks for full explanation, or 1 mark for each correct point.

   *[4 marks]*

2  A portable emergency light can run from a 12V car battery and draws a current of 5A.

   a)  Which of these bulbs should be used?  a) 40W   b) 60W   c) 12W   d) 100W
   *Use P = VI:   P = 12 x 5 = 60W*  2 marks for correct answer, otherwise 1 mark for correctly stating formula.

   *[2 marks]*

   b)  How much charge passes through the bulb in 90 seconds?
   *Use Q = It:   Q = 5 x 90 = 450C*

   *[2 marks]*

   c)  Using the information on voltage and current given at the start of the question, calculate how much energy it takes to run the bulb for 90 seconds.
   *Use E = IVt:   E = 5 x 12 x 90 = 5400J*

   *[2 marks]*

   d)  Using a different formula from the one you used in part c), calculate again how much energy it takes to run the bulb for 90 seconds.  You may wish to make use of your result from part b).  Show that both methods give the same result.
   *Use E = QV:   E = 450 x 12 = 5400J, which is the same as the answer to (c)*

   *[2 marks]*

   e)  What is the resistance of the light?
   *Use V = IR rearranged to R = V/I:   R = 12/5 = 2.4Ω*

   *[2 marks]*

# Exam Questions

1  a)  A careless student is wiring a plug to operate a kettle rated at 2000W for a voltage of 240V. He connects the brown wire to the live terminal, the blue wire to the earth terminal and the green and yellow wire to the neutral terminal.

What was wrong with the way he wired the plug?

........................................................................................................................................

*[2 marks]*

b)  Another student corrects the wiring, and fits a 5A fuse.

Explain what happens when he tries to boil the kettle, and why.
Which would have been a more suitable fuse to use, a 3A one or a 13A one?

........................................................................................................................................

........................................................................................................................................

*[2 marks]*

c)  What is the most important reason for making the casing of the plug from plastic?

........................................................................................................................................

*[1 mark]*

d)  What is the danger from a live wire within an appliance such as a toaster coming into contact with its metal case?

........................................................................................................................................

*[1 mark]*

e)  Explain how this danger can be prevented using the earth wire in combination with a fuse. Start by saying what the earth wire must be connected to. Include an explanation of which wire must be fused, and why it is that one.

........................................................................................................................................

........................................................................................................................................

*[3 marks]*

2  a)  Explain in detail why the National Grid uses a high voltage to transmit electrical energy, despite the danger.

........................................................................................................................................

........................................................................................................................................

*[3 marks]*

b)  Why is alternating current used in mains electricity, rather than direct current?

........................................................................................................................................

*[2 marks]*

c)  What is the role of the step-down transformer in the National Grid?

........................................................................................................................................

*[1 mark]*

# Magnets

There's a proper definition of a <u>magnetic field</u> which you really ought to learn:

> A <u>MAGNETIC FIELD</u> *is a region where* <u>MAGNETIC MATERIALS</u> *(like iron and steel)*
> *and also* <u>WIRES CARRYING CURRENTS</u> *experience* <u>A FORCE</u> *acting on them.*

## Learn all These **Magnetic Field Diagrams**, Arrow-perfect

They're really likely to give you one of these diagrams to do in your Exam. So make sure you know them, especially <u>which way the arrows point</u> — <u>always from North to South</u>.

### *Bar* Magnet

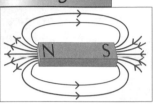

### **Solenoid**

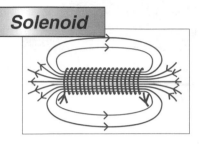

Same field as a bar magnet <u>outside</u>. <u>Strong and uniform</u> field on the <u>inside</u>.

### Two Bar Magnets **Attracting**

<u>Opposite poles attract</u>, as I'm sure you know.

### Two Bar Magnets **Repelling**

<u>Like poles repel</u>, as you must surely know.

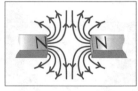

### The **Earth's** Magnetic Field

Note that the <u>magnetic poles</u> are <u>opposite</u> to the <u>Geographic Poles</u>, i.e. the <u>South Pole</u> is at the <u>North Pole</u> — if you see what I mean!

### The Magnetic Field Round a **Current-carrying Wire**

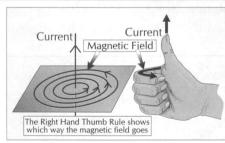

The Right Hand Thumb Rule shows which way the magnetic field goes

## A **Plotting Compass** *is a* **Freely Suspended Magnet**

1) This means it always <u>aligns itself</u> with the <u>magnetic field</u> that it's in.
2) This is great for plotting <u>magnetic field lines</u> like around the <u>bar magnets</u> shown above.
3) Away from any magnets, it will <u>align</u> with the magnetic field of the <u>Earth</u> and point <u>North</u>.
4) <u>Any magnet</u> suspended so it can turn <u>freely</u> will also come to rest pointing <u>North-South</u>.
5) The end of the magnet which points North is called a "<u>North-seeking pole</u>" or "<u>magnetic North</u>".
   The end pointing South will therefore be a "<u>magnetic South pole</u>". This is how they got their names.

---

## *Magnetic fields — there's no getting away from them...*

This is a nice easy page. Learn the definition of a magnetic field and the six field diagrams. Also learn those five details about plotting compasses and which way the poles are compared to the Earth. Then <u>cover the page</u> and <u>jot it all down</u>.

# Electromagnets

## An *Electromagnet* is just a *Coil of Wire* with an *Iron Core*

1) Electromagnets are really simple.
2) They're simply a solenoid (which is just a coil of wire) with a piece of "soft" iron inside.
3) When current flows through the wires of the solenoid it creates a magnetic field around it.
4) The soft iron core has the effect of increasing the magnetic field strength.

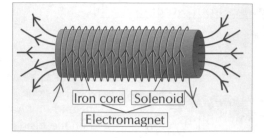

Iron core   Solenoid
Electromagnet

1) The magnetic field around an electromagnet is just like the one round a bar magnet, only stronger.
2) This means that the ends of a solenoid act like the North Pole and South Pole of a bar magnet.
3) Pretty obviously, if the direction of the current is reversed, the N and S poles will swap ends.

4) If you imagine looking directly into one end of a solenoid, the direction of current flow tells you whether it's the N or S pole you're looking at, as shown by the two diagrams opposite. You need to remember those diagrams. They may show you a solenoid in the Exam and ask you which pole it is.

N-Pole   S-Pole

| The STRENGTH of an ELECTROMAGNET depends on THREE FACTORS: | 1) The size of the CURRENT. |
| --- | --- |
| | 2) The number of TURNS the coil has. |
| | 3) What the CORE is made of. |

## Iron is Magnetically "*Soft*" — Ideal for *Electromagnets*

In magnetic terms, "soft" means it changes easily between being magnetised and demagnetised. Iron is "soft" which makes it perfect for electromagnets which need to be turned on and off.

## Steel is Magnetically "*Hard*" — Ideal for *Permanent* Magnets

Magnetically "hard" means that the material retains its magnetism. This would be useless in an electromagnet, but is exactly what's required for permanent magnets.

## To *MAGNETISE* a piece of steel, etc:

Put it in a solenoid with a steady DC supply. Turn off the current, pull it out, and there it is, a permanent magnet.

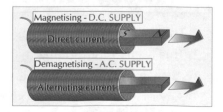

Magnetising - D.C. SUPPLY
Direct current

Demagnetising - A.C. SUPPLY
Alternating current

## To *DEMAGNETISE* a piece of steel, etc:

Put it in a solenoid with an AC supply, and then pull it out with the AC current still going and there it is, demagnetised.

## *Remember* — "hard" magnets can be magnetised or demagnetised

This is all very basic information, and quite memorable I'd have thought. Learn the headings and diagrams first, then cover the page and scribble them down. Then gradually fill in the other details. Keep looking back and checking. Make sure you learn all the points.

# Electromagnets

Electromagnets always have a <u>soft iron core</u>, which <u>increases the strength</u> of the magnet. The core has to be <u>soft</u> (magnetically soft, that is), so that when the <u>current</u> is turned <u>off</u>, the magnetism <u>disappears</u> with it. The four applications below depend on that happening.

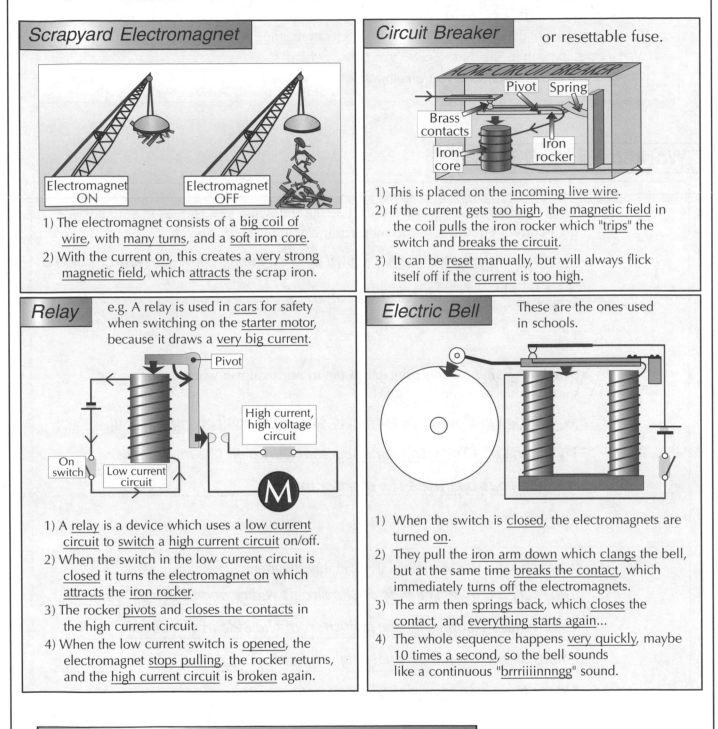

## Scrapyard Electromagnet

Electromagnet ON · Electromagnet OFF

1) The electromagnet consists of a <u>big coil of wire</u>, with <u>many turns</u>, and a <u>soft iron core</u>.
2) With the current <u>on</u>, this creates a <u>very strong magnetic field</u>, which <u>attracts</u> the scrap iron.

## Circuit Breaker — or resettable fuse.

Pivot · Spring · Brass contacts · Iron core · Iron rocker

1) This is placed on the <u>incoming live wire</u>.
2) If the current gets <u>too high</u>, the <u>magnetic field</u> in the coil <u>pulls</u> the iron rocker which "<u>trips</u>" the switch and <u>breaks the circuit</u>.
3) It can be <u>reset</u> manually, but will always flick itself off if the <u>current</u> is <u>too high</u>.

## Relay

e.g. A relay is used in <u>cars</u> for safety when switching on the <u>starter motor</u>, because it draws a <u>very big current</u>.

Pivot · High current, high voltage circuit · On switch · Low current circuit · M

1) A <u>relay</u> is a device which uses a <u>low current circuit</u> to <u>switch</u> a <u>high current circuit</u> on/off.
2) When the switch in the low current circuit is <u>closed</u> it turns the <u>electromagnet on</u> which <u>attracts</u> the <u>iron rocker</u>.
3) The rocker <u>pivots</u> and <u>closes the contacts</u> in the high current circuit.
4) When the low current switch is <u>opened</u>, the electromagnet <u>stops pulling</u>, the rocker returns, and the <u>high current circuit</u> is <u>broken</u> again.

## Electric Bell

These are the ones used in schools.

1) When the switch is <u>closed</u>, the electromagnets are turned <u>on</u>.
2) They pull the <u>iron arm down</u> which <u>clangs</u> the bell, but at the same time <u>breaks the contact</u>, which immediately <u>turns off</u> the electromagnets.
3) The arm then <u>springs back</u>, which <u>closes</u> the contact, and <u>everything starts again</u>...
4) The whole sequence happens <u>very quickly</u>, maybe <u>10 times a second</u>, so the bell sounds like a continuous "<u>brrriiiinnngg</u>" sound.

## Only **Iron, Steel**, **Nickel** and **Cobalt** are **Magnetic**

Don't forget that <u>all</u> other <u>common metals</u> are <u>not magnetic at all</u>. So a magnet <u>won't stick</u> to <u>aluminium ladders</u> or <u>copper kettles</u> or <u>brass trumpets</u> or <u>gold rings</u> or <u>silver spoons</u>.

## Learn the four main uses of electromagnets

They nearly always have one of these in the Exam. Usually it's a circuit diagram of one of them and likely as not they'll ask you to explain exactly how it works. The scrapyard example's pretty easy, but you still need to <u>learn all those details</u> for the three trickier ones.

# Warm-Up and Worked Exam Question

## Warm-up Questions

1  What is a solenoid?
2  State three factors that affect the strength of an electromagnet.
3  What does it mean to say that steel is "magnetically hard"?
4  Name three devices that use electromagnets.

## Worked Exam Question

1  a)  What is an electromagnetic relay?  Use the following key phrases in your answer: high-current circuit, low-current circuit, electromagnetic switch.

*A relay is an electromagnetic switch which uses a low-current circuit to switch a high-current circuit on and off*

*[2 marks]*

b)  Briefly explain how this principle is put to practical use when operating the starter motor of a car.

*When the ignition key is turned it causes a small current to run.*

*By means of a relay this small current turns on the much larger current that operates the starter motor.*

*[2 marks]*

c)  Why is it safer to use a relay circuit in this situation?

*The relay isolates the side of the circuit where people are putting their hands to operate the switch from the side of the circuit where there is a dangerously high current.*

*Hence it reduces the risk of electrocution.*

*[2 marks]*

d)  Why does the core of the electromagnet used in a relay need to be made of soft iron?

*The core of the electromagnet must be soft iron rather than steel because soft iron can be easily magnetised and demagnetised.*

*This means that it loses its magnetism when the switch is turned off.*

*[1 mark]*

# Exam Questions

1   You are provided with a steel rod and a solenoid connected to a 12V power supply, as shown in the diagram.

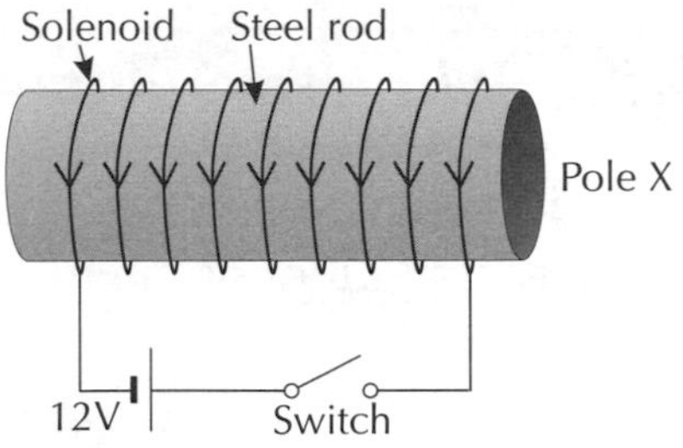

Solenoid   Steel rod

Pole X

12V   Switch

a)   On the diagram, would Pole X be the north or the south pole of the magnetic field formed around the solenoid?

...........................................................................................................................................................

*[1 mark]*

b)   How could you reverse the polarity of Pole X?

...........................................................................................................................................................

*[1 mark]*

c)   What difference would it make to the strength of the magnetic field formed around the solenoid if there were fewer turns in the coil?

...........................................................................................................................................................

*[1 mark]*

d)   How could you make the steel rod into a permanent magnet using the apparatus shown in the diagram?

...........................................................................................................................................................

*[1 mark]*

e)   How could you subsequently demagnetise the same steel rod?

...........................................................................................................................................................

*[2 marks]*

2   The diagram shows how an electric bell works. Study the diagram and put the six sentences A to F in the correct order to describe what happens when the switch is closed.

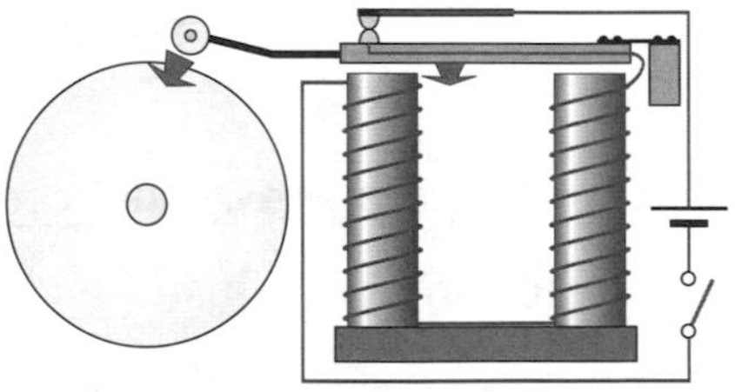

A - The contacts joining the iron arm to the circuit pull apart, breaking the circuit.

B - The iron arm is attracted to the electromagnet, causing the gong to strike the bell.

C - The iron arm, no longer attracted to the electromagnet, springs back.

D - The electromagnet is turned on.

E - The contacts close again, so current again runs in the circuit.

F - The electromagnet is turned off.

*[5 marks]*

# Motors, Generators and Electricity

Anything <u>carrying a current</u> in a <u>magnetic field</u> will experience a <u>force</u>.
There are <u>three important cases</u>, detailed on these two pages.

## *A Current in a Magnetic Field Experiences a Force*

The two tests below demonstrate the <u>force</u> on a <u>current-carrying wire</u> placed in a <u>magnetic field</u>. The <u>force</u> gets <u>bigger</u> if either the <u>current</u> or the <u>magnetic field</u> is made bigger.

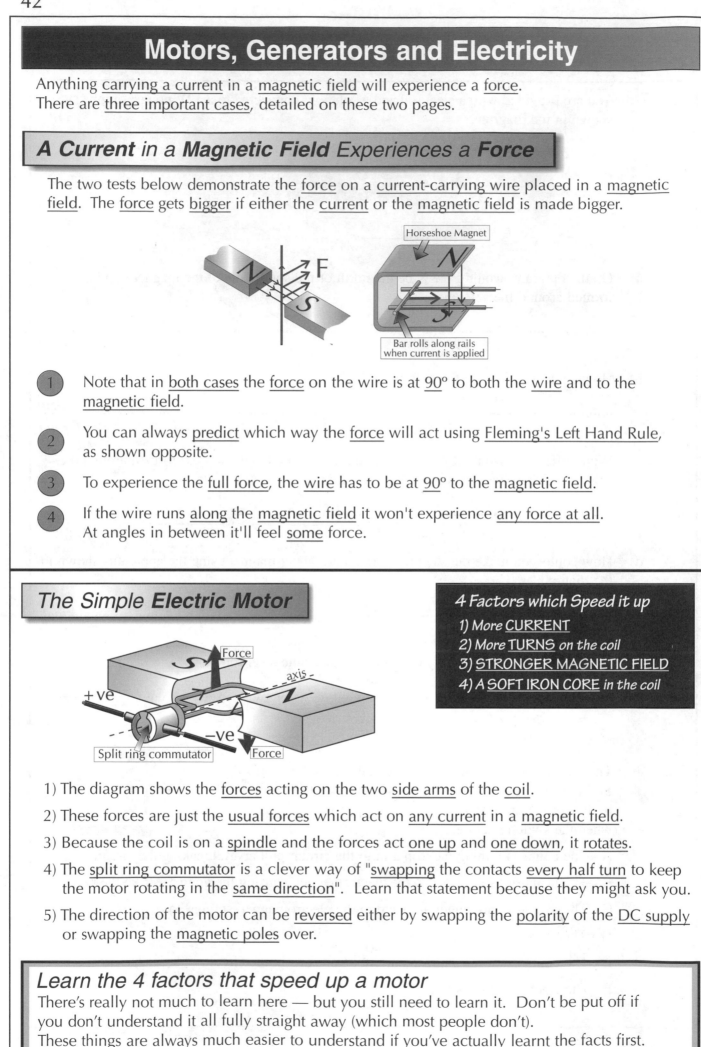

Horseshoe Magnet

Bar rolls along rails
when current is applied

(1) Note that in <u>both cases</u> the <u>force</u> on the wire is at <u>90°</u> to both the <u>wire</u> and to the <u>magnetic field</u>.

(2) You can always <u>predict</u> which way the <u>force</u> will act using <u>Fleming's Left Hand Rule</u>, as shown opposite.

(3) To experience the <u>full force</u>, the <u>wire</u> has to be at <u>90°</u> to the <u>magnetic field</u>.

(4) If the wire runs <u>along</u> the <u>magnetic field</u> it won't experience <u>any force at all</u>. At angles in between it'll feel <u>some</u> force.

## *The Simple Electric Motor*

Force

axis

+ve

−ve

Split ring commutator

Force

### 4 Factors which Speed it up
1) More <u>CURRENT</u>
2) More <u>TURNS</u> on the coil
3) <u>STRONGER MAGNETIC FIELD</u>
4) A <u>SOFT IRON CORE</u> in the coil

1) The diagram shows the <u>forces</u> acting on the two <u>side arms</u> of the <u>coil</u>.

2) These forces are just the <u>usual forces</u> which act on <u>any current</u> in a <u>magnetic field</u>.

3) Because the coil is on a <u>spindle</u> and the forces act <u>one up</u> and <u>one down</u>, it <u>rotates</u>.

4) The <u>split ring commutator</u> is a clever way of "<u>swapping</u> the contacts <u>every half turn</u> to keep the motor rotating in the <u>same direction</u>". Learn that statement because they might ask you.

5) The direction of the motor can be <u>reversed</u> either by swapping the <u>polarity</u> of the <u>DC supply</u> or swapping the <u>magnetic poles</u> over.

---

## *Learn the 4 factors that speed up a motor*

There's really not much to learn here — but you still need to learn it. Don't be put off if you don't understand it all fully straight away (which most people don't).
These things are always much easier to understand if you've actually learnt the facts first.

# Motors, Generators and Electricity

## *Fleming's Left Hand Rule* tells you *which way* the Force Acts

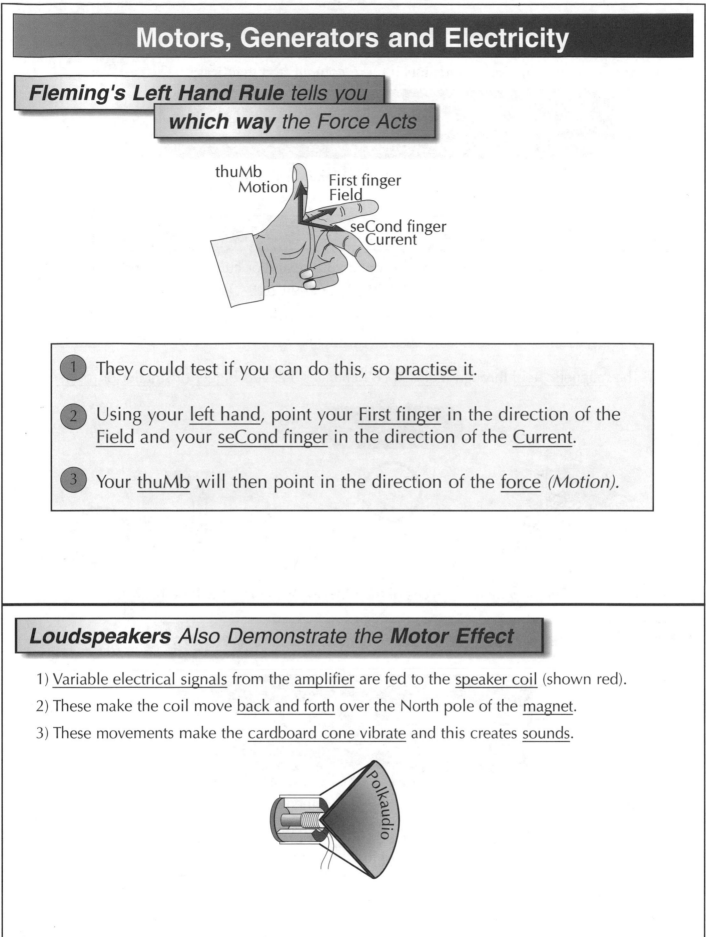

① They could test if you can do this, so <u>practise it</u>.

② Using your <u>left hand</u>, point your <u>First finger</u> in the direction of the <u>Field</u> and your <u>seCond finger</u> in the direction of the <u>Current</u>.

③ Your <u>thuMb</u> will then point in the direction of the <u>force</u> *(Motion)*.

## *Loudspeakers* Also Demonstrate the *Motor Effect*

1) <u>Variable electrical signals</u> from the <u>amplifier</u> are fed to the <u>speaker coil</u> (shown red).

2) These make the coil move <u>back and forth</u> over the North pole of the <u>magnet</u>.

3) These movements make the <u>cardboard cone vibrate</u> and this creates <u>sounds</u>.

## *Learn Fleming's Rule — it's really useful*

Same routine here. <u>Learn all the details</u>, along with the diagrams, then <u>cover up the page</u> and <u>write it all down</u> again <u>from memory</u>. It doesn't have to be neat — all you're trying to do is make sure that you really do <u>know it all</u>.

# Motors, Generators and Electricity

Electromagnetic induction sounds much more complicated than it is:

> **ELECTROMAGNETIC INDUCTION:** *The creation of a* <u>VOLTAGE</u>
> *(and maybe current) in a wire which is experiencing a* <u>CHANGE IN</u>
> <u>MAGNETIC FIELD</u>.

It's called "<u>induction</u>" rather than "<u>creation</u>", but it amounts to the <u>same thing</u>.

## EM Induction — a) *Flux cutting*   b) *Field Through a Coil*

<u>Electromagnetic induction</u> is the <u>induction</u> of a <u>voltage</u> and/or <u>current</u> in a conductor. There are <u>two different situations</u> where you get <u>EM induction</u>. You need to know about <u>both</u> of them:

1) The <u>conductor</u> moves across a <u>magnetic field</u> and "<u>cuts</u>" through the lines of <u>magnetic flux</u>.

2) The <u>magnetic field</u> through a <u>closed coil changes</u>, i.e. gets <u>bigger</u> or <u>smaller</u> or <u>reverses</u>.

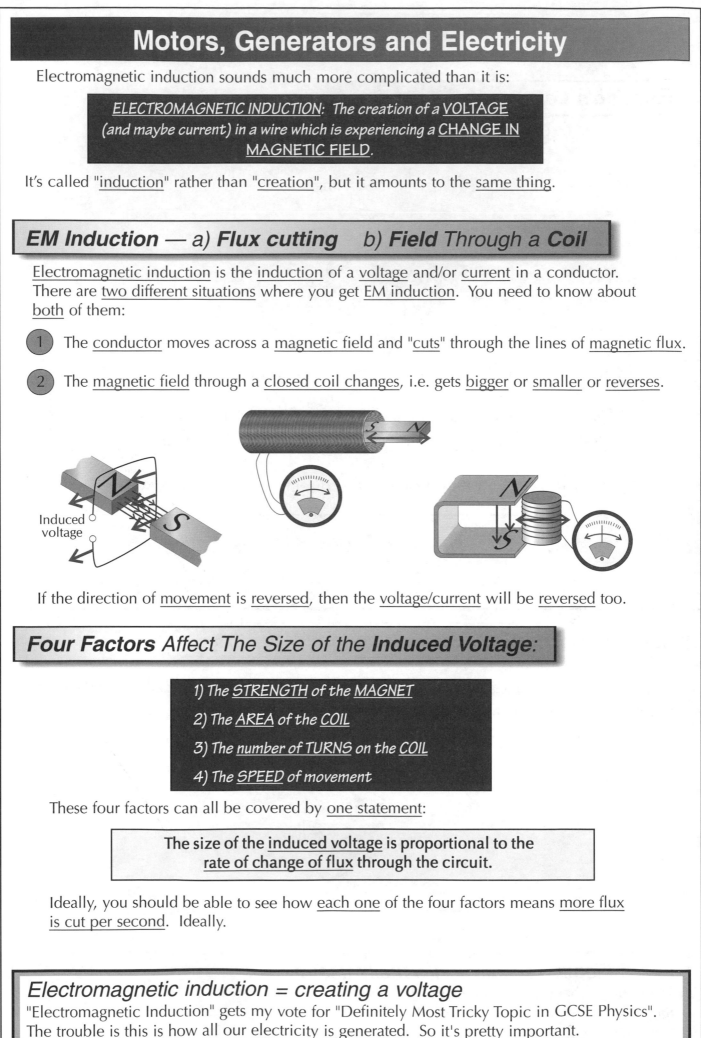

If the direction of <u>movement</u> is <u>reversed</u>, then the <u>voltage/current</u> will be <u>reversed</u> too.

## *Four Factors* Affect The Size of the *Induced Voltage*:

> 1) *The* <u>STRENGTH</u> *of the* <u>MAGNET</u>
>
> 2) *The* <u>AREA</u> *of the* <u>COIL</u>
>
> 3) *The* <u>number of TURNS</u> *on the* <u>COIL</u>
>
> 4) *The* <u>SPEED</u> *of movement*

These four factors can all be covered by <u>one statement</u>:

> **The size of the <u>induced voltage</u> is proportional to the
> <u>rate of change of flux</u> through the circuit.**

Ideally, you should be able to see how <u>each one</u> of the four factors means <u>more flux is cut per second</u>. Ideally.

---

### *Electromagnetic induction = creating a voltage*

"Electromagnetic Induction" gets my vote for "Definitely Most Tricky Topic in GCSE Physics". The trouble is this is how all our electricity is generated. So it's pretty important.

# Motors, Generators and Electricity

## Generators and Dynamos

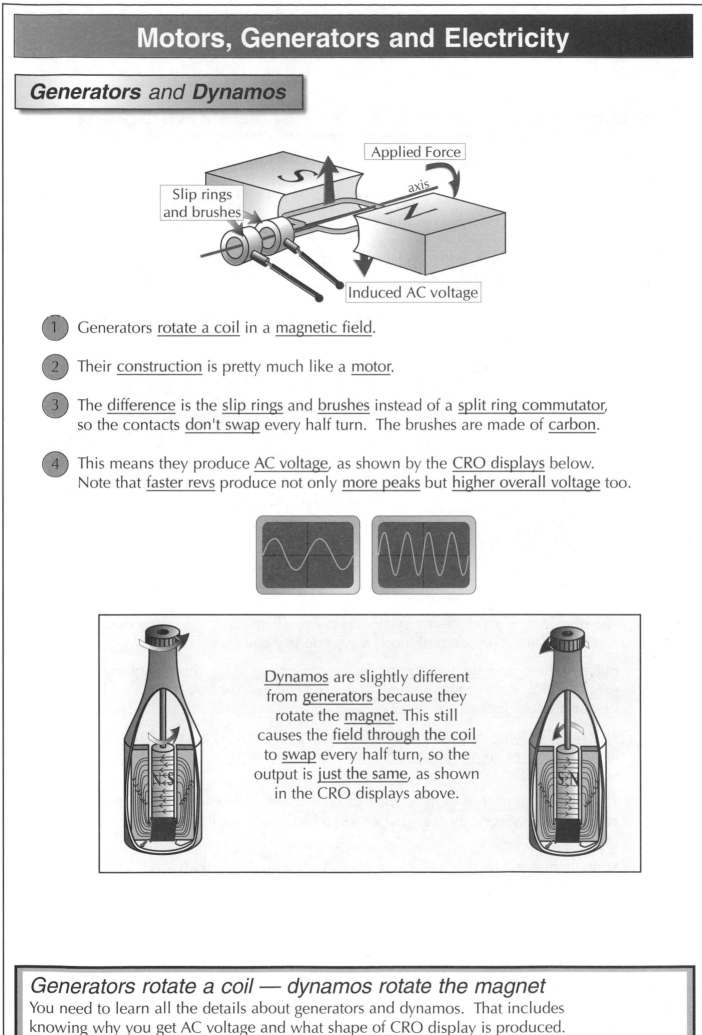

1. Generators <u>rotate a coil</u> in a <u>magnetic field</u>.

2. Their <u>construction</u> is pretty much like a <u>motor</u>.

3. The <u>difference</u> is the <u>slip rings</u> and <u>brushes</u> instead of a <u>split ring commutator</u>, so the contacts <u>don't swap</u> every half turn. The brushes are made of <u>carbon</u>.

4. This means they produce <u>AC voltage</u>, as shown by the <u>CRO displays</u> below. Note that <u>faster revs</u> produce not only <u>more peaks</u> but <u>higher overall voltage</u> too.

<u>Dynamos</u> are slightly different from <u>generators</u> because they rotate the <u>magnet</u>. This still causes the <u>field through the coil</u> to <u>swap</u> every half turn, so the output is <u>just the same</u>, as shown in the CRO displays above.

## *Generators rotate a coil — dynamos rotate the magnet*

You need to learn all the details about generators and dynamos. That includes knowing why you get AC voltage and what shape of CRO display is produced.

# Motors, Generators and Electricity

Transformers use <u>electromagnetic induction</u>.  So they will <u>only</u> work on <u>AC</u>.

## Transformers *Change the* Voltage *— but only* AC Voltages

1   The <u>laminated iron core</u> is purely for transferring the <u>magnetic flux</u> from the primary coil to the secondary.

2   No <u>electricity</u> flows round the <u>iron core</u>, only <u>magnetic flux</u>.

3   The iron core is <u>laminated</u> with <u>layers of insulation</u> to reduce the <u>eddy currents</u> which <u>heat it up</u>, and therefore <u>waste energy</u>.

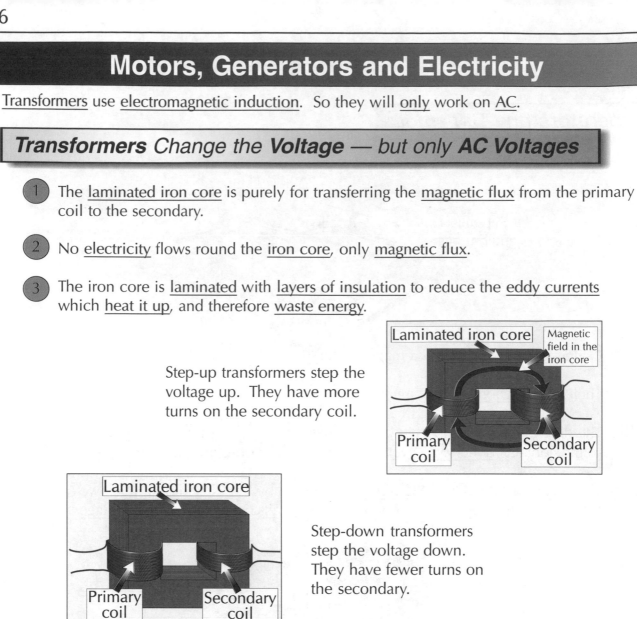

Step-up transformers step the voltage up.  They have more turns on the secondary coil.

Step-down transformers step the voltage down. They have fewer turns on the secondary.

1)   The primary coil <u>produces magnetic flux</u> (field) which stays <u>within the iron core</u> and this means it <u>all</u> passes through the <u>secondary coil</u>.

2)   Because there is <u>alternating current</u> (AC) in the <u>primary coil</u>, this means that the magnetic flux in the iron core is <u>reversing</u> (100 times a second, if it's at 50 Hz) — i.e. it is a <u>changing</u> flux.

3)   This <u>rapidly changing</u> magnetic flux is then experienced by the <u>secondary coil</u> and this <u>induces</u> an <u>alternating voltage</u> in it — <u>electromagnetic induction</u> of a voltage in fact.

4)   The <u>relative number of turns</u> on the two coils determines whether the voltage created in the secondary is <u>greater</u> or <u>less</u> than the voltage in the primary.

5)   If you supplied DC to the primary, you'd get <u>nothing</u> out of the secondary at all. Sure, there'd still be flux in the iron core, but it wouldn't be <u>constantly changing</u> so there'd be no <u>induction</u> in the secondary because you need a <u>changing flux</u> to induce a voltage.  So don't forget it — transformers only work with <u>AC</u>. They won't work with DC <u>at all</u>.

---

## *Transformers need a constantly changing flux — so it's got to be AC*

Besides their iron core, transformers have lots of other <u>important</u> details which also need to be <u>learnt</u>.  Like all that complicated business about why DC doesn't work.

# Motors, Generators and Electricity

## The *Transformer Equation* — use it *Either Way Up*

In words:

> The RATIO OF TURNS on the two coils equals the RATIO OF THEIR VOLTAGES.

$$\frac{Primary\ Voltage}{Secondary\ Voltage} = \frac{Number\ of\ turns\ on\ Primary}{Number\ of\ turns\ on\ Secondary}$$

$$\frac{V_P}{V_S} = \frac{N_P}{N_S}$$

*or*

$$\frac{V_S}{V_P} = \frac{N_S}{N_P}$$

It's just another formula. You put in the numbers you've got and work out the one that's left. It's really useful to remember you can write it either way up — this example's much trickier algebra-wise if you start with $V_S$ on the bottom...

Example: *A transformer has 40 turns on the primary and 800 on the secondary. If the input voltage is 1000 V find the output voltage.*

ANSWER: $V_S/V_P = N_S/N_P$ so $V_S/1000 = 800/40$ $V_S = 1000 \times (800/40) = \underline{20\ 000\ V}$

## There's also "*Power In = Power Out*" which gives "$V_P I_P = V_S I_S$"

This formula is true because transformers are nearly 100% efficient. But don't worry, it's just another formula — you stick numbers in and work out the bit that's left, and that's all there is to it. End of story.

## *Imagine a revision method as efficient as a transformer*

You'll need to practise with these tricky equations. They're unusual because they can't be put into formula triangles but other than that the method is the same. Just practise.

# Warm-Up and Worked Exam Question

## Warm-up Questions

1  In Fleming's Left Hand Rule, write down what quantities have their direction shown by the thumb, first finger and second finger.
2  Write down four factors that can speed up an electric motor.
3  Explain what electromagnetic induction is.
4  A generator consists of a wire coil rotating in a magnetic field. Write down the factors which significantly affect the size of the induced voltage in the coil.
5  A transformer has 20 turns on the primary coil and 100 turns on the secondary coil. If 4V a.c. is supplied to the primary coil, what voltage would there be across the secondary coil, assuming the transformer was 100% efficient?

## Worked Exam Question

1  A loop of current-carrying wire runs between the poles of a magnet as shown.

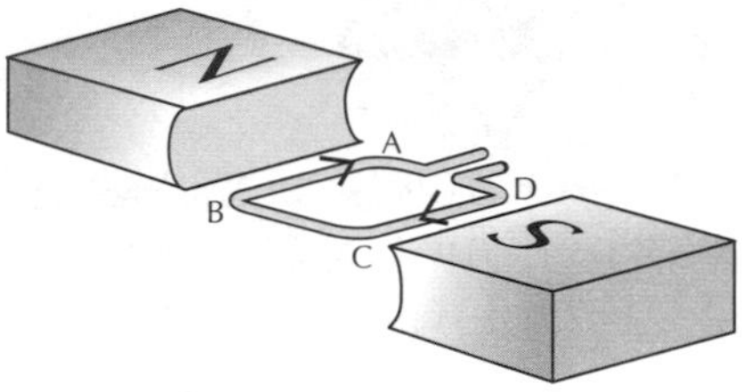

a)  In what direction does a force act on section AB of the loop?

Downwards  (using Fleming's left hand rule)

[1 mark]

b)  In what direction does a force act on section CD of the loop?

Upwards  (Fleming's left hand rule again)

[1 mark]

c)  What will these two forces cause the loop to do?

Rotate

[1 mark]

d)  Is the force acting on section BC of the loop:

(i) less than the force on AB but not zero   (ii) greater than the force on AB
(iii) equal to the force on AB              (iv) zero?

(iv)   A wire running along the magnetic field doesn't experience a force.

[1 mark]

e)  The single loop of wire is now replaced by a coil with many turns. What effect will this have on the force acting on the side arms of the coil?

It will increase the force.

[1 mark]

# Exam Questions

1  a) A motor and a generator are both devices that transform one type of energy into another type. Fill in the table below showing what type of energy forms the input and output of both devices.

| | Input | Output |
|---|---|---|
| Motor | | |
| Generator | | |

*[2 marks]*

b) A simple motor and a simple generator can both be constructed by having a coil rotate between the poles of a magnet. What component or components does a generator have where a motor would have a split ring commutator?

*[1 mark]*

c) What is the difference between a generator and a dynamo?

*[2 marks]*

d) Is the induced voltage in a dynamo the result of:
(i) a conductor "cutting" lines of magnetic flux
(ii) a conductor "cutting" the electric field
(iii) a change in the resistance of a coil
(iv) a change in the magnetic field through a coil?

*[1 mark]*

Mary connects an a.c. generator to a CRO in order to display the output voltage obtained when the coil rotates steadily.

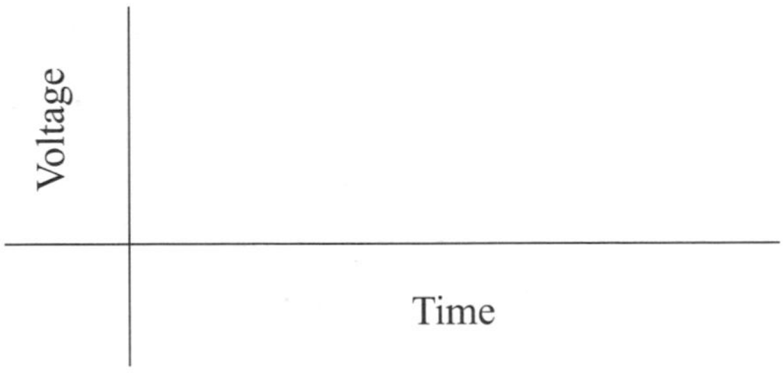

Voltage

Time

e) On the axes above, sketch the display she sees on the screen, labelling your trace "$V_1$".

*[2 marks]*

f) On the same diagram sketch what the CRO display looks like when Mary doubles the speed of rotation of the coil. Show both the frequency and the overall strength of the voltage. Your trace should be to the same scale as the previous one, and should be labelled "$V_2$".

*[2 marks]*

# Exam Questions

2  a)  On the diagram of a d.c. motor on the right,
       label the split ring commutator, the
       current direction in the coil and show the
       direction of the forces on the
       two side arms of the coil.

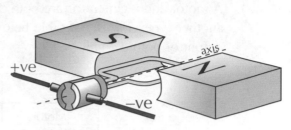

*[4 marks]*

   b)  The split ring commutator in a d.c. motor:
       (i) reverses the voltage across the cell twice in each revolution
       (ii) reverses the direction of rotation of the coil twice in each revolution
       (iii) reverses the direction of the current in the coil twice in each revolution
       (iv) reverses the polarity of the magnet twice in each revolution?

*[1 mark]*

   c)  Why must a d.c. motor have a split ring commutator to keep working continuously?

*[2 marks]*

   d)  What would be the effect of swapping <u>both</u> the direction of the magnetic field and the
       direction of the current?

*[1 mark]*

3      A power station transformer puts out a voltage of 20 000 V. Then a step-up transformer
       is used to step up this voltage to 400 000 V so that the power can be transmitted at a high
       voltage across the National Grid. Assume the transformer is perfectly efficient.

   a)  Is the ratio of the number of turns in the primary coil to the number of turns in the
       secondary coil:  A)  1:20     B) 2000:1     C) 1:20 000     D) 40:1

*[1 mark]*

   b)  If the current in the National Grid cables is 0.1A, what is the current going into the
       transformer?

*[2 marks]*

   c)  What role do transformers play in the National Grid? (Write no more than 3 lines.)

*[3 marks]*

   d)  In real life no transformer is 100% efficient, although they often come close. Explain
       one way in which energy is wasted in a transformer and how it can be constructed to
       make this wastage as small as possible.

*[2 marks]*

   e)  What flows round the iron core of the transformer? Is it:

       A) electric current     B) magnetic flux     C) voltage     D) power?

*[1 mark]*

# Revision Summary for Section One

Electricity and magnetism. This is definitely Physics at its most tricky. The big problem with Physics in general is that usually there's nothing to "see". You're told that there's a current flowing or a magnetic field somewhere, but there's nothing you can actually see with your eyes. That's what makes it so difficult. To get to grips with Physics you have to get used to learning about things which you can't see. Try these questions and see how well you're doing.

1) Look at the table on page 1. Cover the last three columns and write down all the hidden details.
2) Explain how formula triangles work. What are the three rules for using any formula?
3) What are the two rules to remember about units? Give an example of each.
   a) Find the current when a resistance of 96 Ω is connected to a battery of 12V.
   b) Find the charge passed when a current of 2 A flows for 2 minutes.
   c) Find the power output of a heater which provides 77 kJ of heat energy in 4 mins.
   d) Find the resistance of a hairdrier which draws 10 A and gives out 1.4 kW.
4) What carries current in metals? What's "conventional current" and what's the problem?
5) Sketch CRO traces for DC and AC currents. Label the key features.
6) Sketch out the standard test circuit with all the details. Describe how it's used.
7) Sketch the four standard I-V graphs and explain their shapes. How do you get R from them?
8) Sketch 18 circuit symbols that you know, with their names.
9) Write down two facts about: a) variable resistors  b) diodes  c) LEDs  d) LDRs  e) thermistors.
10) Sketch a typical series circuit and say why it is a series circuit, not a parallel one.
11) State five rules about the current, voltage and resistance in a series circuit.
12) Give examples of lights wired in series and wired in parallel and explain the main differences.
13) Sketch a typical parallel circuit, showing voltmeter and ammeter positions.
14) State five rules about the current, voltage and resistance in a parallel circuit.
15) Draw a circuit diagram of part of a car's electrics, and explain why they are in parallel.
16) What is static electricity? What is nearly always the cause of it building up?
17) Which particles move when static builds up, and which ones don't?
18) Explain how charging by induction works. Also explain what causes sparks to happen.
19) Give two examples each of static being: a) helpful  b) shocking  c) dangerous. Write all the details.
20) What are the four types of energy that electricity can easily be converted into?
21) Sketch a circuit showing four devices converting energy. Describe all the energy changes.
22) Sketch a view of a circuit to explain the formula "E = QV". Which definitions go with it?
23) Explain what the number on an electricity meter represents.
24) What's a kilowatt-hour? What are the two easy formulas for finding the cost of electricity?
25) Sketch a properly wired plug. Explain fully how fuses work.
26) Describe what earthing and double insulation are. Why are they useful?
27) Sketch a typical power station, and the national grid and explain why it's at 400 kV.
28) Sketch magnetic fields for: a) a bar magnet,  b) a solenoid,  c) two magnets attracting,
   d) two magnets repelling,  e) the Earth's magnetic field,  f) a current-carrying wire.
29) What is an electromagnet made of? Explain how to decide on the polarity of the ends.
30) What is meant by magnetically hard and soft? How do you magnetise and demagnetise?
31) Sketch and give details of: a) a scrapyard magnet,  b) a circuit breaker,  c) a relay, d) an electric bell.
32) Sketch two demos of the motor effect. Sketch a motor and list the 4 factors which speed it up.
33) What do you use Fleming's Left Hand Rule for? Which direction do your fingers point in?
34) Give a definition of electromagnetic induction. Sketch three cases where it happens.
35) List the four factors which affect the size of the induced voltage.
36) Sketch a generator, labelling all the parts. Describe how it works and what all the bits do.
37) Write down how a dynamo works.
38) Sketch the two types of transformer, and highlight the main details. Explain how they work.
39) Write down the transformer equation. Do your own worked example — it's all good practice.

# Gravity

## Gravity *is the* Force of Attraction *Between* All Masses

Gravity attracts all masses, but you only notice it when one of the masses is really really big, e.g. a planet. Anything near a planet or star is attracted to it very strongly. This has three important effects:

 It makes all things accelerate towards the ground
(all with the same acceleration, g, which ≈ 10 m/s² on Earth).

 It gives everything a weight.

3  It keeps planets, moons and satellites in their orbits. The orbit is a balance between the forward motion of the object and the force of gravity pulling it inwards.

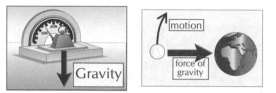

## Weight *and* Mass *are* Not the Same

To understand this you must learn all these facts about mass and weight.

1) Mass is the amount of matter in an object.
   For any given object this will have the same value anywhere in the Universe.

2) Weight is caused by the pull of gravity. In most questions the weight of an object is just the force of gravity pulling it towards the centre of the Earth.

3) An object has the same mass whether it's on Earth or on the Moon — but its weight will be different. A 1 kg mass will weigh less on the Moon (1.6N) than it does on Earth (10N), simply because the force of gravity pulling on it is less.

4) Weight is a force measured in newtons.
   It must be measured using a spring balance or Newton meter.

5) Mass is not a force.
   It's measured in kilograms with a mass balance (never a spring balance).

## Remember — "g" is "an acceleration of about 10 m/s²"
Very often the only way to "understand" something is to learn all the facts about it. That's certainly true here. "Understanding" the difference between mass and weight is no more than learning all those facts about them. When you've learnt all those facts, you'll understand it.

# Weight and Moments

## The *Very Important Formula* relating *Mass*, *Weight* and *Gravity*

$$W = m \times g \quad \text{(Weight = mass} \times g\text{)}$$

1) Remember, <u>weight and mass are not the same</u>. Mass is in <u>kg</u>, weight is in <u>newtons</u>.

2) The letter "g" represents the <u>strength of the gravity</u> and its value is <u>different</u> for <u>different planets</u>. <u>On Earth</u> g = 10 N/kg.
On the <u>Moon</u>, where the gravity is weaker, g is just 1.6 N/kg.

3) This formula is <u>very easy</u> to use:

<u>Example</u>: *What is the weight, in Newtons, of a 5kg mass, both on Earth and on the Moon?*
<u>Answer</u>: *"W = m × g". On Earth: W = 5 × 10 = <u>50N</u> (The weight of the 5kg mass is 50N)*
*On the Moon: W = 5 × 1.6 = <u>8N</u> (The weight of the 5kg mass is 8N)*

Easy — as long as you've learnt what all the letters mean.

## A *Moment* is a *Turning Force* — Force × Perpendicular Distance

When a force acts on something which has a <u>pivot</u>, it creates a <u>turning force</u> called a "<u>moment</u>". <u>Moments</u> are <u>calculated</u> using this formula:

> moment = force × perpendicular distance from pivot

<u>Also</u>, for the system to be in <u>equilibrium</u> (i.e. all <u>balanced</u> and <u>not moving</u>), then <u>this must be true</u> too:

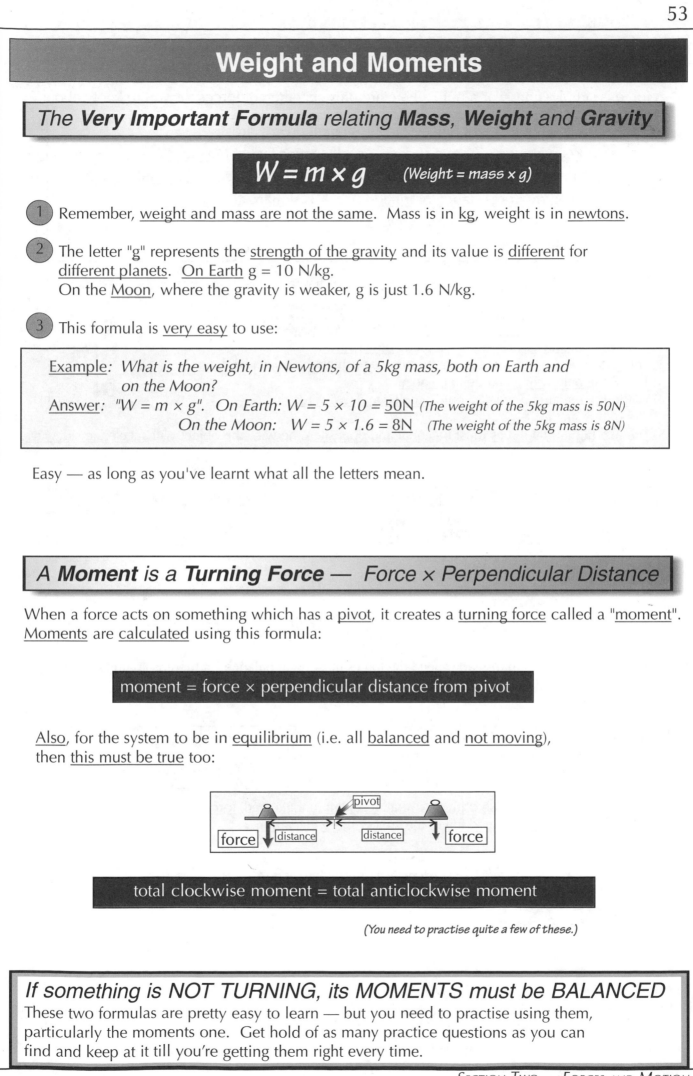

> total clockwise moment = total anticlockwise moment

*(You need to practise quite a few of these.)*

## If something is NOT TURNING, its MOMENTS must be BALANCED

These two formulas are pretty easy to learn — but you need to practise using them, particularly the moments one. Get hold of as many practice questions as you can find and keep at it till you're getting them right every time.

# Balanced and Unbalanced Forces

A <u>force</u> is simply a <u>push</u> or a <u>pull</u>.  There are only <u>six different forces</u> for you to know about:

> 1) <u>Gravity</u> or <u>weight</u>, always acting <u>straight downwards</u>.
>
> 2) <u>Reaction force</u> from a <u>surface</u>, usually acting <u>straight upwards</u>.
>
> 3) <u>Thrust</u> or <u>push</u> or <u>pull</u> due to an engine or rocket <u>speeding something up</u>.
>
> 4) <u>Drag</u> or <u>air resistance</u> or <u>friction</u> which is <u>slowing the thing down</u>.
>
> 5) <u>Lift</u> due to an <u>aeroplane wing</u>.
>
> 6) <u>Tension</u> in a <u>rope</u> or <u>cable</u>.

And there are basically only <u>five different force diagrams</u> you can get:

## 1) *Stationary Object* — *All Forces in Balance*

1) The force of <u>Gravity</u> (or weight) is acting <u>downwards</u>.
2) This causes a <u>reaction force</u> from the surface <u>pushing</u> the object <u>back up</u>.
3) This is the <u>only way</u> it can be in <u>balance</u>.
4) <u>Without</u> a reaction force, it would <u>accelerate downwards</u> due to the pull of gravity.
5) The two <u>horizontal forces</u> must be <u>equal and opposite</u> otherwise the object will <u>accelerate sideways</u>.

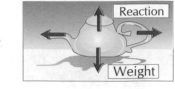

## 2) *Steady Horizontal Velocity* — *All Forces in Balance*

## 3) *Steady Vertical Velocity* — *All Forces in Balance*

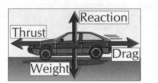

*This skydiver is free-falling at 'terminal velocity' — see page 68.*

<u>Take note</u> — to move with a <u>steady speed</u> the forces must be in <u>balance</u>.  If there is an <u>unbalanced force</u> then you get <u>acceleration</u>, not steady speed.  That's <u>really important</u> so don't forget it.

## 4) *Horizontal Acceleration* — *Unbalanced Forces*

## 5) *Vertical Acceleration* — *Unbalanced Forces*

1) You only get <u>acceleration</u> with an overall <u>resultant</u> (unbalanced) <u>force</u>.
2) The <u>bigger</u> this <u>unbalanced force</u>, the <u>greater</u> the <u>acceleration</u>.

Note that the forces in the <u>other direction</u> are still <u>balanced</u>.

*Just after dropping out of the plane, the skydiver accelerates — see page 68.*

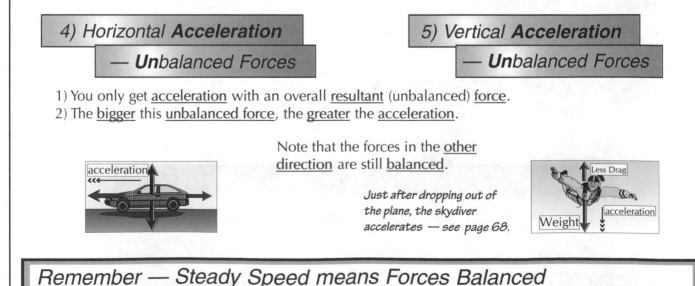

## *Remember — Steady Speed means Forces Balanced*

Make sure you learn those five different force diagrams.  You'll almost certainly get one of them in your Exam.  All you really need to remember is how the relative sizes of the arrows relate to the type of motion.  It's pretty simple so long as you make the effort to <u>learn it</u>.

*SECTION TWO — FORCES AND MOTION*

# Friction

## *Friction* is Always There to *Slow things Down*

1. If an object has <u>no force</u> propelling it along, it will always <u>slow down and stop</u> because of <u>friction</u>.

2. To travel at a <u>steady speed</u>, things always need a <u>driving force</u> to counteract the friction.

3. Friction occurs in <u>three main ways</u>:

### a) *FRICTION* BETWEEN *SOLID SURFACES* WHICH ARE *GRIPPING*

For example between <u>tyres and the road</u>.  There's always a <u>limit</u> as to how far two surfaces can <u>grip</u> each other, and if you demand <u>more force of friction</u> than they can manage, then they start to <u>slide</u> past each other instead, e.g. if you try to brake <u>too hard</u>, you'll <u>skid</u>.

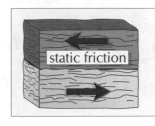

### b) *FRICTION* BETWEEN *SOLID SURFACES* WHICH ARE *SLIDING PAST EACH OTHER*

For example between <u>brake pads and brake discs</u>. There's just as much force of <u>friction</u> here as between the tyres and the road.  In fact in the end, if you brake hard enough, the friction here becomes <u>greater</u> than at the tyres, and then the wheel <u>skids</u>.

### c) *RESISTANCE* OR *"DRAG"* FROM *FLUIDS (AIR OR LIQUID)*

The most important factor <u>by far</u> in <u>reducing drag in fluids</u> is keeping the shape of the object <u>streamlined</u>, like fish bodies or boat hulls or bird wings/bodies.
The <u>opposite extreme</u> is a <u>parachute</u> which is about as <u>high drag</u> as you can get — which is, of course, <u>the whole idea</u>.

## *Learn the three main types of Friction*

To check you've really absorbed all this, write a mini essay about a car stopping, a) in normal conditions, and b) when it's icy and the car skids instead of stopping.  Then write a paragraph explaining why jumping out of a plane without a parachute is a really bad idea.

# Friction

## *Friction* Always *Increases* as the *Speed Increases*

A car has much more friction to work against when travelling at 60mph compared to 30mph. So at 60mph the engine has to work much harder just to maintain a steady speed. It therefore uses more petrol than it would going just as far at 30mph.

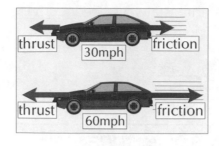

## *But We Also* **Need Friction** to **Move** and to **Stop**

It's easy to think of friction as generally a nuisance because we always seem to be working against it, but don't forget that without it we wouldn't be able to drive or walk or run or go sky-diving. It also holds nuts and bolts together.

## *Friction* Causes *Wear* and *Heating*

1) Friction always acts between surfaces that are sliding over each other, e.g. in machinery.

2) Friction always produces heat and wearing of the surfaces.

3) Lubricants are used to keep the friction as low as possible.

4) These make the machinery run more freely so it needs less power, and it also reduces wear.

5) The heating effect of friction can be enormous. For example the brakes on Grand Prix racing cars can often glow red hot. Another example is if an engine runs without oil it will quickly seize up as the moving parts get red hot through friction and eventually weld themselves together.

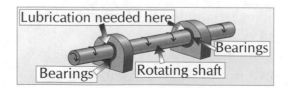

## *Friction causes energy to be lost — usually as heat and sound*

I would never have thought there was so much to say about friction. Nevertheless, it's all mentioned in the specifications, and is very likely to come up in your Exam.

# Warm-Up and Worked Exam Questions

## Worked Exam Questions

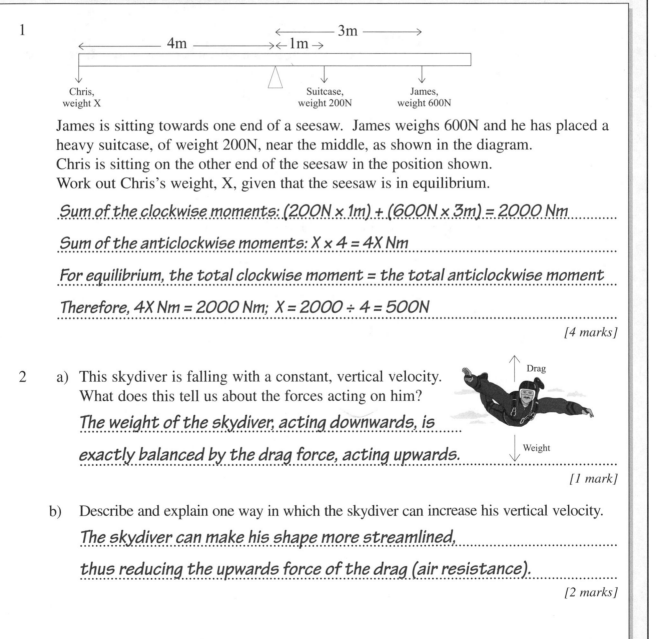

1

James is sitting towards one end of a seesaw.  James weighs 600N and he has placed a heavy suitcase, of weight 200N, near the middle, as shown in the diagram.
Chris is sitting on the other end of the seesaw in the position shown.
Work out Chris's weight, X, given that the seesaw is in equilibrium.

*Sum of the clockwise moments: (200N x 1m) + (600N x 3m) = 2000 Nm*

*Sum of the anticlockwise moments: X x 4 = 4X Nm*

*For equilibrium, the total clockwise moment = the total anticlockwise moment*

*Therefore, 4X Nm = 2000 Nm; X = 2000 ÷ 4 = 500N*

[4 marks]

2  a)  This skydiver is falling with a constant, vertical velocity.
        What does this tell us about the forces acting on him?

*The weight of the skydiver, acting downwards, is*

*exactly balanced by the drag force, acting upwards.*

[1 mark]

   b)  Describe and explain one way in which the skydiver can increase his vertical velocity.

*The skydiver can make his shape more streamlined,*

*thus reducing the upwards force of the drag (air resistance).*

[2 marks]

# Exam Questions

1   a)   Explain why a car driver must use less braking force in icy conditions than in dry
         conditions.

_because the if you break hard there is more friction
the the tyer will slid_

[2 marks]

    b)   In recent years, most lorries have been fitted with
         aerofoils, as shown in the diagram.  By thinking
         about the forces acting on the lorry, explain what
         the aerofoil does and why it is fitted.

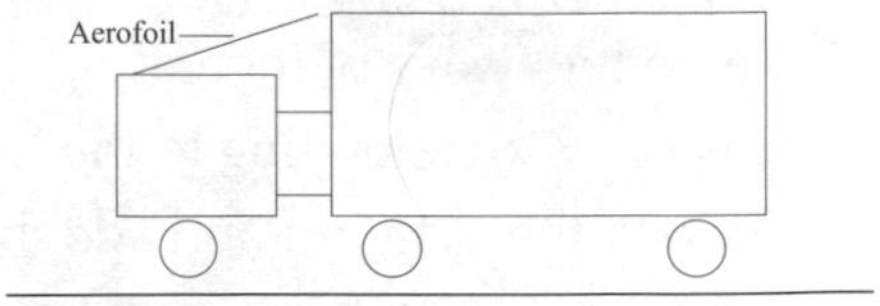

Aerofoil

_it is reducing the friction and drag and it has
to be strecam lined_

[3 marks]

    c)   The driver receives extra pay if she can use less petrol on her journey.
         Explain why she decides to go no faster than 65 km/h, even on the motorway.

_If you go in a higher speed the is more friction
so the use of petol is high_

[2 marks]

2        A satellite has a mass of 85kg on the Earth's surface, where the gravitational field
         strength, g = 10N/kg.

    a)   What is the weight of the satellite when it is on the Earth's surface?

$W = M \times g = 85 \times 10 = 850n$

[1 mark]

    b)   This satellite is put into orbit around the Earth at a place where the gravitational field
         strength, g = 6N/kg.  Find:

         i) The mass of the satellite in orbit and       ii) the weight of the satellite in orbit

_the mass it always be same 85 kg          W = 85.86 = 510
$M = \frac{W}{g}$                                         /2 × 6_

[2 marks]

    c)   When the satellite was designed, explain why little money was spent in making it
         aerodynamic.

[2 marks]

3   a)   Jenny and her father are sitting on a symmetrical seesaw as shown.
         The seesaw is moving anticlockwise.

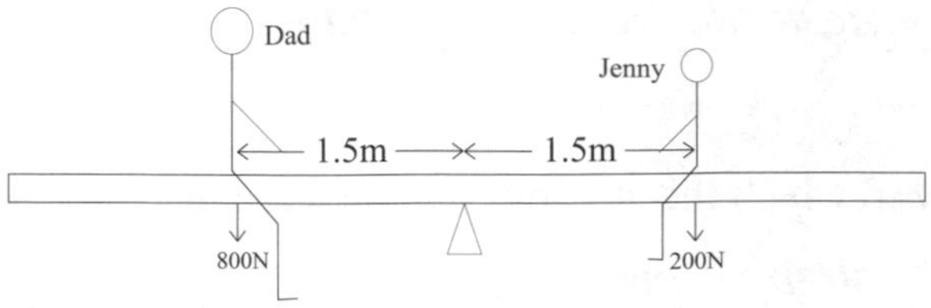

         Dad
                              Jenny
         ←— 1.5m —→|←— 1.5m —→

         800N            △            200N

         Name two things that they can do to balance the seesaw.

_decrease the distance from the Pivot and not moving
Jenny moves further_

[2 marks]

b)

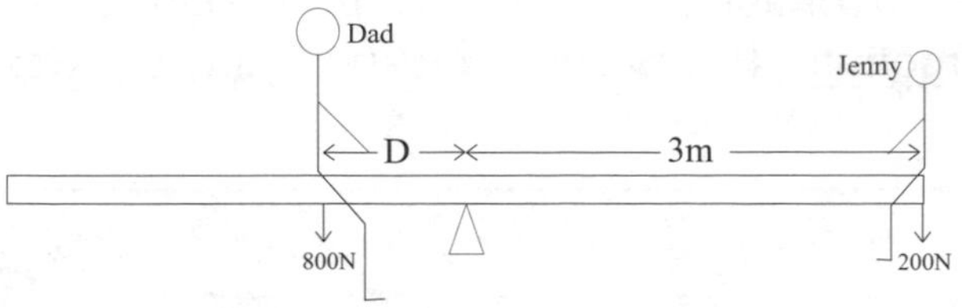

Work out how far away from the pivot Jenny's father must sit in order to balance the seesaw. Explain the principle that you use to find the answer.

Moment is Force x Depth =

W1.d1 = W2.d2

800 d? = 200 x 3.m = 600/800 = 0.75 m

*[4 marks]*

c)

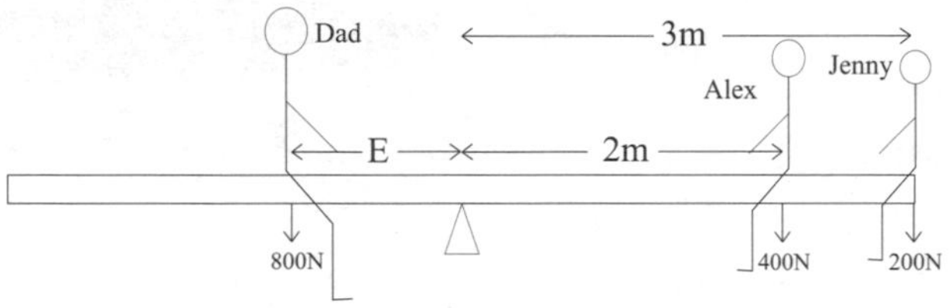

Jenny's brother, Alex, who weighs 400N, climbs onto the seesaw and sits 2m from the pivot and 1m in front of Jenny as shown. Where should their father sit now to restore the seesaw to equilibrium?

400 x 2 = 800          800 + 600 = 1400

200 x 3 = 600

E = -clockwise - anticlockwise

E = 800          = 1400/600 = 1.25

*[5 marks]*

d) If the gravitational field strength at the Earth's surface, g = 10N/kg, calculate Jenny's mass.

200 ÷ 10 = 20 kg

*[1 mark]*

e) If Jenny, Alex and their father could use the same seesaw on the moon, where g = 1.6N/kg, would they need to sit in different places in order to balance the seesaw?

No because the mass stay the same every where in the universe

*[2 marks]*

# Laws of Motion

Around about the time of the Great Plague in the 1660s, Isaac Newton worked out The Three Laws of Motion. They're really important and unless you understand them you'll never fully understand forces and motion. This page covers the first two laws:

## First Law — *Balanced Forces* mean *No Change* in *Velocity*

*So long as the forces on an object are all UNDERLINE BALANCED, then it'll just STAY STILL, or else if it's already moving it'll just carry on at the SAME VELOCITY — so long as the forces are all BALANCED.*

1. When a train or car or bus or anything else is moving at a constant velocity then the forces on it must all be balanced.

2. Never let yourself think that things need a constant overall force to keep them moving.

3. To keep going at a steady speed, there must be zero resultant force — you must remember that.

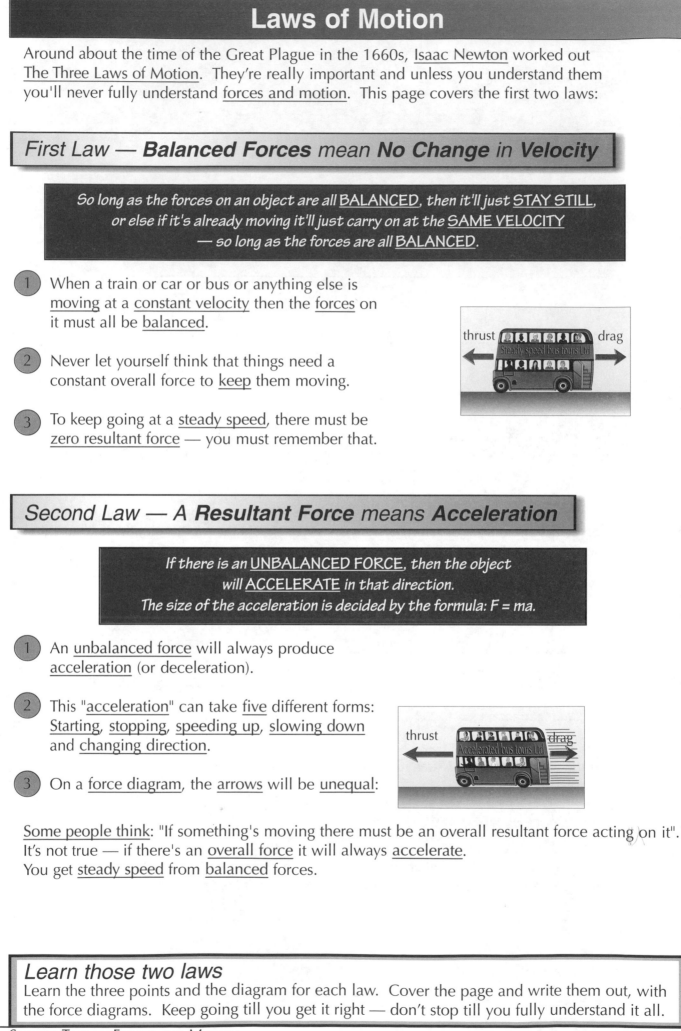

## Second Law — *A Resultant Force* means *Acceleration*

*If there is an UNBALANCED FORCE, then the object will ACCELERATE in that direction. The size of the acceleration is decided by the formula: F = ma.*

1. An unbalanced force will always produce acceleration (or deceleration).

2. This "acceleration" can take five different forms: Starting, stopping, speeding up, slowing down and changing direction.

3. On a force diagram, the arrows will be unequal:

Some people think: "If something's moving there must be an overall resultant force acting on it". It's not true — if there's an overall force it will always accelerate. You get steady speed from balanced forces.

## *Learn those two laws*
Learn the three points and the diagram for each law. Cover the page and write them out, with the force diagrams. Keep going till you get it right — don't stop till you fully understand it all.

# Laws of Motion

## The Overall **Unbalanced Force** is often called The **Resultant Force**

Any resultant force will produce acceleration and this is the formula for it:

$$F = ma \quad \text{or} \quad a = F/m$$

m = mass,  a = acceleration,  F is always the resultant force (see pages 2-3 on using formulas).

## Three Points to Note:

1. The bigger the force, the greater the acceleration or deceleration.

2. The bigger the mass, the smaller the acceleration.

3. To get a big mass to accelerate as much as a small mass, it needs a bigger force. Just think about pushing heavy trolleys.

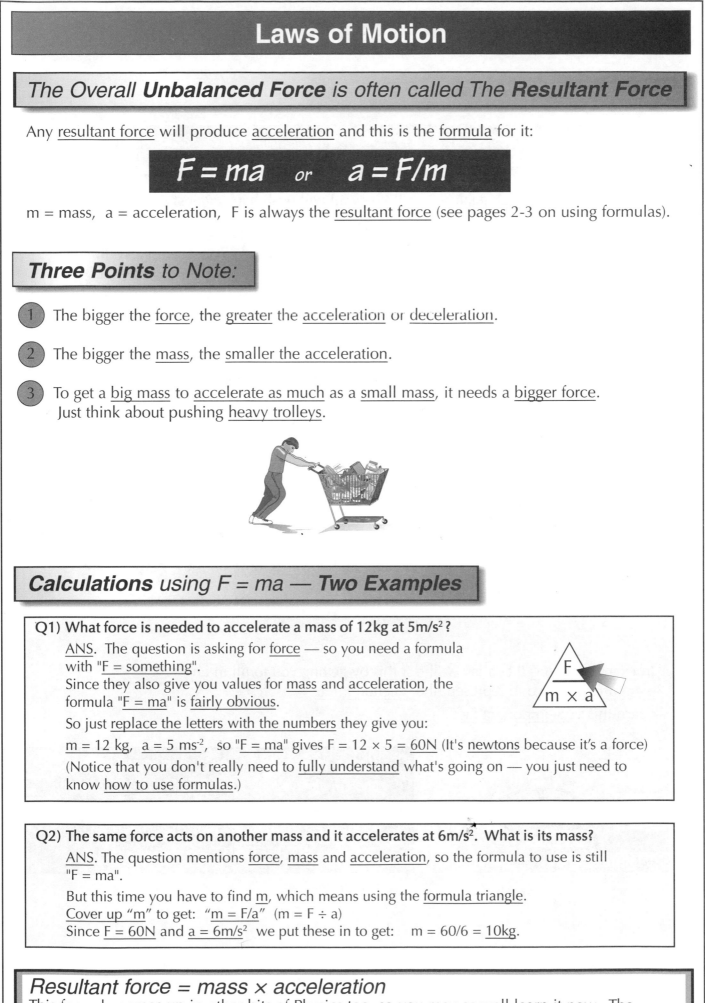

## Calculations using F = ma — Two Examples

**Q1) What force is needed to accelerate a mass of 12kg at 5m/s²?**

ANS. The question is asking for force — so you need a formula with "F = something".

Since they also give you values for mass and acceleration, the formula "F = ma" is fairly obvious.

So just replace the letters with the numbers they give you:

m = 12 kg,  a = 5 ms⁻², so "F = ma" gives F = 12 × 5 = 60N (It's newtons because it's a force)

(Notice that you don't really need to fully understand what's going on — you just need to know how to use formulas.)

**Q2) The same force acts on another mass and it accelerates at 6m/s². What is its mass?**

ANS. The question mentions force, mass and acceleration, so the formula to use is still "F = ma".

But this time you have to find m, which means using the formula triangle.
Cover up "m" to get:  "m = F/a"  (m = F ÷ a)
Since F = 60N and a = 6m/s² we put these in to get:    m = 60/6 = 10kg.

## Resultant force = mass × acceleration
This formula comes up in other bits of Physics too, so you may as well learn it now. The calculations are easy enough with a formula triangle — just make sure you know the formula.

# Laws of Motion

## The Third Law — *Reaction Forces*

> *If object A* <u>EXERTS A FORCE</u> *on object B then object B exerts* <u>THE EXACT OPPOSITE FORCE</u> *on object A*

1) That means if you <u>push against a wall</u>, the wall will <u>push back</u> against you, <u>just as hard</u>.

2) And as soon as you <u>stop</u> pushing, <u>so does the wall</u>.

3) It makes sense — there must be an <u>opposing force</u> when you lean against a wall, otherwise you (and the wall) would <u>fall over</u>.

4) If you <u>pull a cart</u>, whatever force <u>you exert</u> on the rope, the rope exerts the <u>exact opposite</u> pull on <u>you</u>.

5) If you put a book on a table, the <u>weight</u> of the book acts <u>downwards</u> on the table — and the table exerts an <u>equal and opposite</u> force <u>upwards</u> on the book.

6) If you support a book on your <u>hand</u>, the book exerts its <u>weight</u> downwards on you, and you provide an equal <u>upwards</u> force on the book and it all stays <u>in balance</u>.

In <u>Exam questions</u> they may well <u>test this</u> by getting you to fill in an <u>extra arrow</u> or two to represent the <u>reaction force</u>.

Learn this <u>very important fact</u>:

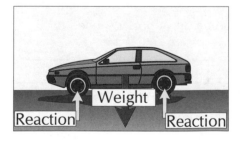

> Whenever an object is on a horizontal <u>SURFACE</u>, there'll always be a <u>REACTION FORCE</u> pushing <u>UPWARDS</u>, supporting the object. The total <u>REACTION FORCE</u> will be <u>EQUAL AND OPPOSITE</u> to the weight.

## *The 3rd Law — equal and opposite reaction force to stop you moving*

The wording of this law is a bit complicated, but the concept's not that hard so make sure you understand it. In this topic there really is <u>no substitute</u> for fully understanding <u>The Three Laws</u>.

# Warm-Up and Worked Exam Question

This might feel like very theoretical stuff, but understanding forces is an essential part of engineering. Before you can start designing bridges you need to understand these basic facts.

## Warm-up Questions

1   In each of the following situations, say whether the forces are balanced or unbalanced:

   a) A car travelling in a straight line at a constant speed of 15 m/s

   b) The Moon orbiting the Earth

   c) A ballet dancer standing on the points of her toes

   d) A coin dropping from a high building

2   a) A car has a mass of 900kg.  If it accelerates at 3 m/s$^2$, what is the resultant force on it?

   b) A force of 18N acts on a mass of 3kg.  What is the size of the acceleration?

3   A tractor is towing a trailer.  If the tractor exerts a force of 100N on the trailer, how much force will the trailer exert on the tractor?

## Worked Exam Question

Take a look at this worked exam question.  It's not too hard but it should give you a good idea of what to write.  You'll usually get at least one balancing forces question in the exam.

1   a)   A car has a mass of 800kg.  The thrust from the engine is 2500N and the drag force is 580N.  Calculate the acceleration of the car.

   *First find the resultant force = 2500N – 580N = 1920N*

   *Then find the acceleration using a = F/m = 1920N/800kg = 2.4m/s$^2$*

   *[3 marks]*

   b)   Two extra passengers with total mass 160kg get into the car.  Assuming that the forces on the car remain constant, find the new acceleration of the car.

   *The new total mass of the car and passengers = 800 + 160 = 960kg.*

   *Therefore, the new acceleration, a = F/m = 1920N/960kg = 2.0m/s$^2$*

   *[2 marks]*

   c)   The car has a top speed of 45 m/s.  If the thrust from the engine has remained at 2500N, how big is the drag force?  Explain your answer.

   *The drag force = 2500N. At top speed, there is no acceleration &*

   *the resultant force is zero. Therefore, force acting forwards from*

   *the engine is exactly balanced by the drag force acting backwards.*

   *[3 marks]*

# Exam Questions

1    The Saturn V rocket was used to launch the Apollo spacecraft. On the launch pad, the total mass of the rocket and spacecraft was 2 700 000 kg ($2.7 \times 10^6$kg). The rocket had five engines, each producing a thrust of 7 000 000N ($7.0 \times 10^6$N).

a)    If the gravitational field strength at the Earth's surface, g, = 10N/kg, calculate the weight of the combined Saturn V rocket and Apollo spacecraft.

........................................................................................................................................................

*[1 mark]*

b)    At take-off, when each engine was producing its maximum thrust, calculate the resultant force on the rocket.

........................................................................................................................................................

*[2 marks]*

c)    What would the acceleration of the rocket have been as it left the Earth's surface?

........................................................................................................................................................

*[2 marks]*

d)    Assuming that the thrust of the engines remained constant, describe and explain what would have happened to the acceleration in the first few minutes of the flight.

........................................................................................................................................................

*[2 marks]*

e)    Once out of the Earth's atmosphere, the rocket continued to accelerate.
By considering the forces involved, explain how this was possible.

........................................................................................................................................................

*[2 marks]*

f)    Later in the flight, well away from any gravitational fields, the engines were shut down. What would have happened to the motion of the rocket?

........................................................................................................................................................

*[3 marks]*

2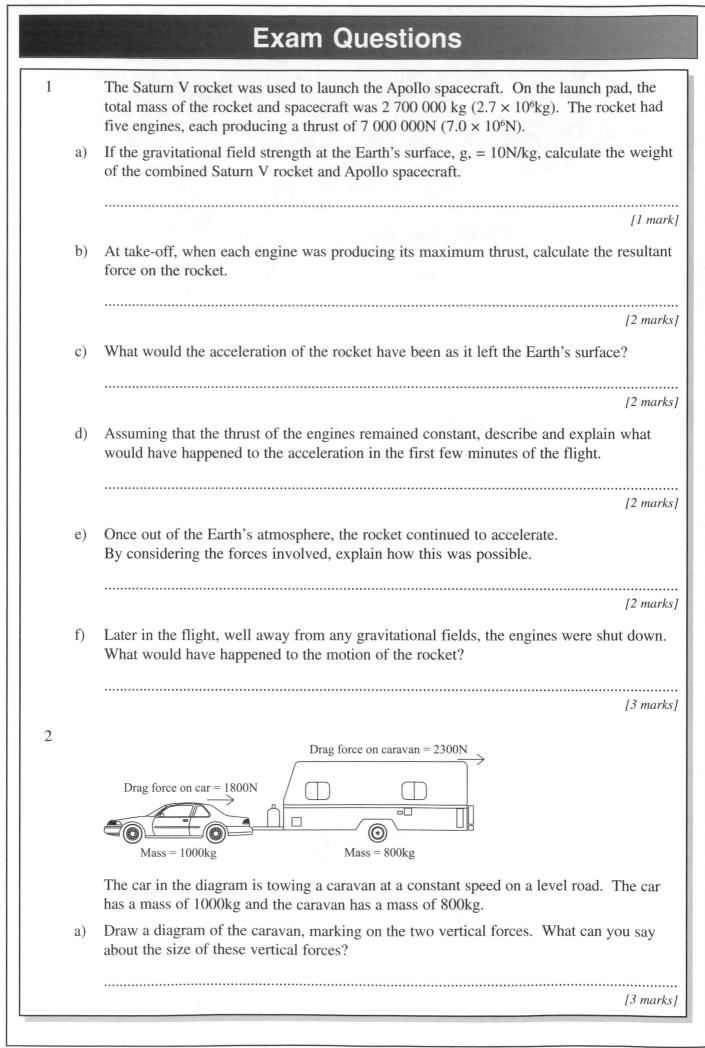

Drag force on caravan = 2300N

Drag force on car = 1800N

Mass = 1000kg                    Mass = 800kg

The car in the diagram is towing a caravan at a constant speed on a level road. The car has a mass of 1000kg and the caravan has a mass of 800kg.

a)    Draw a diagram of the caravan, marking on the two vertical forces. What can you say about the size of these vertical forces?

........................................................................................................................................................

*[3 marks]*

# Exam Questions

b)  Calculate the force exerted by the car on the caravan.

......................................................................................................................................

......................................................................................................................................
*[2 marks]*

c)  What will be the thrust exerted by the car's engine?

......................................................................................................................................

......................................................................................................................................
*[2 marks]*

d)  The driver wishes to accelerate the car and caravan at 2 m/s². Assuming that the drag forces stay constant, what will be the new thrust force from the engine?

......................................................................................................................................
*[2 marks]*

3  a)  The diagram shows the Earth orbiting the Sun.

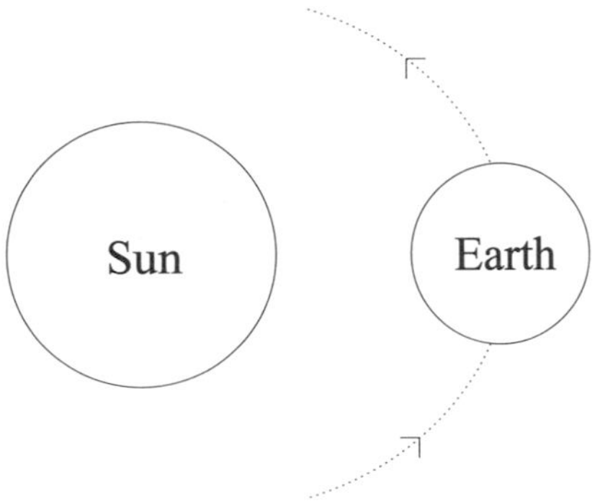

i) Draw arrows on the diagram to show the force acting on each body.

......................................................................................................................................
*[2 marks]*

ii) What can you say about the size of these two forces?

......................................................................................................................................
*[1 mark]*

b)  i) Explain why a football moves when you kick it.

......................................................................................................................................
*[2 marks]*

ii) Explain why you can feel the football when you kick it.

......................................................................................................................................
*[2 marks]*

# Speed, Velocity and Acceleration

## Speed and Velocity both mean how fast you're going

Speed and velocity are both measured in m/s (or km/h or mph). They both simply say how fast you're going, but there's a subtle difference between them which you need to know:

> SPEED is just how fast you're going (e.g. 30mph or 20m/s) with no regard to the direction.
> VELOCITY however must also have the DIRECTION specified, e.g. 30mph north or 20m/s, 060°

They expect you to remember that distinction, so learn it.

### Speed, Distance and Time — the Formula:

$$\text{Speed} = \frac{\text{Distance}}{\text{Time}}$$

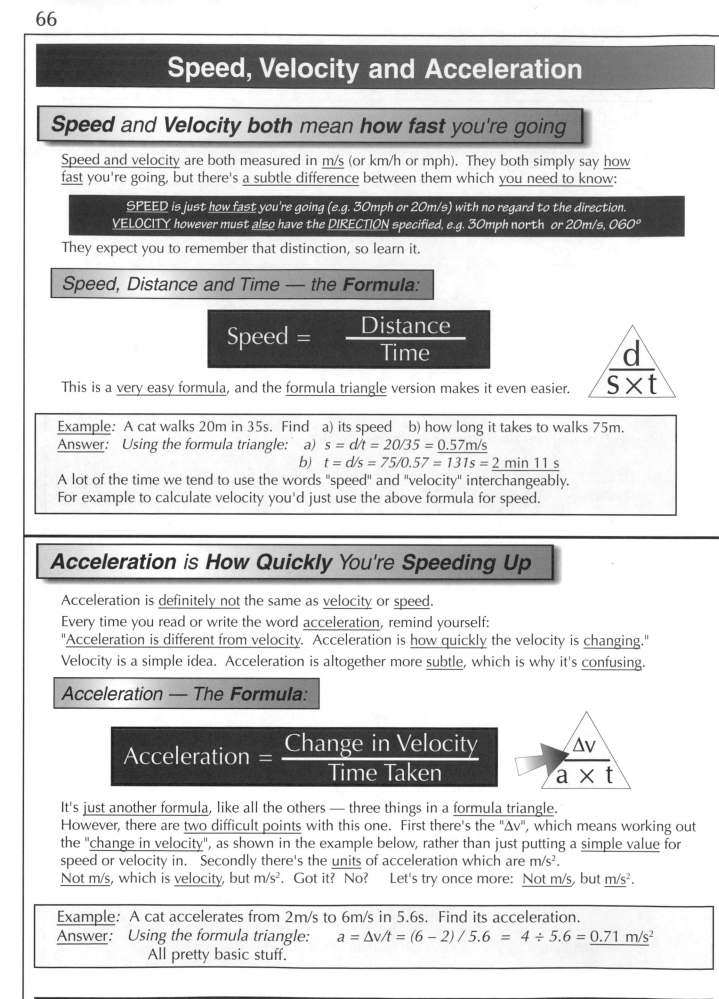

This is a very easy formula, and the formula triangle version makes it even easier.

Example: A cat walks 20m in 35s. Find   a) its speed   b) how long it takes to walks 75m.
Answer:   Using the formula triangle:   a)  s = d/t = 20/35 = 0.57m/s
                               b)  t = d/s = 75/0.57 = 131s = 2 min 11 s
A lot of the time we tend to use the words "speed" and "velocity" interchangeably.
For example to calculate velocity you'd just use the above formula for speed.

## Acceleration is How Quickly You're Speeding Up

Acceleration is definitely not the same as velocity or speed.

Every time you read or write the word acceleration, remind yourself:
"Acceleration is different from velocity.  Acceleration is how quickly the velocity is changing."
Velocity is a simple idea.  Acceleration is altogether more subtle, which is why it's confusing.

### Acceleration — The Formula:

$$\text{Acceleration} = \frac{\text{Change in Velocity}}{\text{Time Taken}}$$

It's just another formula, like all the others — three things in a formula triangle.
However, there are two difficult points with this one. First there's the "Δv", which means working out the "change in velocity", as shown in the example below, rather than just putting a simple value for speed or velocity in.   Secondly there's the units of acceleration which are $m/s^2$.
Not m/s, which is velocity, but $m/s^2$. Got it? No?    Let's try once more:  Not m/s, but $m/s^2$.

Example: A cat accelerates from 2m/s to 6m/s in 5.6s.  Find its acceleration.
Answer:   Using the formula triangle:    a = Δv/t = (6 − 2) / 5.6  =  4 ÷ 5.6 = 0.71 m/s²
                All pretty basic stuff.

## Velocity and Acceleration — learn the difference

A lot of people don't realise that velocity and acceleration are totally different things.  Make sure you do.  Learn the definitions and the formulae, cover the page and write it all down.

# Distance-Time and Velocity-Time Graphs

Make sure you learn all these details — and make sure you can <u>distinguish</u> between the two.

## Distance-Time Graphs

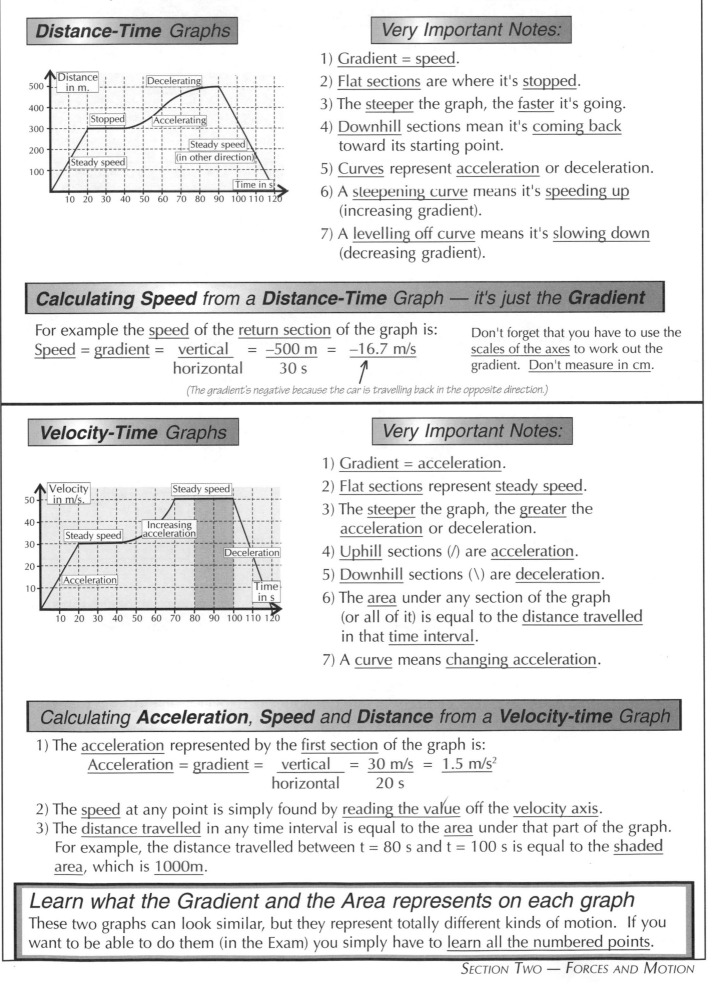

### Very Important Notes:

1) <u>Gradient = speed</u>.
2) <u>Flat sections</u> are where it's <u>stopped</u>.
3) The <u>steeper</u> the graph, the <u>faster</u> it's going.
4) <u>Downhill</u> sections mean it's <u>coming back</u> toward its starting point.
5) <u>Curves</u> represent <u>acceleration</u> or deceleration.
6) A <u>steepening curve</u> means it's <u>speeding up</u> (increasing gradient).
7) A <u>levelling off curve</u> means it's <u>slowing down</u> (decreasing gradient).

## Calculating Speed from a Distance-Time Graph — it's just the Gradient

For example the <u>speed</u> of the <u>return section</u> of the graph is:

Speed = gradient = $\frac{vertical}{horizontal}$ = $\frac{-500\ m}{30\ s}$ = <u>−16.7 m/s</u> ↑

Don't forget that you have to use the scales of the axes to work out the gradient. <u>Don't measure in cm</u>.

*(The gradient's negative because the car is travelling back in the opposite direction.)*

## Velocity-Time Graphs

### Very Important Notes:

1) <u>Gradient = acceleration</u>.
2) <u>Flat sections</u> represent <u>steady speed</u>.
3) The <u>steeper</u> the graph, the <u>greater</u> the <u>acceleration</u> or deceleration.
4) <u>Uphill</u> sections (/) are <u>acceleration</u>.
5) <u>Downhill</u> sections (\) are <u>deceleration</u>.
6) The <u>area</u> under any section of the graph (or all of it) is equal to the <u>distance travelled</u> in that <u>time interval</u>.
7) A <u>curve</u> means <u>changing acceleration</u>.

## Calculating Acceleration, Speed and Distance from a Velocity-time Graph

1) The <u>acceleration</u> represented by the <u>first section</u> of the graph is:

   Acceleration = gradient = $\frac{vertical}{horizontal}$ = $\frac{30\ m/s}{20\ s}$ = <u>1.5 m/s²</u>

2) The <u>speed</u> at any point is simply found by <u>reading the value</u> off the <u>velocity axis</u>.
3) The <u>distance travelled</u> in any time interval is equal to the <u>area</u> under that part of the graph. For example, the distance travelled between t = 80 s and t = 100 s is equal to the <u>shaded area</u>, which is <u>1000m</u>.

## Learn what the Gradient and the Area represents on each graph

These two graphs can look similar, but they represent totally different kinds of motion. If you want to be able to do them (in the Exam) you simply have to <u>learn all the numbered points</u>.

# Resultant Force and Terminal Velocity

## *Resultant Force* is Really Important — Especially for "*F = ma*"

The notion of <u>resultant force</u> is a really important one. In most <u>real situations</u> there are <u>at least two</u> <u>forces</u> acting on an object along any direction. The <u>overall effect</u> of these forces will decide the <u>motion</u> of the object — whether it will <u>accelerate</u>, <u>decelerate</u> or stay at a <u>steady speed</u>. The "overall effect" is found by just <u>adding or subtracting</u> the forces which point along the <u>same</u> direction. The overall force you get is called the <u>resultant force</u>. And when you use the <u>formula "F = ma"</u>, F must always be the <u>resultant force</u>.

<u>Example</u>:  A car of mass of 1750kg has an engine which provides a driving force of 5200N. At 70mph the drag force acting on the car is 5150N. Find its acceleration  a) when first setting off from rest  b) at 70mph.

<u>Answer</u>:  1) First draw a force diagram for both cases (no need to show the vertical forces):

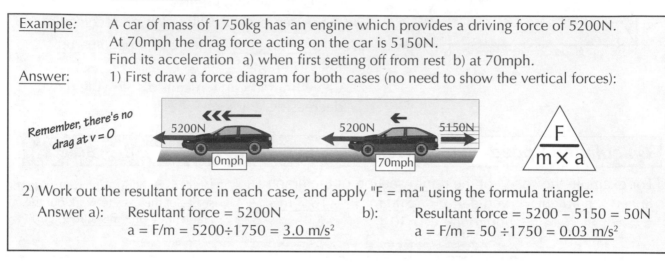

*Remember, there's no drag at v = 0*

2) Work out the resultant force in each case, and apply "F = ma" using the formula triangle:

Answer a):  Resultant force = 5200N  
a = F/m = 5200÷1750 = <u>3.0 m/s²</u>

b):  Resultant force = 5200 − 5150 = 50N  
a = F/m = 50 ÷1750 = <u>0.03 m/s²</u>

## *Cars* and *Free-Fallers* all Reach a *Terminal Velocity*

When cars and free-falling objects <u>first set off</u> they have <u>much more</u> force <u>accelerating</u> them than <u>resistance</u> slowing them down. As the <u>speed</u> increases the resistance <u>builds up</u>.

This gradually <u>reduces</u> the <u>acceleration</u> until eventually the <u>resistance force</u> is <u>equal</u> to the <u>accelerating force</u> and then it won't be able to accelerate any more. It will have reached its maximum speed, or <u>terminal velocity</u>.

## The *Terminal Velocity* of *Falling Objects* depends on their *Shape and Area*

The <u>accelerating force</u> acting on <u>all falling objects</u> is <u>gravity</u>, and it would make them all fall at the <u>same rate</u> if it wasn't for <u>air resistance</u>.

To prove this: on the Moon, where there's <u>no air</u>, a stone and a feather dropped simultaneously will <u>hit the ground together</u>. However, on Earth, <u>air resistance</u> causes things to fall at <u>different speeds</u>, and the <u>terminal velocity</u> of any object is determined by its <u>drag</u> in <u>comparison</u> to its <u>weight</u>. The drag depends on its <u>shape and area</u>.

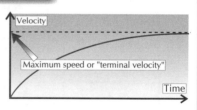

The most important example is the <u>human skydiver</u>. Without his parachute open he has quite a <u>small area</u> and a force of "<u>W=mg</u>" pulling him down. He reaches a <u>terminal velocity</u> of about <u>120mph</u>. But with the parachute <u>open</u>, there's much more <u>air resistance</u> (at any given speed) and still only the same force "<u>W=mg</u>" pulling him down.

This means his <u>terminal velocity</u> comes right down to about <u>15mph</u>, which is a <u>safe speed</u> to hit the ground at.

*In <u>both</u> cases <u>R = W</u>.*

*The difference is the <u>speed</u> at which that happens.*

## *Terminal velocity is all about air resistance*

The best way of checking how much you know is to <u>write a quick mini-essay</u> for each of the three sections. Then <u>check back</u> and see what you <u>missed</u>. Then try again. <u>And keep trying</u>.

# Stopping Distances for Cars

This often comes up in Exam questions, so make sure you learn it properly.

## The **Many Factors** Which Affect Your Total **Stopping Distance**

The distance it takes to stop a car is divided into the thinking distance and the braking distance.

### 1) Thinking Distance

"The distance the car travels in the split-second between a hazard appearing and the driver applying the brakes." It's affected by three main factors:

a) **How fast you're going** — Whatever your reaction time, the faster you're going, the further you'll go.

b) **How alert you are** — This is affected by tiredness, drugs, alcohol, old-age, etc.

c) **How bad the visibility is** — lashing rain and oncoming lights, etc. make hazards harder to spot.

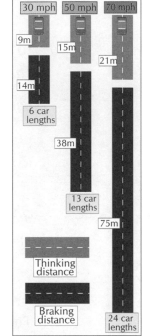

*The figures below for typical stopping distances are from the Highway Code. Note how far it takes to stop when you're going at 70mph.*

### 2) Braking Distance

"The distance the car travels during its deceleration whilst the brakes are being applied." It's affected by four main factors:

a) **How fast you're going** — obviously. The faster you're going the further it takes to stop (see right).

b) **How heavily loaded the vehicle is** — with the same brakes, a heavily-laden vehicle takes longer to stop. A car won't stop as quickly when it's full of people and luggage and towing a caravan.

c) **How good your brakes are** — all brakes must be checked and maintained regularly. Worn or faulty brakes will let you down catastrophically just when you need them the most, i.e. in an emergency.

d) **How good the grip is** — this depends on three things:
1) road surface, 2) weather conditions, 3) tyres.

Leaves and diesel spills and mud on the road are serious hazards because they're unexpected. Wet or icy roads are always much more slippy than dry roads, but often you only discover this when you try to brake hard. Tyres should have a minimum tread depth of 1.6mm. This is essential for getting rid of the water in wet conditions. Without tread, a tyre will simply ride on a layer of water and skid very easily. This is called "aquaplaning" and isn't nearly as cool as it sounds.

## Stopping Distances **Increase Alarmingly** with **Extra Speed**

### — Mainly Because of the **v²** bit in **KE=½mv²**

To stop a car, the kinetic energy, ½mv², has to be converted to heat energy at the brakes and tyres:

$$\text{Kinetic Energy Transferred} = \text{Work Done by Brakes}$$
$$\tfrac{1}{2}mv^2 = F \times d$$

*(See pages 157 and 159 on Energy and Work)*

v = speed of car    F = maximum braking force    d = braking distance

Learn this: if you double the speed, you double the value of $v$, but the $v^2$ means that the KE is then increased by a factor of four. However, "F" is always the maximum possible braking force which can't be increased, so d must increase by a factor of four to make the equation balance. E.g. if you go twice as fast, the braking distance "d" must increase by a factor of four to dissipate the extra KE.

## Thinking Distance + Braking Distance = Stopping Distance

They mention this specifically in the syllabus and are likely to test you on it since it involves safety. Learn all the details and write yourself a mini-essay to see how much you really know.

# Warm-Up and Worked Exam Question

## Warm-up Questions

1    A train travels 280m in 8s.  Calculate its speed.
2    A motorbike starts from rest at traffic lights and reaches 15m/s after 3s.  Find its acceleration.
3    a) A car is travelling at 16m/s.  How far does it travel in 6s?
     b) If the same car has a mass of 900kg, find its kinetic energy.
4    Name two factors that affect:    a) thinking distance    b) braking distance

## Worked Exam Question

1    As part of an aerobatic display, a skydiver jumps out of an aeroplane and falls towards the ground.  She does not open her parachute until she reaches her terminal velocity.

a)    (i)  Explain why she reaches a terminal velocity.

*The force downwards, due to her weight, is equal to the force upwards, caused by air resistance or drag.*

*[1 mark]*

(ii)  Given that she cannot change her weight during the fall, explain how she could reach a higher terminal velocity.

*She could make her shape more streamlined in order to reduce her air resistance.*

(Note: Since she can't increase her weight, any comments about weight would not receive any marks.)

*[1 mark]*

b)    The graph shows the motion of the skydiver.

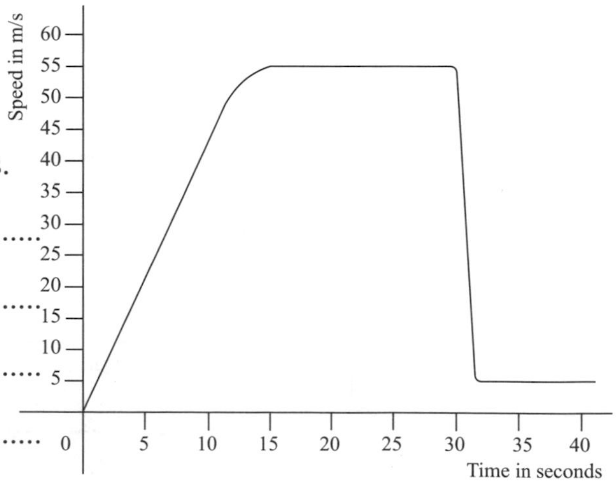

(i)  Calculate the distance travelled   between 15s and 30s.

*= the area under the graph between 15s and 30s (1 mark)*

*= (30s - 15s) × 55m/s*

*= 15s × 55m/s (1 mark)*

*= 825m (1 mark)*

*[3 marks]*

(ii)  Describe and explain what is happening between 30s and 35s.

*She opens her parachute, which unbalances the forces so she slows right down then reaches a new terminal velocity when the forces balance again.*

*[2 marks]*

# Exam Questions

1  a)  This graph shows the motion
of a car from the top of a hill
to the bottom of the hill.
After 12s the car reaches
the bottom of the hill.

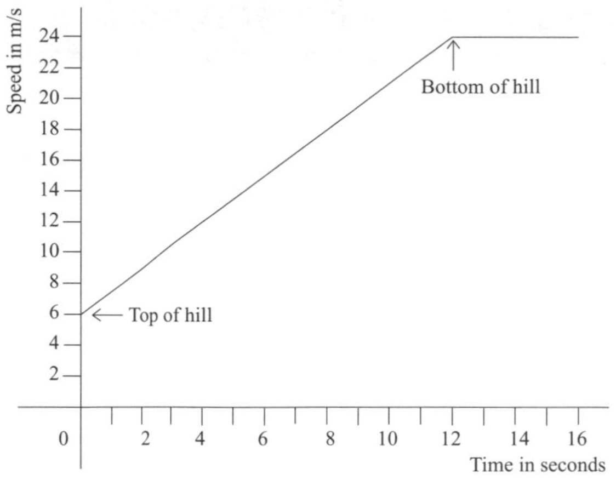

(i)  Calculate the acceleration
of the car.

..............................................................................................................................

*[3 marks]*

(ii)  If the car has a mass of 1200kg, what is the resultant force acting on the car?

..............................................................................................................................

*[2 marks]*

(iii)  Use the graph to find the distance travelled by the car down the hill.

..............................................................................................................................

*[3 marks]*

b)  Once the car reaches the bottom of the hill, it travels on a level road at a constant
velocity of 24m/s.

(i)  What is the resultant force on the car now?  Explain your answer.

..............................................................................................................................

*[2 marks]*

(ii)  Calculate the kinetic energy of the car at the bottom of the hill.

..............................................................................................................................

*[3 marks]*

(iii)  By considering the work done in stopping the car, find the average braking force
exerted on the car to bring it to rest in a distance of 50m.

..............................................................................................................................

*[3 marks]*

(iv)  The actual braking force will be less than your answer to part (iii).  Explain why.

..............................................................................................................................

*[2 marks]*

(v)  If three extra people sat in the car, how would its braking distance at this speed be
affected, assuming that the same braking force was used?  Explain your answer.

..............................................................................................................................

*[2 marks]*

# Hooke's Law and Extension

## *Hooke's Law* — *Extension* is Proportional to *Load*

Hooke's Law says:

> If you <u>STRETCH</u> something with a <u>STEADILY INCREASING</u>
> <u>FORCE</u>, then the <u>LENGTH</u> will <u>INCREASE STEADILY</u> too.

The important thing to measure in a Hooke's Law experiment is not so much the total length as the <u>extension</u>.

> <u>EXTENSION</u> is the <u>INCREASE IN LENGTH</u>
> compared to the original length
> with <u>no force applied</u>.

For most materials, you'll find that the <u>extension is directly proportional to the load, which</u> just means if you <u>double</u> the load, the <u>extension is doubled too</u>.

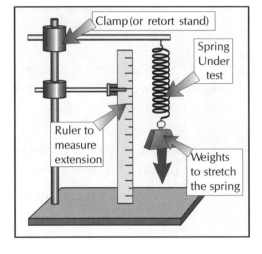

Clamp (or retort stand)

Spring Under test

Ruler to measure extension

Weights to stretch the spring

You should LEARN that this always gives a <u>straight line</u> graph through the <u>origin</u>, as shown here. This is the graph you get for the <u>two important cases</u>: a <u>metal wire</u> and a <u>spring</u>.

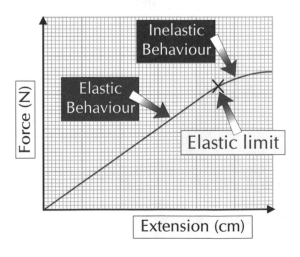

Inelastic Behaviour

Elastic Behaviour

Elastic limit

Force (N)

Extension (cm)

Notice that for both the <u>wire</u> and the <u>spring</u>, there's an <u>elastic limit</u>. For extensions <u>less</u> than this, the wire or spring <u>returns to its original shape</u>, but if stretched <u>beyond</u> the elastic limit, it behaves <u>inelastically</u>, which means it <u>doesn't</u> follow Hooke's Law and that it also <u>won't return</u> to its original shape.

## *Hooke's Law — it can stretch you to the limit...*

Hooke's Law is pretty standard stuff, so make sure you know all the little details.
You've got to be able to draw the graph, label it right and <u>explain</u> the ideas behind the straight bit and the curved bit. Find out what you know: <u>cover, write it out, check</u>, etc.

# Boyle's Law and Pressure

<u>Boyle's Law</u> sounds a lot more confusing than it actually is.  This is the proper definition:

> When the <u>PRESSURE IS INCREASED</u> on a *fixed mass of gas* kept at <u>constant temperature</u>, the <u>VOLUME WILL DECREASE</u>.  *The changes in pressure and volume are in* <u>INVERSE PROPORTION</u>.

A simpler version would be:

> *If you squash a gas into a smaller space, the pressure goes up in proportion to how much you squash it.  E.g. if you squash it to half the amount of space, it'll end up at twice the pressure it was before (so long as you don't let it get hotter or colder, or let any escape).*

It can work <u>both ways</u> too.  If you <u>increase the pressure</u>, the <u>volume must decrease</u>.
If you <u>increase the volume</u>, the <u>pressure must decrease</u>.  That's all pretty obvious though isn't it?

## *Gas Syringe* Experiments Are Good For Showing *Boyle's Law*

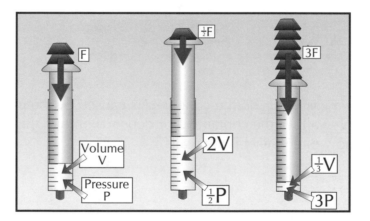

1. A <u>gas syringe</u> makes a pretty good <u>airtight seal</u> and is great for demonstrating <u>Boyle's Law</u>.

2. You put <u>weights on the top</u> to give a <u>definite</u> known force pushing down on the piston.

3. If you <u>double the weight</u>, you also <u>double the force</u> which <u>doubles the pressure</u>.

4. You can then measure the <u>volume change</u> using the <u>scale</u> on the side of the syringe.

---

## *Less space, more collisions, more pressure*

This is another topic that can seem a lot more confusing than it really is.  The basic principle of Boyle's Law is simple enough, and so is the Gas Syringe demo.  It might look complicated but really there's nothing to it.  In the end it's just stuff that needs <u>learning</u>, that's all.

# Boyle's Law and Pressure

## Using the Formula "**PV = Constant**" or "**$P_1V_1 = P_2V_2$**"

It's another formula, but unfortunately you can't put this in a formula triangle.
It's still the same idea though: <u>put in the numbers</u> they give you,
and <u>work out the value</u> for the remaining letter.

Remember that you don't need to <u>fully understand</u> the Physics, you just need a bit of
"common sense" about <u>formulas</u>.  Understanding always helps, but you can still get the right
answer without it.

You've just got to identify the values for each letter — the rest is <u>very routine</u>.

---

<u>Example</u>:  *A gas is compressed from a volume of 300cm³ at a pressure of 2.5 atmospheres*
*down to a volume of 175cm³.  Find the new pressure, in atmospheres.*

<u>Answer</u>:  "$P_1V_1 = P_2V_2$"  gives:    $2.5 \times 300 = P_2 \times 175$,  so $P_2 = (2.5 \times 300) \div 175 = 4.3$ atm.
        NB:  For <u>this formula</u>, always keep the units <u>the same</u> as they give them
        (in this case, it's in <u>atmospheres</u>).

---

## *Kinetic Theory Explains it all Very Nicely*

(1)  The <u>pressure</u> which a gas <u>exerts</u> on the <u>container</u> is caused by the particles moving
about and <u>hitting the walls</u> of the container.   It depends on <u>two things</u>: how <u>fast</u>
they're going and <u>how often</u> they hit the walls.

(2)  <u>How often</u> they hit the walls depends on how <u>squashed up</u> they are.  When the
<u>volume is reduced</u>, the particles become <u>more squashed up</u> and so they hit the walls
<u>more often</u>, and hence the <u>pressure increases</u>.   The <u>speed</u> of the particles <u>won't</u>
<u>change</u> so long as the <u>temperature</u> doesn't change.

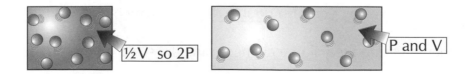

## *Pressure can be measured in pascals (Pa) or atmospheres (atm)*
This is actually quite common sense stuff — it makes more sense than a lot of Physics.
But you still need to learn it of course — cover the book and write out a calculation using
the Boyle's Law equation, then write a mini essay explaining it all using kinetic theory.

# Warm-Up and Worked Exam Question

## Warm-up Questions

1) A spring extends by 5mm when a force of 10N is applied. How far will it extend when a force of (a) 20N and (b) 30 N is applied, assuming that the elastic limit is not exceeded?

2) A syringe contains 80cm³ of gas at a pressure of 2 atmospheres. If the temperature is kept the same, what will be the volume of gas in the syringe at a pressure of (a) 4 atmospheres and (b) 1 atmosphere?

3) A spring is 10cm long when unloaded. A force of 40N stretches it to a length of 15cm. What force would be needed to stretch the spring to (a) 12cm and (b) 13cm?

4) When some gas, trapped in a syringe, expands at a constant temperature, the pressure of the gas falls. Two explanations are given below. Choose the correct one.

   a) The pressure falls because the particles slow down and don't hit the walls as hard.

   b) The pressure falls because the particles have to travel further and don't hit the walls as often.

5) A sealed can contains 120cm³ of gas at a pressure of 100kPa. What will be the volume of gas in the container when the pressure is increased to 250kPa at a constant temperature?

## Worked Exam Question

1   A student measured the volume of gas in a syringe at different pressures. She recorded the results in the table. Each time she changed the pressure, she waited a few minutes before taking the readings of pressure and volume.

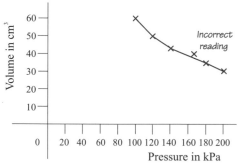

| Pressure/kPa | Volume/cm³ |
|---|---|
| 100 | 60 |
| 120 | 50 |
| 140 | 43 |
| 160 | 40 |
| 180 | 33 |
| 200 | 30 |

a)   Plot a graph of pressure against volume, but do not draw a line through the points yet.
        *Graph: 1 mark each for sensible choice of scale on each axis (2 marks total);*      *[7 marks]*
    *1 mark each for each axis being labelled correctly and having the correct unit (2 marks total);*
    *3 marks for accurate plotting of points (1 mark deducted for each error) (3 marks total).*

b)   (i) One reading of the volume seems to be wrong.
        Label on the graph the incorrect reading.

   *Incorrect reading occurs at a pressure of 160kPa*      *[1 mark]*

   (ii) Draw a line of best fit and use it to find out and estimate what the correct volume should be.

   *Volume should be 38cm³.*
        *One mark for good line of best fit, one for good estimate.*      *[2 marks]*

c)   Explain why the student waited a few minutes between each set of readings.

   *When the pressure of a gas changes, its temperature changes so she*

   *needs to allow it return to room temperature before taking each reading.*

                                                                        *[2 marks]*

# Exam Questions

1   A student used the apparatus in the diagram to investigate the extension of a spring. His results are show in the table.

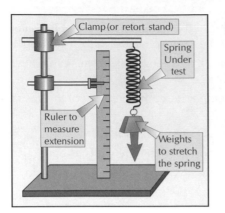

| Force in N | Ruler reading in mm | Extension in mm |
|---|---|---|
| 0 | 120 | |
| 2 | 126 | |
| 4 | 132 | |
| 6 | 138 | |
| 8 | 144 | |

a)   Copy the table, completing it to show the extension in millimetres.

*[1 mark]*

b)   Plot the graph of extension against force.

*[8 marks]*

c)   Use your graph to find the force that produces an extension of 15mm.

.......................................................................................................................................

*[1 mark]*

The student decided to make a weighing machine. He took off the existing spring and replaced it with a different one. He observed that the reading on the ruler was 55mm with a weight of 10N and 90mm with a weight of 80N.

d)   When he put on another weight, the ruler reading was 75mm. How big was this weight?

.......................................................................................................................................

*[4 marks]*

e)   The student made sure that he did not exceed the elastic limit of the spring when he used his weighing machine.

(i)   What is meant by the term "elastic limit"?

.......................................................................................................................................

*[2 marks]*

(ii)   Why is it important not to go above the elastic limit of the spring in a weighing machine?

.......................................................................................................................................

*[2 marks]*

# Exam Questions

2    The diagram shows a common piece of laboratory apparatus which is used to investigate the variation in volume with the pressure of a gas. A fixed mass of air is trapped inside a glass tube.
Pumping the oil higher up the tube can increase the pressure on the air and decrease its volume.

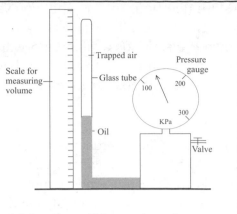

A student started the experiment by pumping the oil as high as possible up the tube. She took readings of the volume and pressure of the air. Then she opened a valve to allow the oil to drop a little way down the tube. After a short delay, she took further readings of the pressure and the volume of the air. She repeated this process several times until the pressure on the air fell to the normal atmospheric pressure.
Her results are given in the table:

| Pressure in kPa | Volume in m³ | 1/Volume in m³ |
|---|---|---|
| 300 | 0.000 17 | 6000 |
| 250 | 0.000 20 | |
| 200 | 0.000 25 | |
| 150 | 0.000 33 | 3000 |
| 100 | 0.000 50 | |

a)    Complete the table by calculating the three missing values of 1/volume.

*[3 marks]*

b)    Plot a graph of pressure on the y-axis against 1/volume on the x-axis.

*[6 marks]*

c)    Explain how this graph shows that the gas obeys Boyle's law.

......................................................................................................................................................

*[3 marks]*

d)    In what other way could the student have used these results to show that this gas obeys Boyle's law?

......................................................................................................................................................

*[2 marks]*

e)    Use the graph to find out what the volume of the gas would be at a pressure of 75 kPa.

......................................................................................................................................................

*[2 marks]*

f)    Why did the student pause between each set of readings?

......................................................................................................................................................

*[2 marks]*

g)    Calculate the pressure of this gas when the volume is:  (i) 0.00080 m³    (ii) 0.00120 m³

......................................................................................................................................................

*[4 marks]*

# Revision Summary for Section Two

More questions to test what you've learned.  There are lots of facts about forces and motion which you definitely need to know.  Some bits are certainly quite tricky to understand, but there's also loads of straightforward stuff which just needs to be learned, ready for instant regurgitation in the Exam.  You have to practise these questions over and over and over again, until you can answer them all really easily.

Burgundy questions are for Edexcel students only.

1) What is gravity?  List three main effects that gravity produces.
2) Explain the difference between mass and weight.  What units are they measured in?
3) What's the formula for weight?  Illustrate it with a worked example of your own.
4) List six different kinds of force.  Sketch diagrams to illustrate them all.
5) Sketch each of the five standard force diagrams, showing the forces and the type of motion.
6) List the three types of friction with a sketch to illustrate each one.
7) Describe how friction is affected by speed.  What 2 effects does friction have on machinery?
8) Is friction at all useful?  Describe five problems we would have if there was no friction.
9) Write down the First Law of Motion.  Illustrate with a diagram.
10) If an object has zero resultant force on it, can it be moving?  Can it be accelerating?
11) Write down the Second Law of Motion.  Illustrate with a diagram.   What's the formula  for it?
12) A force of 30N pushes on a trolley of mass 4kg.  What will be its acceleration?
13) What's the mass of a cat which accelerates at 9.8 m/s$^2$ when acted on by a force of 56N?
14) Write down the Third Law of Motion.  Illustrate it with four diagrams.
15) Explain what *reaction force* is and where it occurs.  Is it important to know about it?
16) What's the difference between speed and velocity?  Give an example of each.
17) Write down the formula for working out speed.   Find the speed of a partly chewed mouse which hobbles 3.2m in 35s.  Find how far he would get in 25 minutes.
18) What's acceleration?  Is it the same thing as speed or velocity?  What are its units?
19) Write down the formula for acceleration.
    What's the acceleration of a ball, kicked from rest to a speed of 14 m/s in 0.4s?
20) Sketch a typical distance-time graph and point out all the important parts of it.
21) Sketch a typical velocity-time graph and point out all the important parts of it.
22) Write down seven important points relating to each of these graphs.
23) Explain how to calculate velocity from a distance-time graph.
24) Explain how to find speed, distance and acceleration from a velocity-time graph.
25) Explain what "resultant force" is.  Illustrate with a diagram.  When do you most need it?
26) What is "terminal velocity"?  Is it the same thing as maximum speed?
27) What are the two main factors affecting the terminal velocity of a falling object?
28) What are the two different parts of the overall stopping distance of a car?
29) List three or four factors which affect each of the two parts of a car's stopping distance.
30) Which formula explains why the stopping distance increases so much?  Explain why it does.
31) What is Hooke's Law?  Sketch the usual apparatus.  Explain what you must measure.
32) Sketch the Hooke's Law graph for a spring and explain its shape.
    Explain "elastic" and "inelastic".
33) What is Boyle's Law?  Sketch an experiment which demonstrates it.  What's the formula?
34) A fixed amount of gas at 5,000 Pa is compressed down to 60cm$^3$, and in the process its pressure rises to 260,000 Pa.  What was the volume before it got compressed?
35) What does Kinetic Theory say affects the pressure on the walls of a container of gas?
36) What happens to particles in a gas if you reduce the volume?
    What happens to the pressure?

# Waves — Basic Principles

Waves are <u>different</u> from anything else.  They have various features which <u>only waves have</u>:

## *Amplitude*, *Wavelength*, *Frequency* and *Period*

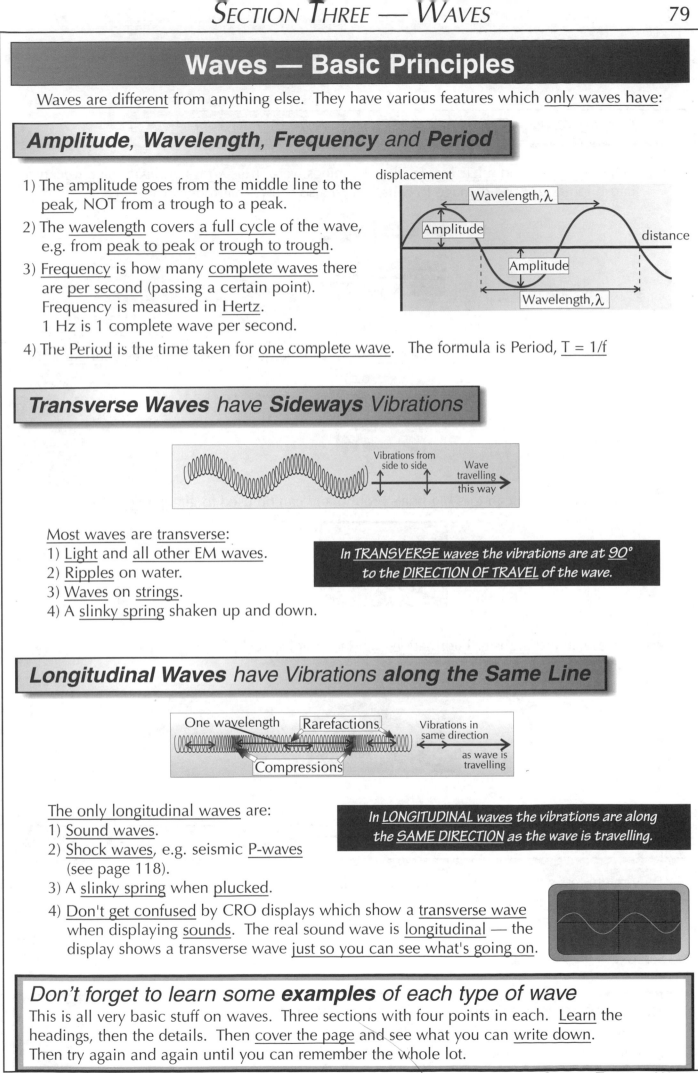

1) The <u>amplitude</u> goes from the <u>middle line</u> to the <u>peak</u>, NOT from a trough to a peak.

2) The <u>wavelength</u> covers <u>a full cycle</u> of the wave, e.g. from <u>peak to peak</u> or <u>trough to trough</u>.

3) <u>Frequency</u> is how many <u>complete waves</u> there are <u>per second</u> (passing a certain point). Frequency is measured in <u>Hertz</u>. 1 Hz is 1 complete wave per second.

4) The <u>Period</u> is the time taken for <u>one complete wave</u>.   The formula is Period, <u>T = 1/f</u>

## *Transverse Waves* have *Sideways* Vibrations

Most waves are <u>transverse</u>:
1) <u>Light</u> and <u>all other EM waves</u>.
2) <u>Ripples</u> on water.
3) <u>Waves</u> on <u>strings</u>.
4) A <u>slinky spring</u> shaken up and down.

> *In <u>TRANSVERSE waves</u> the vibrations are at <u>90°</u> to the <u>DIRECTION OF TRAVEL</u> of the wave.*

## *Longitudinal Waves* have Vibrations *along the Same Line*

The only longitudinal waves are:
1) <u>Sound waves</u>.
2) <u>Shock waves</u>, e.g. seismic <u>P-waves</u> (see page 118).
3) A <u>slinky spring</u> when <u>plucked</u>.
4) <u>Don't get confused</u> by CRO displays which show a <u>transverse wave</u> when displaying <u>sounds</u>.  The real sound wave is <u>longitudinal</u> — the display shows a transverse wave <u>just so you can see what's going on</u>.

> *In <u>LONGITUDINAL waves</u> the vibrations are along the <u>SAME DIRECTION</u> as the wave is travelling.*

---

## *Don't forget to learn some **examples** of each type of wave*

This is all very basic stuff on waves.  Three sections with four points in each.  <u>Learn</u> the headings, then the details.  Then <u>cover the page</u> and see what you can <u>write down</u>. Then try again and again until you can remember the whole lot.

# Waves — Basic Principles

## All Waves *Carry Energy* — Without Transferring *Matter*

1) Light, infra red, and microwaves all make things warm up. X-rays and gamma rays can cause ionisation and damage to cells, which also shows that they carry energy.

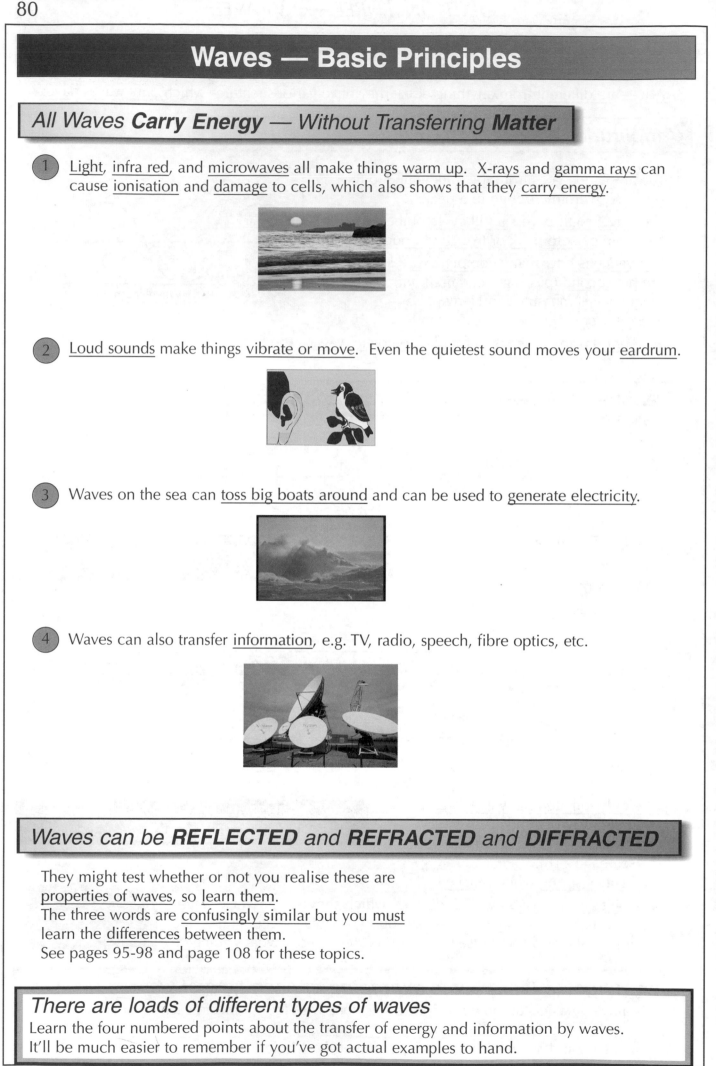

2) Loud sounds make things vibrate or move. Even the quietest sound moves your eardrum.

3) Waves on the sea can toss big boats around and can be used to generate electricity.

4) Waves can also transfer information, e.g. TV, radio, speech, fibre optics, etc.

## Waves can be **REFLECTED** and **REFRACTED** and **DIFFRACTED**

They might test whether or not you realise these are properties of waves, so learn them.
The three words are confusingly similar but you must learn the differences between them.
See pages 95-98 and page 108 for these topics.

## *There are loads of different types of waves*

Learn the four numbered points about the transfer of energy and information by waves.
It'll be much easier to remember if you've got actual examples to hand.

# Warm-Up and Worked Exam Question

Make sure you've got a handle on the basics of waves by doing these warm-up questions.

## Warm-up Questions

1) Name the features A and B shown on the diagram.

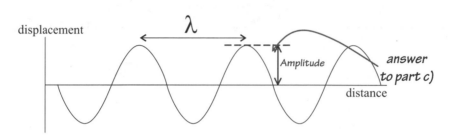

2) Name the two types of wave and give one example of each type.
3) Explain how we can tell that gamma rays carry energy.

## Worked Exam Question

Here's the kind of question you'll get in the real exam:

1    The diagram shows waves travelling at sea.

*displacement*

$\lambda$

*Amplitude*

*answer to part c)*

*distance*

a)   Name the feature of the wave labelled $\lambda$.

   *wavelength*
   .............................................................................................................
                                                                      *(1 mark)*

b)   A sailor counts the number of waves passing a buoy every second.

   i)   What is this quantity called?

       *frequency*          *Fairly obvious really.*
       .....................................................................................................
                                                                  *(1 mark)*

   ii)   What are its SI units?

       *hertz (Hz)*
       .....................................................................................................
                                                                  *(1 mark)*

c)   Mark the amplitude of the wave on the diagram.
                                     *answer on diagram*        *(1 mark)*

d)   Are the waves transverse or longitudinal?  Explain how you know.
   *Transverse — because the vibrations are at right angles*
   .............................................................................................................
   *to the wave direction.*          *Remember — transverse = across,*
   ..............................................        *longitudinal = along*          ............
                                                                      *(2 marks)*

# Exam Questions

1) Here is a diagram of a transverse wave.

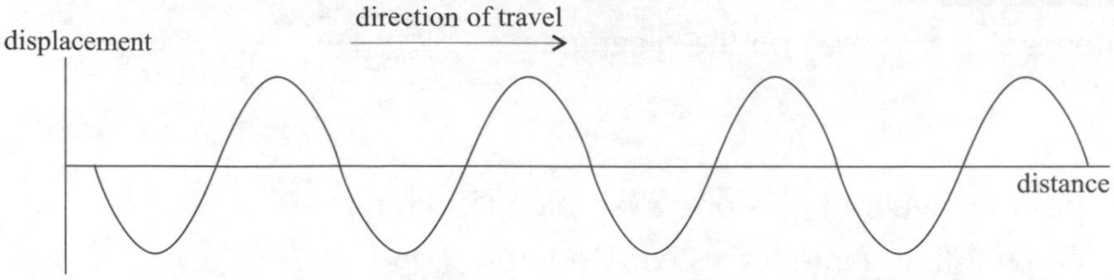

a) The wave has a frequency of 20 Hz. Explain what this means.

......................................................................................................................................

*(1 mark)*

b) The period is the time for one complete wave. Calculate the period of this wave.

......................................................................................................................................

......................................................................................................................................

*(2 marks)*

c) Give two examples of transverse waves.

......................................................................................................................................

......................................................................................................................................

*(2 marks)*

2) The diagram shows a longitudinal wave travelling along a slinky spring.

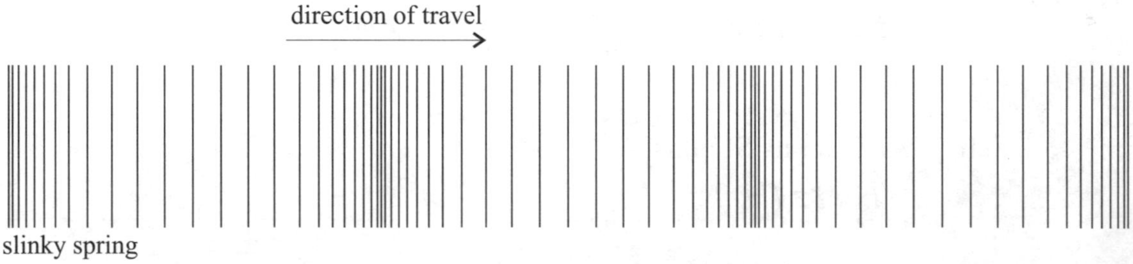

a) Label a <u>wavelength</u> on the diagram and areas of <u>rarefaction</u> and <u>compression</u>.

*(3 marks)*

b) Give another example of a longitudinal wave.

......................................................................................................................................

*(1 mark)*

# How Sound Travels

## Sound Travels at *Various Speeds* in Different Substances

① Sound Waves are caused by vibrating objects.

② Sound waves are longitudinal waves, which travel at fixed speeds in particular media, as shown in the table underneath.

③ As you can see, the denser the medium, the faster sound travels through it, generally speaking anyway.

④ Sound generally travels faster in solids than in liquids, and faster in liquids than in gases.

| Substance | Density | Speed of Sound |
|-----------|---------|----------------|
| Iron | 7.9 g/cm$^3$ | 5000 m/s |
| Rubber | 0.9 g/cm$^3$ | 1600 m/s |
| Water | 1.0 g/cm$^3$ | 1400 m/s |
| Cork | 0.3 g/cm$^3$ | 500 m/s |
| Air | 0.001 g/cm$^3$ | 330 m/s |

## Sound *Doesn't Travel* In A *Vacuum*

① Sound waves can be reflected, refracted and diffracted.

② But one thing they can't do is travel through a vacuum.

③ This is nicely demonstrated by the bell jar experiment.

④ As the air is sucked out by the vacuum pump, the sound gets quieter and quieter.

⑤ The bell has to be mounted on something like foam to stop the sound from travelling through the solid surface and making the bench vibrate, because you'd hear that instead.

Ringing bell — very quiet brrriiiinngg — Glass bell jar — Foam — air → To vacuum pump

## Sound works by vibrating particles: no particles = no sound.

It's hard to imagine sound not travelling through a vacuum — but the glass bell jar experiment proves it. It's one of those "classic" experiments that you ought to be able to draw and explain without thinking.

# Noise Pollution

## Hearing can be Damaged by Excessive Noise

1) The normal range of human hearing is 20 Hz to 20 000 Hz, but the upper limit decreases with age. Sounds with frequencies above 20 000 Hz just can't be heard by humans.

2) Dogs however can hear up to about 40 000 Hz so dog whistles are between 20kHz and 40 kHz so we can't hear them but the dogs can.

3) Too much loud noise will damage your hearing. The higher end of the frequency range is affected more. Personal stereos and loud machinery are the main culprits for damaging people's hearing.

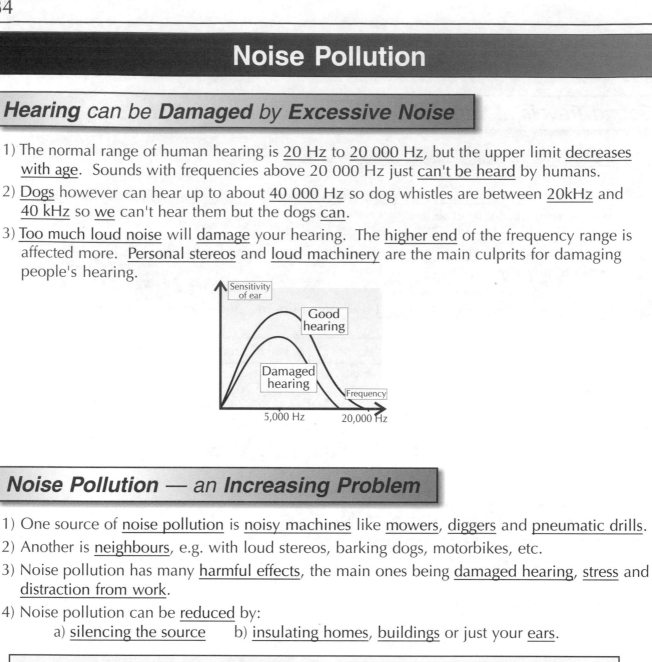

## Noise Pollution — an Increasing Problem

1) One source of noise pollution is noisy machines like mowers, diggers and pneumatic drills.

2) Another is neighbours, e.g. with loud stereos, barking dogs, motorbikes, etc.

3) Noise pollution has many harmful effects, the main ones being damaged hearing, stress and distraction from work.

4) Noise pollution can be reduced by:
    a) silencing the source    b) insulating homes, buildings or just your ears.

> Specific ways of reducing noise pollution:
> 1) Fitting silencers to engines and some sort of muffler to any other machinery.
> 2) Sound insulation in buildings: acoustic tiles, curtains, carpets and double glazing.
> 3) Wearing ear plugs.

## Echoes and Reverberation are due to Sound Being Reflected

1) Sound will only be reflected from hard flat surfaces. Things like carpets and curtains act as absorbing surfaces which will absorb sounds rather than reflect them.

2) This is very noticeable in the reverberation in an empty room. A big empty room sounds completely different once you've put carpet, curtains and furniture in, because these things absorb the sound quickly and stop it echoing (reverberating) around the room.

## Learn all the problems and all the solutions

Once again the page is broken up into sections with important numbered points for each. All those numbered points are important. They're all mentioned specifically in the syllabuses so you should expect them to test exactly this stuff in the Exams.

# Sound Waves

## *Amplitude is a Measure of the Energy Carried by Any Wave*

1) The greater the amplitude, the more energy the wave carries.
2) In sound this means it'll be louder.
3) With light, a bigger amplitude means it'll be brighter.

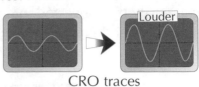

CRO traces

## *The Frequency of a Sound Wave Determines its Pitch*

1) High frequency sound waves sound high pitched like a squeaking mouse.
2) Low frequency sound waves sound low pitched like a mooing cow.
3) Frequency is the number of complete vibrations each second.
4) Common units are kHz (1000 Hz) and MHz (1 000 000 Hz).
5) High frequency (or high pitch) also means shorter wavelength.
6) These CRO traces are very important so make sure you know all about them:

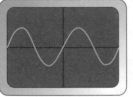 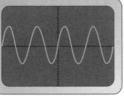  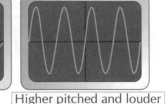

| Original Sound | Higher pitched | Lower pitched | Higher pitched and louder |

## *Microphones Turn Sound Waves into Electrical Signals*

1) The microphone changes the sound wave into a varying electrical current.
2) The variations in the current carry the information.
3) The currents from a microphone are very small and are amplified into much bigger signals by an amplifier.
4) These signals from the microphone can be recorded and played back through speakers.
5) Speakers turn electrical signals into sound waves — exactly the opposite of what a microphone does.

There's more about signals on pages 106-107.

## *More technical terms to learn*

There really is a lot of stuff on sound in the Physics syllabuses — so the more of it you really learn properly, the more marks you'll get in the Exam. Most Exam questions, even in Physics, only test whether you've learned the basic facts.

# Ultrasound

## Ultrasound is Sound with a Higher Frequency than We Can Hear

Electrical devices can be made which produce underlying electrical oscillations of any frequency. These can easily be converted into mechanical vibrations to produce sound waves beyond the range of human hearing (i.e. frequencies above 20 kHz). This is called ultrasound and it has loads of uses:

## 1) Industrial Cleaning

Ultrasound can be used to clean delicate mechanisms without them having to be dismantled. The ultrasound waves can be directed on very precise areas and are extremely effective at removing dirt and other deposits which form on delicate equipment. The alternatives would either damage the equipment or else would require it to be dismantled first.

The same technique is used for cleaning teeth. Dentists use ultrasonic tools to easily and painlessly remove hard deposits of tartar which build up on teeth and which would lead to gum disease.

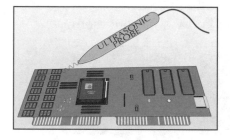
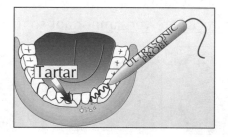

## 2) Breaking Down Kidney Stones

This works like the cleaning method above. An ultrasound beam concentrates high energy shockwaves at the kidney stone and turns it into sand-like particles. These particles then pass out of the body in urine. It's a good method because the patient doesn't need surgery and it's relatively painless.

## 3) Industrial Quality Control

Ultrasound waves can pass through something like a metal casting and whenever they reach a boundary between two different media (like metal and air) some of the wave is reflected back and detected. The exact timing and distribution of these echoes give detailed information about the internal structure.

The echoes are usually processed by computer to produce a visual display of what the object must be like inside. If there are cracks or holes where there shouldn't be they'll show up.

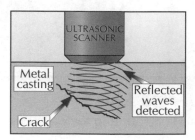

# Ultrasound

## 4) For **Pre-Natal Scanning** of a **Foetus**

This follows the <u>same principle</u> as the industrial quality control.  As the ultrasound hits <u>different media</u> some of the sound wave is <u>reflected</u> and these reflected waves are <u>processed by computer</u> to produce a <u>video image</u> of the foetus.  No one knows for sure whether ultrasound is safe in all cases but <u>X-rays</u> would definitely be dangerous to the foetus.

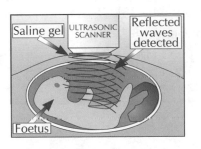

## 5) Range and Direction Finding — **SONAR**

<u>Bats</u> send out <u>high-pitched squeaks</u> (ultrasound) and pick up the <u>reflections</u> with their <u>big ears</u>.  Their brains are able to <u>process</u> the reflected signal and turn it into a <u>picture</u> of what's around.

So the bats basically "<u>see</u>" with <u>sound waves</u>, well enough in fact to <u>catch moths</u> in <u>mid-flight</u> in <u>complete darkness</u>.

The same technique is used for <u>sonar</u> which uses sound waves <u>underwater</u> to detect features in the water and on the seabed.  The <u>pattern</u> of the reflections indicates the <u>depth</u> and basic features.

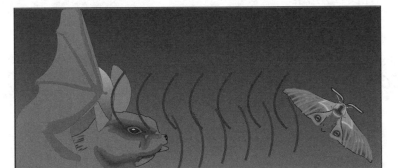

## *Learn the 5 main uses of ultrasound*

Ultrasound has a lot of different uses — it can also be used to speed up the breaking up of internal scar tissue.  But you only need to learn the uses on these two pages, because they're the ones that'll come up in the exam.

# Warm-Up and Worked Exam Question

## Warm-up Questions

1) Finish the sentences by picking the right word from each pair.

   Sound is caused by a vibrating/stationary object. Sound is a transverse/longitudinal wave.

   Sound travels faster in a solid/gas than a solid/gas. Sound cannot travel through a vacuum/pipe.

2) A dog whistle produces a sound with a frequency of 35kHz. a) What does "frequency of 35kHz" mean? b) Give two reasons why 35kHz is an ideal frequency for a dog whistle.

3) Match these sounds to their frequencies:

   deep bass drum           8000 Hz

   high-pitched squeak      100 Hz

   ultrasound scan          40 000 Hz

## Worked Exam Question

1) An electric guitar is plugged into an amplifier, and a microphone is placed in front of the speaker so the sound produced can be shown on an oscilloscope.

a) Which piece of equipment converts sound energy into electrical energy?

   *the microphone*

   *(1 mark)*

b) The guitarist plays a note with a frequency of 550 Hz. What does this mean?

   *550 waves each second*

   *(1 mark)*

c) The 550 Hz note produced this trace on the oscilloscope:

On the grid below, draw a trace that might be produced:

(i) if the same note was played louder.        *(2 marks)*

(ii) if a lower pitched and quieter note were played.        *(2 marks)*

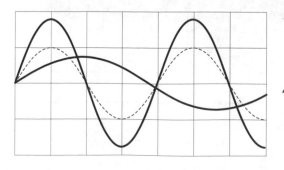

*Answer (i) — same time period but greater amplitude*

*Answer (ii) — longer time period and smaller amplitude*

# Exam Questions

1    Diagrams A, B, C and D show oscilloscope traces for four different sounds.

The oscilloscope settings are the same for each trace.

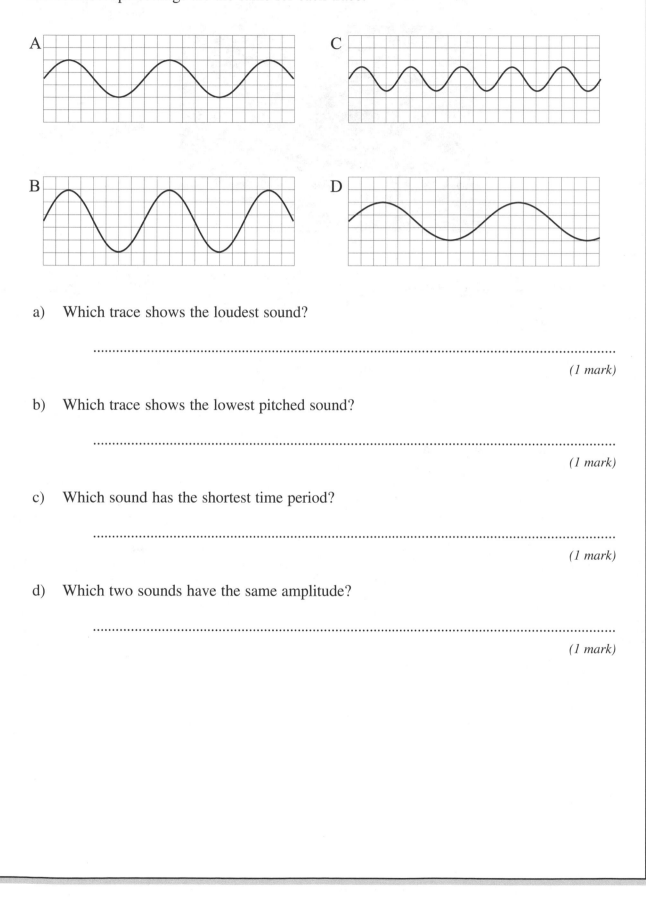

a)    Which trace shows the loudest sound?

    ....................................................................................................................................

    *(1 mark)*

b)    Which trace shows the lowest pitched sound?

    ....................................................................................................................................

    *(1 mark)*

c)    Which sound has the shortest time period?

    ....................................................................................................................................

    *(1 mark)*

d)    Which two sounds have the same amplitude?

    ....................................................................................................................................

    *(1 mark)*

## Exam Questions

2   The photograph shows a lady having an ultrasound scan.
    Ultrasonic waves are used to examine the foetus inside the mother's womb.

a)   What are ultrasonic waves?

    ..............................................................................................................................
    *(2 marks)*

b)   Why are ultrasonic waves used instead of X-rays?

    ..............................................................................................................................
    *(1 mark)*

c)   Describe how ultrasound is used to produce an image of the foetus.

    ..............................................................................................................................

    ..............................................................................................................................

    ..............................................................................................................................
    *(2 marks)*

d)   Name two other uses of ultrasonic waves.

    ..............................................................................................................................

    ..............................................................................................................................
    *(2 marks)*

e)   Which of the following frequencies could be ultrasonic waves?
     16 000 Hz, 4000 Hz, 35 kHz, 18 kHz.

    ..............................................................................................................................
    *(1 mark)*

# Speed of Waves

These are formulae, <u>just like all the other formulae</u>, and the <u>same rules apply</u> (see pages 2-3). There are a <u>few extra details</u> that go with these wave formulas though.  Learn them now:

## The **First Rule**:  *Try and Choose the Right Formula*

1) People have <u>difficulty</u> deciding <u>which formula</u> to use.
2) Often the question starts with "*A wave is travelling...*", so people immediately go for "v = f$\lambda$".
3) To choose the <u>right formula</u> you have to look for the <u>three quantities</u> mentioned in the question.
4) If the question mentions <u>speed</u>, <u>frequency</u> and <u>wavelength</u> then "v = f$\lambda$" <u>is</u> the one to use.
5) But if it has <u>speed</u>, <u>time</u> and <u>distance</u> then you need "s = d/t" instead.

### Example 1 — *Water Ripples*

a) *Some ripples travel 55 cm in 5 seconds.  Find their speed in cm/s.*
   <u>Answer</u>:  Speed, distance and time are mentioned in the question,
   so you must use "s=d/t":        s = d/t = 55/5 = <u>11 cm/s</u>
b) *The wavelength of these waves is found to be 2.2 cm.  What is their frequency?*
   <u>Answer</u>:  This time f and $\lambda$ mentioned, so use "v = f$\lambda$", and you'll need this:
   which tells you that f = v/$\lambda$ = 11 cm/s ÷ 2.2 cm. = <u>5 Hz</u> (It's fine to use cm/s with cm, s and Hz.)

## The **Second Rule**:  *Watch those* **Units**

1) The <u>standard (SI) units</u> involved with waves are:  <u>metres</u>, <u>seconds</u>, <u>m/s</u> and <u>Hertz</u> (Hz).
   Always <u>convert into SI units</u> (m, s, Hz, m/s) before you work anything out.
2) The trouble is waves often have <u>high frequencies</u>, given in <u>kHz</u> or <u>MHz</u>, so make sure you <u>learn</u>
   <u>this</u> too:      1 kHz (kiloHertz) = 1000 Hz            1 MHz (1 MegaHertz) = 1 000 000 Hz
3) <u>Wavelengths</u> can also be given in <u>different units</u>, e.g. <u>km</u> for long wave radio, or <u>cm</u> for sound.
4) Note — the <u>speed of light</u> is $3 \times 10^8$m/s = <u>300 000 000 m/s</u>.
   This, along with numbers like <u>900 MHz = 900 000 000 Hz</u> won't fit into a lot of calculators.
   That leaves you <u>three choices</u>:
       1) Enter the numbers as <u>standard form</u> ($3 \times 10^8$ and $9 \times 10^8$), or...
       2) <u>Cancel</u> three or six <u>noughts</u> off both numbers, (if you're <u>dividing</u> them) or...
       3) Do it entirely <u>without a calculator</u>.

### Example 2 — *Sound*

Q) *A sound wave travelling in a solid has a frequency of 19 kHz and a wavelength of 12cm.*
   *Find its speed.*
<u>Answer</u>:  You've got f and $\lambda$ mentioned, so use "v = f$\lambda$".  But you must convert the units into SI:
   So,  v = f×$\lambda$  = 19 000 Hz × 0.12 m = <u>2280 m/s</u> — convert the units and there's <u>no problem</u>.

### Example 3 — *EM radiation:*

Q) *A radio wave has a frequency of 92.2 MHz.  Find its wavelength.*
   *(The speed of all EM waves is $3 \times 10^8$ m/s.)*
<u>Answer</u>:  f and $\lambda$ are mentioned, so use "v = f$\lambda$".  Radio waves travel at the speed of light of course.
   Once again, convert the units into SI, but you'll also have to use standard form:
   $\lambda$ = v/f  = $3 \times 10^8$ / 92 200 000  = $3 \times 10^8$ / $9.22 \times 10^7$  = 3.25m

## And finally:  **Frequency = 1/ Time Period**        *(Watch out — there are a few bits that could go wrong here.)*

Q) *A wave does 40 complete cycles in 8 seconds.  Find its time period, T, and its frequency, f, in Hz.*
<u>Answer</u>:  T = time taken for one cycle = 8 s ÷ 40  = 0.2 s      f = 1/T = 1 ÷ 0.2 = 5 Hz

---

## *There are just* **two formulas** *to learn, that's all*
<u>Sift out</u> the main rules on this page, then <u>cover it up</u> and <u>write them down</u>.  You need to know
all the units, both the formulas, and an example of when you should use each formula.

# Speed of Waves

## *Relative Speeds of Sound and Light*

1) <u>Light</u> travels about <u>a million times faster</u> than <u>sound</u>, so you never need to calculate how long it takes compared to sound. You only work out the time taken for the <u>sound</u> to travel.

2) The <u>formula</u> needed is always the <u>s = d/t</u> one for <u>speed, distance and time</u> (see page 66).

3) When something makes a sound more than about <u>100 m away</u> and you can actually <u>see</u> the action which makes the sound then the effect is quite <u>noticeable</u>. Good examples are:

   a) <u>Live cricket</u> — you hear the "<u>knock</u>" a while after seeing the ball being struck.

   b) <u>Hammering</u> — you hear the "<u>clang</u>" when the hammer is back up <u>in mid air</u>.

   c) <u>Starting pistol</u> — you <u>see the smoke</u> and then <u>hear the bang</u>.

   d) <u>Jet aircraft</u> — they're always <u>ahead</u> of where it sounds like they are.

   e) <u>Thunder and lightning</u> — the flash of lightning causes the sound of the thunder, and the <u>time interval</u> between the <u>flash</u> and the <u>rumble</u> tells you how far away the lightning is. There's approximately <u>five seconds delay for every mile</u>. [1 mile = 1600 m, ÷ 330 (speed of sound) = 4.8 s]

<u>Example</u>: *Looking out from his office across the school yard, the Headmaster saw five students destroying something with a hammer. Before acting swiftly, he did take the time to notice that there was a delay of exactly 0.4 seconds between the hammer striking and the sound reaching his ear. How far away were the children? (Sound travels at 330 m/s in air.)*

<u>Answer</u>: The formula we want is "Speed = Distance/Time" or "s=d/t".
We want to find the distance, d. We already know the time is 0.4 s
and the speed of sound in air = 330 m/s. So we need d=s×t (from the triangle).
This gives: d = 330×0.4 = <u>132m</u>. (That's how far the sound travels in 0.4 s.)

## *Echo Questions — Don't Forget the Factor of Two*

1) The <u>big thing</u> to remember with <u>echo questions</u> is that because the sound has to travel <u>both ways</u>, then to get the <u>right answer</u> you'll need to either <u>double something</u> or <u>halve something</u>.

2) Make sure you remember: sound travels at about <u>330 m/s in air</u> and <u>1400 m/s in water</u>. Any echo question is likely to be in air or water and if you have to work out the speed of the sound it's really useful to know what sort of number you should be getting.
So for example, if you get 170 m/s for the speed of sound in air then you should realise you've <u>forgotten the factor of two</u> somewhere, and then you can <u>easily go back and sort it out</u>.

<u>Example</u>: *Having expelled five students from his school, the Headmaster popped open a bottle of Champagne and heard the echo 0.6 s later from the other side of his office. How big was his office?*

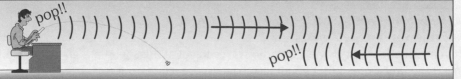

<u>Answer</u>: The formula is "Speed = Distance/Time" or "s=d/t" again. We want to find the distance, d.
We already know the time, 0.6 s and the speed (of sound in air), hence d=s×t (from triangle)
This gives: d = 330×0.6 = <u>198 m</u>
But watch out — <u>don't forget the factor of two for echo questions</u>:
The 0.6 s is for <u>there and back</u>, so the office is only <u>half</u> that distance, <u>99 m long</u>.

---

*Remember Echoes have a Factor of Two...Factor of Two...Factor of Two...*
<u>Learn</u> the details on this page, then <u>cover it up</u> and <u>write them all down from memory</u>.

# Warm-Up and Worked Exam Question

## Warm-up Questions

1) A sound wave has a frequency of 2500 Hz and a wavelength of 13.2 cm.
   Calculate its speed.

2) The radio waves for Radio 4 have a wavelength of 1.5 km.
   Find their frequency.

3) A cricketer hits a ball and a man hears the knock 0.8 s later.
   How far away is the man?  (Assume the speed of sound to be 330 m/s.)

4) A ship sends a sonar signal to the sea bed and detects the echo 0.7 s later.
   How deep is the sea at that point?  (The speed of sound in water is 1400 m/s.)

## Worked Exam Question

1   The diagram shows waves at sea passing a buoy.

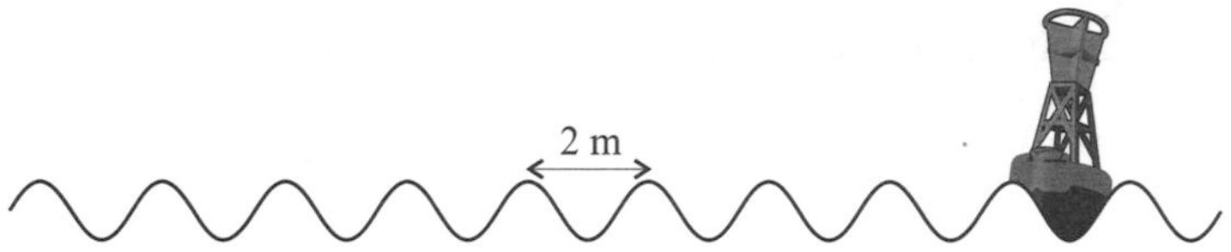

2 m

The distance between consecutive peaks is measured and is found to be 2 m.

a)   What name is given to this distance?

*wavelength*

*(1 mark)*

20 waves are counted passing the buoy every 10 seconds.

b)   Calculate the frequency of the waves.

*20 ÷ 10 = 2 waves per second = 2 Hz*

*(1 mark)*

c)   Calculate the speed of the waves.

$v = f\lambda$   *1 mark*

*= 2 × 2 = 4 m/s   1 mark*

*(2 marks)*

d)   Nearer the shore, the waves travel slower, at 3 m/s.
     Calculate the wavelength of waves near the shore if their frequency is 2 Hz.

$\lambda = v/f$   *1 mark*

*= 3 ÷ 2 = 1.5 m   1 mark*

*(2 marks)*

## Exam Questions

1)    The speed of sound in water is 1400 m/s.

     a)    An ultrasonic wave has a frequency of 28 000Hz.
         Calculate its wavelength in water.

......................................................................................................................................

......................................................................................................................................

*(2 marks)*

     b)    A boat uses ultrasonic waves to calculate the depth of the sea.
         The diagram shows how.

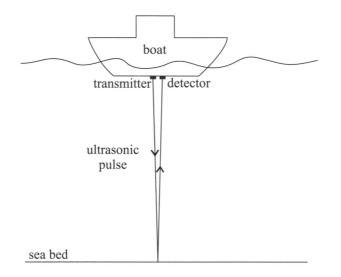

The pulse takes 0.2s to travel from the transmitter to the seabed and back to the detector.  Calculate the distance to the seabed.

......................................................................................................................................

......................................................................................................................................

......................................................................................................................................

*(3 marks)*

# Reflection of Waves

## *The Ripple Tank is Really Good for Displaying Waves*

Learn all these diagrams showing <u>reflection of waves</u>. They could ask you to complete <u>any</u> <u>one of them</u> in the Exam. It can be quite a bit <u>trickier</u> than you think unless you've <u>practised</u> them really well <u>beforehand</u>.

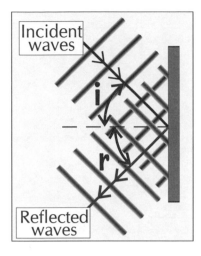

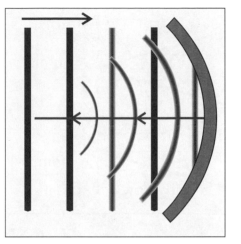

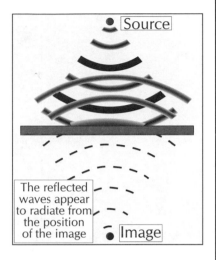

## *Reflection of Light*

(1) <u>Reflection of light</u> is what allows us to <u>see</u> objects.

(2) When light reflects from an <u>uneven surface</u> such as a <u>piece of paper</u> the light reflects off <u>at all different angles</u> and you get a DIFFUSE REFLECTION.

(3) When light reflects from an <u>even surface</u> (<u>smooth and shiny</u> like a <u>mirror</u>) then it's all reflected at the <u>same angle</u> and you get a <u>clear reflection</u>.

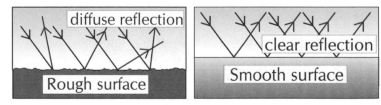

(4) But don't forget, the <u>LAW OF REFLECTION</u> applies to <u>every reflected ray</u>:

### Angle of *INCIDENCE (i)* = Angle of *REFLECTION (r)*

## *These diagrams are as important as the facts*

Describing what's happening with waves would be pretty hard if you didn't have a diagram. It's really important you get to grips with drawing both ripple tank diagrams and ray diagrams.

# Reflections in a Mirror

## Reflection In a Plane Mirror — How to Locate The Image

You need to be able to <u>reproduce</u> this entire diagram of <u>how an image is formed</u> in a <u>PLANE MIRROR</u>.

Learn these <u>three important points</u>:

1. The <u>image</u> is the <u>same size</u> as the <u>object</u>.

2. It is <u>AS FAR BEHIND</u> the mirror as the object is <u>in front</u>.

3. It's formed from <u>diverging rays</u>, which means it's a <u>virtual image</u>.

## How to Draw a Reflected Ray

1. <u>To draw any reflected ray</u>, just make sure the <u>angle of reflection</u>, r, equals the <u>angle of incidence</u>, i.

2. Note that these two angles are <u>ALWAYS</u> defined between the ray itself and the <u>dotted NORMAL</u>.

3. <u>Don't ever</u> label them as the angle between the ray and the <u>surface</u>.

---

*The angles are always with the NORMAL, not with the mirror*
You need to know this well enough to answer typical Exam questions like: "<u>Label the angles of reflection and incidence on this diagram</u>" and "<u>Why is the image in a plane mirror virtual?</u>"

# Refraction

1) _Refraction_ is when waves _change direction_ as they _enter a different medium_.

2) This is caused _entirely_ by the _change in speed_ of the waves.

3) It also causes the _wavelength_ to change, but remember that the _frequency_ does _not_ change.

## 1) **Refraction** is Shown by **Waves** in a Ripple Tank **Slowing Down**

1) The waves travel <u>slower</u> in <u>shallower water</u>, causing <u>refraction</u> as shown.

2) There's a <u>change in direction</u> and a <u>change in wavelength</u>, but <u>no change in frequency</u>.

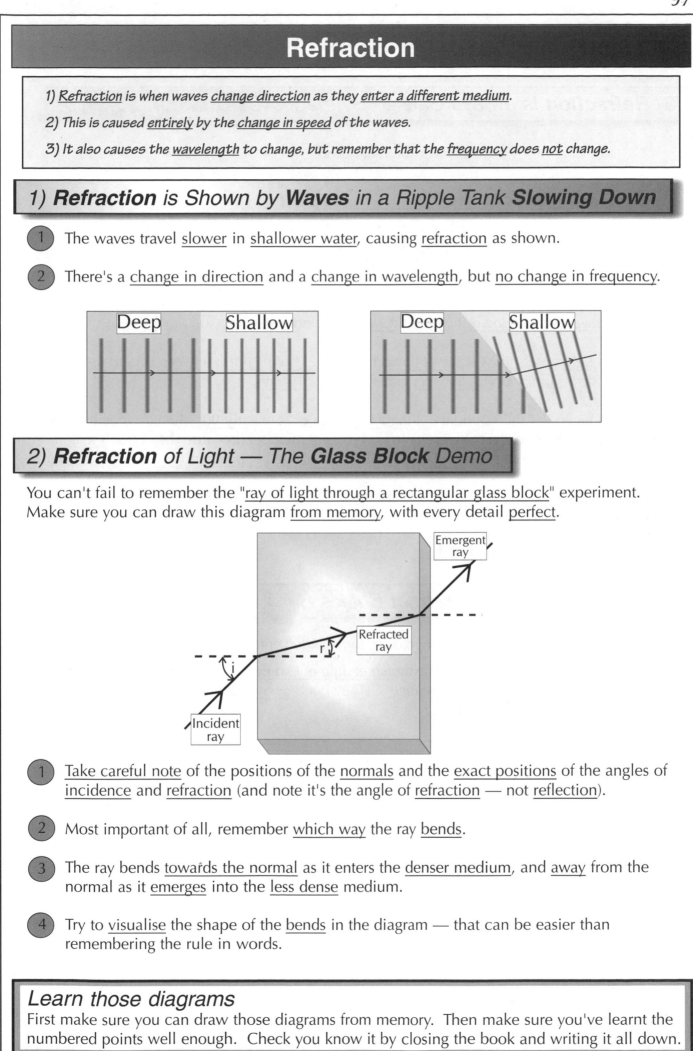

## 2) **Refraction** of Light — The **Glass Block** Demo

You can't fail to remember the "<u>ray of light through a rectangular glass block</u>" experiment. Make sure you can draw this diagram <u>from memory</u>, with every detail <u>perfect</u>.

1) <u>Take careful note</u> of the positions of the <u>normals</u> and the <u>exact positions</u> of the angles of <u>incidence</u> and <u>refraction</u> (and note it's the angle of <u>refraction</u> — not <u>reflection</u>).

2) Most important of all, remember <u>which way</u> the ray <u>bends</u>.

3) The ray bends <u>towards the normal</u> as it enters the <u>denser medium</u>, and <u>away</u> from the normal as it <u>emerges</u> into the <u>less dense</u> medium.

4) Try to <u>visualise</u> the shape of the <u>bends</u> in the diagram — that can be easier than remembering the rule in words.

## Learn those diagrams

First make sure you can draw those diagrams from memory. Then make sure you've learnt the numbered points well enough. Check you know it by closing the book and writing it all down.

# Refraction

## 3) Refraction Is always Caused By the Waves Changing Speed

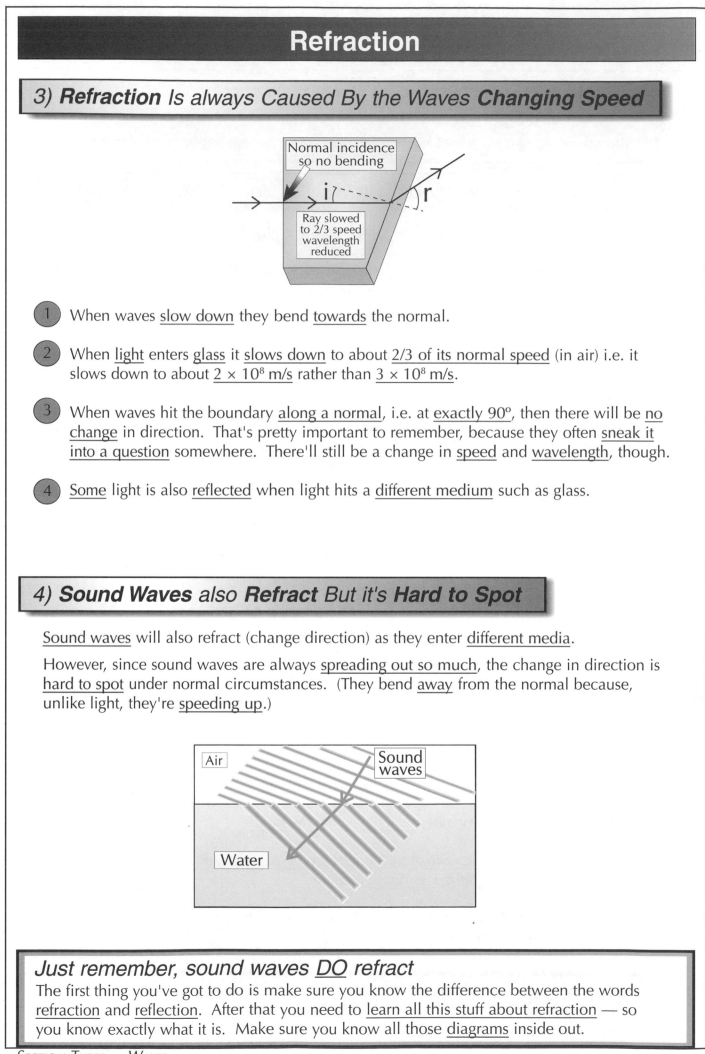

Normal incidence so no bending

Ray slowed to 2/3 speed wavelength reduced

i  r

① When waves <u>slow down</u> they bend <u>towards</u> the normal.

② When <u>light</u> enters <u>glass</u> it <u>slows down</u> to about <u>2/3 of its normal speed</u> (in air) i.e. it slows down to about $2 \times 10^8$ m/s rather than $3 \times 10^8$ m/s.

③ When waves hit the boundary <u>along a normal</u>, i.e. at <u>exactly 90º</u>, then there will be <u>no change</u> in direction.  That's pretty important to remember, because they often <u>sneak it into a question</u> somewhere.  There'll still be a change in <u>speed</u> and <u>wavelength</u>, though.

④ <u>Some</u> light is also <u>reflected</u> when light hits a <u>different medium</u> such as glass.

## 4) Sound Waves also Refract But it's Hard to Spot

<u>Sound waves</u> will also refract (change direction) as they enter <u>different media</u>.

However, since sound waves are always <u>spreading out so much</u>, the change in direction is <u>hard to spot</u> under normal circumstances.  (They bend <u>away</u> from the normal because, unlike light, they're <u>speeding up</u>.)

Air          Sound waves

Water

## Just remember, sound waves <u>DO</u> refract

The first thing you've got to do is make sure you know the difference between the words <u>refraction</u> and <u>reflection</u>.  After that you need to <u>learn all this stuff about refraction</u> — so you know exactly what it is.  Make sure you know all those <u>diagrams</u> inside out.

# Dispersion

## *Dispersion* Produces *Rainbows*

① Different colours of light are refracted by different amounts.

② This is because they travel at slightly different speeds in any given medium (but not in a vacuum).

③ White light is a mixture of lots of colours. A prism can be used to make the different colours of white light emerge at different angles.

④ This produces a spectrum showing all the colours of the rainbow. This effect is called dispersion.

⑤ You need to know that red light is refracted the least — and violet is refracted the most.

⑥ Also know the order of colours in between:
Red Orange Yellow Green Blue Indigo Violet
— which is remembered by:
Richard Of York Gave Battle In Vain
They may well test whether you can put them correctly into the diagram.

⑦ Also learn where infrared and ultraviolet light would appear if you could detect them.

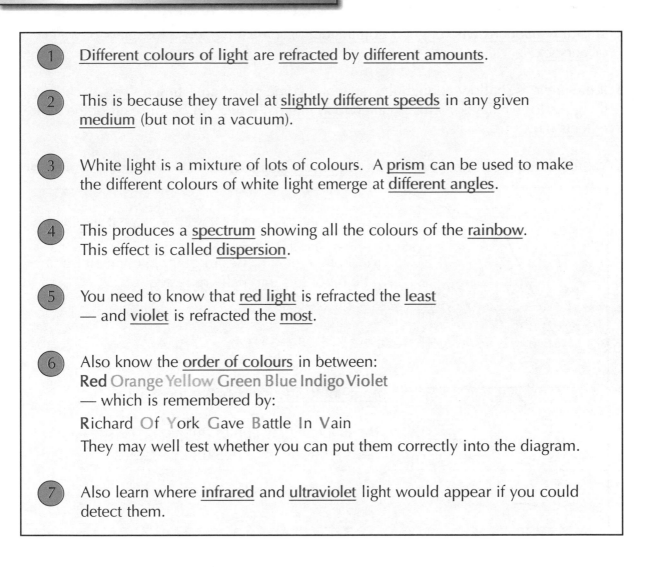

Prism

White light

Angle of deviation

A spectrum

infrared
red
orange
yellow
green
blue
indigo
violet
ultraviolet

Violet is bent the most

## *The Prism effect is called Dispersion*

Rainbows themselves are of course caused in exactly the same way. If it's raining while the sun's shining, the raindrops effectively act as prisms, and the light from the sun is dispersed.

# Total Internal Reflection

## Total Internal Reflection and The Critical Angle

1. This <u>only happens</u> when <u>light</u> is <u>coming out</u> of something <u>dense</u> like <u>glass</u> or <u>water</u> or <u>perspex</u>.

2. If the <u>angle</u> is <u>shallow enough</u> the ray <u>won't come out at all</u>, but it <u>reflects</u> back into the glass (or whatever). This is called <u>total internal reflection</u> because <u>all</u> of the light <u>reflects back in</u>.

3. You definitely need to learn this set of <u>three diagrams</u> which show the three conditions:

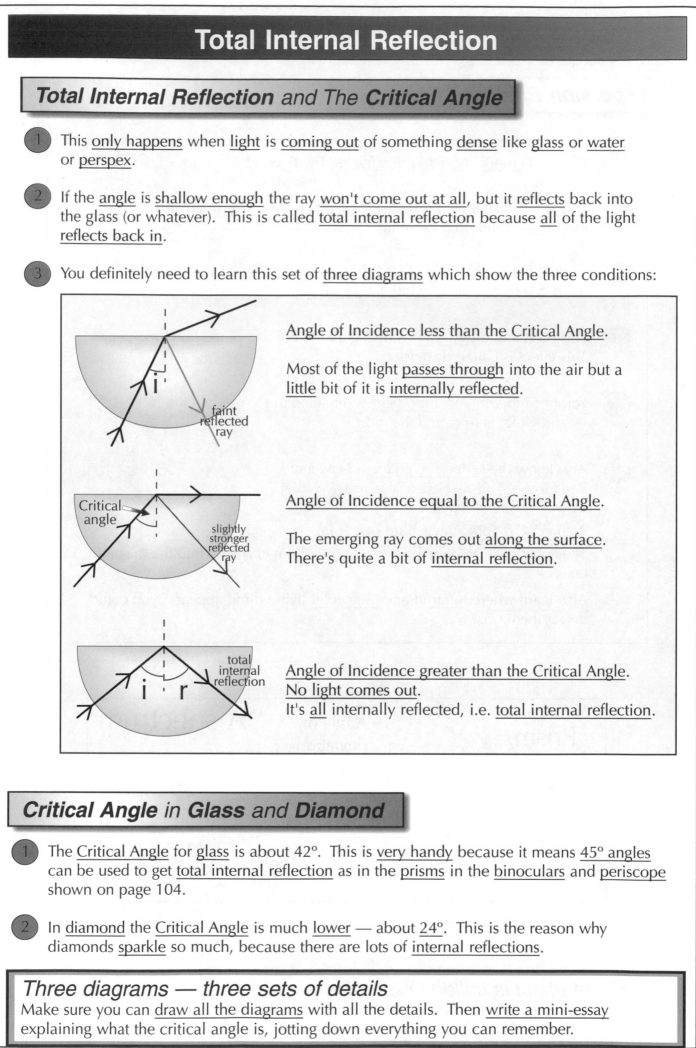

<u>Angle of Incidence less than the Critical Angle.</u>

Most of the light <u>passes through</u> into the air but a <u>little</u> bit of it is <u>internally reflected</u>.

faint reflected ray

Critical angle

slightly stronger reflected ray

<u>Angle of Incidence equal to the Critical Angle.</u>

The emerging ray comes out <u>along the surface</u>. There's quite a bit of <u>internal reflection</u>.

total internal reflection

<u>Angle of Incidence greater than the Critical Angle.</u>
<u>No light comes out.</u>
It's <u>all</u> internally reflected, i.e. <u>total internal reflection</u>.

## Critical Angle in Glass and Diamond

1. The <u>Critical Angle</u> for <u>glass</u> is about 42°. This is <u>very handy</u> because it means <u>45° angles</u> can be used to get <u>total internal reflection</u> as in the <u>prisms</u> in the <u>binoculars</u> and <u>periscope</u> shown on page 104.

2. In <u>diamond</u> the <u>Critical Angle</u> is much <u>lower</u> — about <u>24°</u>. This is the reason why diamonds <u>sparkle</u> so much, because there are lots of <u>internal reflections</u>.

---

### Three diagrams — three sets of details
Make sure you can <u>draw all the diagrams</u> with all the details. Then <u>write a mini-essay</u> explaining what the critical angle is, jotting down everything you can remember.

# Warm-Up and Worked Exam Question

## Warm-up Questions

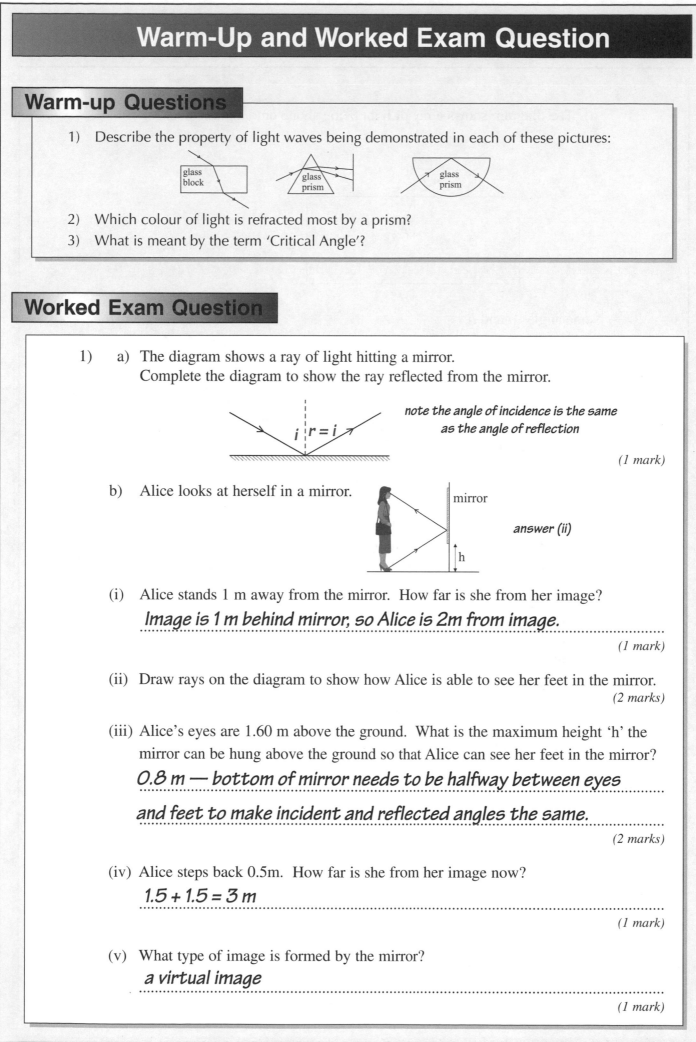

1) Describe the property of light waves being demonstrated in each of these pictures:

glass block

glass prism

glass prism

2) Which colour of light is refracted most by a prism?

3) What is meant by the term 'Critical Angle'?

## Worked Exam Question

1)  a)  The diagram shows a ray of light hitting a mirror.
Complete the diagram to show the ray reflected from the mirror.

$i$  $r = i$

*note the angle of incidence is the same as the angle of reflection*

*(1 mark)*

b)  Alice looks at herself in a mirror.

mirror

*answer (ii)*

h

(i)  Alice stands 1 m away from the mirror.  How far is she from her image?

*Image is 1 m behind mirror, so Alice is 2m from image.*

*(1 mark)*

(ii)  Draw rays on the diagram to show how Alice is able to see her feet in the mirror.

*(2 marks)*

(iii) Alice's eyes are 1.60 m above the ground.  What is the maximum height 'h' the mirror can be hung above the ground so that Alice can see her feet in the mirror?

*0.8 m — bottom of mirror needs to be halfway between eyes and feet to make incident and reflected angles the same.*

*(2 marks)*

(iv) Alice steps back 0.5m.  How far is she from her image now?

*1.5 + 1.5 = 3 m*

*(1 mark)*

(v)  What type of image is formed by the mirror?

*a virtual image*

*(1 mark)*

# Exam Questions

1   a) The diagram shows a ray of light being shone onto a glass block.

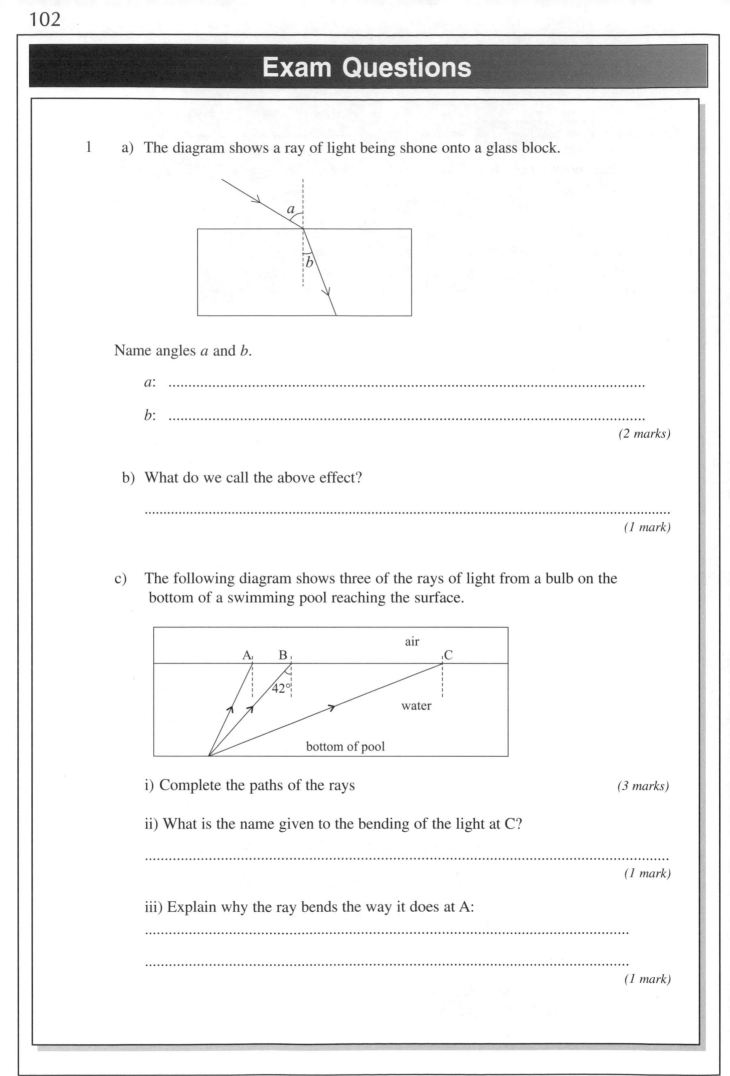

Name angles *a* and *b*.

*a*: ......................................................................................................................

*b*: ......................................................................................................................

*(2 marks)*

b) What do we call the above effect?

.............................................................................................................................

*(1 mark)*

c)  The following diagram shows three of the rays of light from a bulb on the
    bottom of a swimming pool reaching the surface.

i) Complete the paths of the rays                                        *(3 marks)*

ii) What is the name given to the bending of the light at C?

.............................................................................................................................

*(1 mark)*

iii) Explain why the ray bends the way it does at A:

.............................................................................................................................

.............................................................................................................................

*(1 mark)*

# Exam Questions

2)  a)  Name two properties of the wave that change as water waves travel from deep to shallower water.

.......................................................................................................................................................
*(2 marks)*

  b)  Describe how light waves bend when they travel from:

  (i)     air into glass;

.......................................................................................................................................................
*(1 mark)*

  (ii)    glass into air.

.......................................................................................................................................................
*(1 mark)*

  c)  What happens to the frequency of these waves as they cross the boundary?

.......................................................................................................................................................
*(1 mark)*

3)  a)  When white light is shone through a prism it disperses to form a spectrum. Complete the following diagram by inserting the names of the colours.

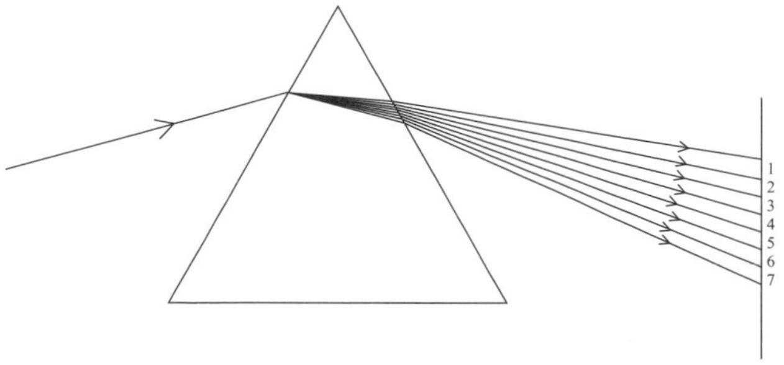

*(2 marks)*

  b)  Why do the different colours refract by different amounts?

.......................................................................................................................................................
*(1 mark)*

  c)  Draw a diagram to show how a prism can be used to make a ray of light change direction by 90°.

*If you get stuck on this, have a look at the next mini-section.*

*(2 marks)*

# Uses of Total Internal Reflection

Total Internal Reflection is used in binoculars, periscopes and bicycle reflectors.
All three use 45° prisms.

## *Binoculars*

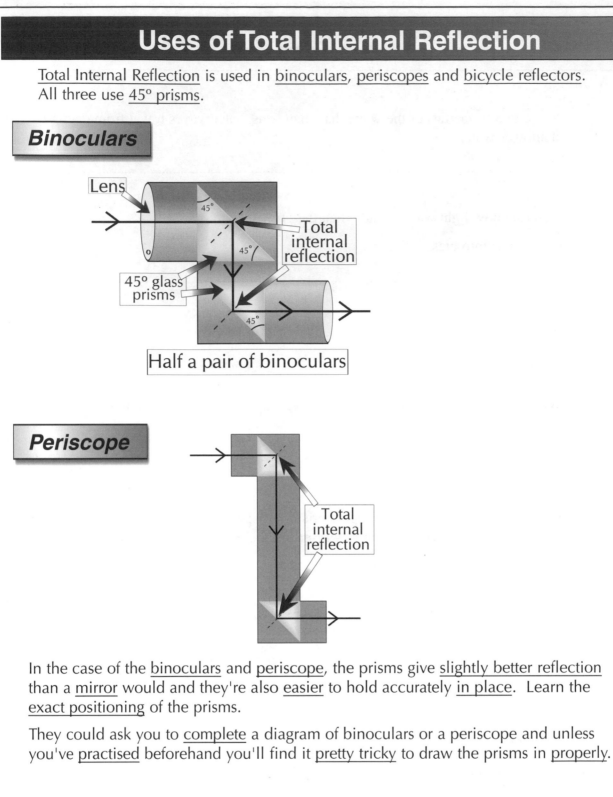

Half a pair of binoculars

## *Periscope*

In the case of the binoculars and periscope, the prisms give slightly better reflection
than a mirror would and they're also easier to hold accurately in place.  Learn the
exact positioning of the prisms.

They could ask you to complete a diagram of binoculars or a periscope and unless
you've practised beforehand you'll find it pretty tricky to draw the prisms in properly.

## *Reflectors*

In the bicycle reflectors the prisms work cleverly
by sending the light back in the opposite direction
that it came from (as shown in the diagram). This
means that whoever shines the light gets a strong
reflection straight back at their eyes.

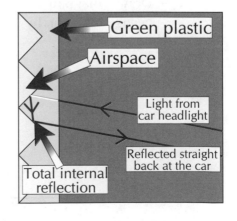

# Uses of Total Internal Reflection

## Optical Fibres — Communications and Endoscopes

(1) Optical fibres can carry information over long distances by repeated total internal reflections.

(2) Optical communications have several advantages over electrical signals in wires:
   a) the signal doesn't need boosting as often.
   b) a cable of the same diameter can carry a lot more information.
   c) the signals cannot be tapped into, or suffer interference from electrical sources.

(3) Normally no light whatever would be lost at each reflection. However some light is lost due to imperfections in the fibre, so it still needs boosting every few km.

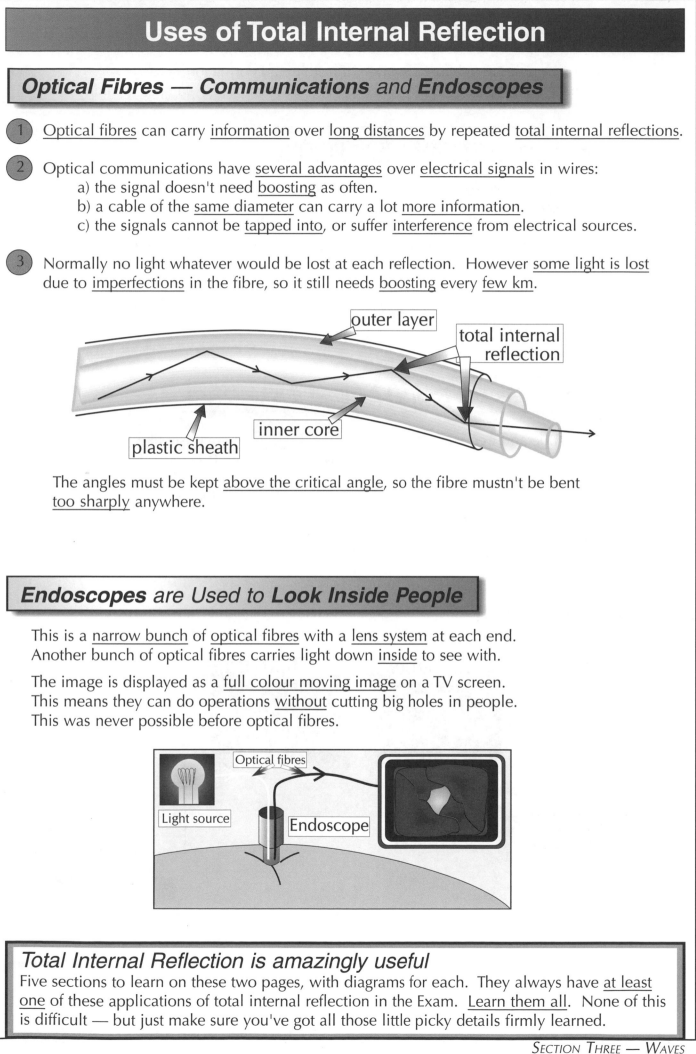

The angles must be kept above the critical angle, so the fibre mustn't be bent too sharply anywhere.

## Endoscopes are Used to Look Inside People

This is a narrow bunch of optical fibres with a lens system at each end. Another bunch of optical fibres carries light down inside to see with.

The image is displayed as a full colour moving image on a TV screen. This means they can do operations without cutting big holes in people. This was never possible before optical fibres.

## Total Internal Reflection is amazingly useful

Five sections to learn on these two pages, with diagrams for each. They always have at least one of these applications of total internal reflection in the Exam. Learn them all. None of this is difficult — but just make sure you've got all those little picky details firmly learned.

# Digital and Analogue

You've got to learn the <u>two</u> different ways of transmitting information.
Without signals there'd be no phones, no computers... even digital watches wouldn't exist.

## *Information* is Converted Into *Signals*

(1) Information (e.g. sound, speech, pictures) is converted into <u>electrical signals</u> before it's transmitted.

(2) It's then sent long distances down <u>cables</u>, like with telephone calls or the internet, or it's carried on <u>EM waves</u>, like radio or TV.

(3) Information can also be sent down <u>optical fibres</u> by first converting it into <u>visible light</u> or <u>infrared</u> signals.

## *Analogue* Varies But *Digital's* Either *On* or *Off*

(1) The <u>amplitude</u> and <u>frequency</u> of analogue signals <u>vary continuously</u> like in sound waves. An analogue signal can have <u>any</u> value in its range.

(2) Dimmer switches, thermometers, speedometers and old fashioned watches are all <u>analogue</u> devices.

(3) Digital signals are <u>coded pulses</u> — they have <u>one</u> of only <u>two</u> values: on or off, true or false, 0 or 1...

(4) On/off switches, digital clocks and digital meters are <u>digital</u> devices.

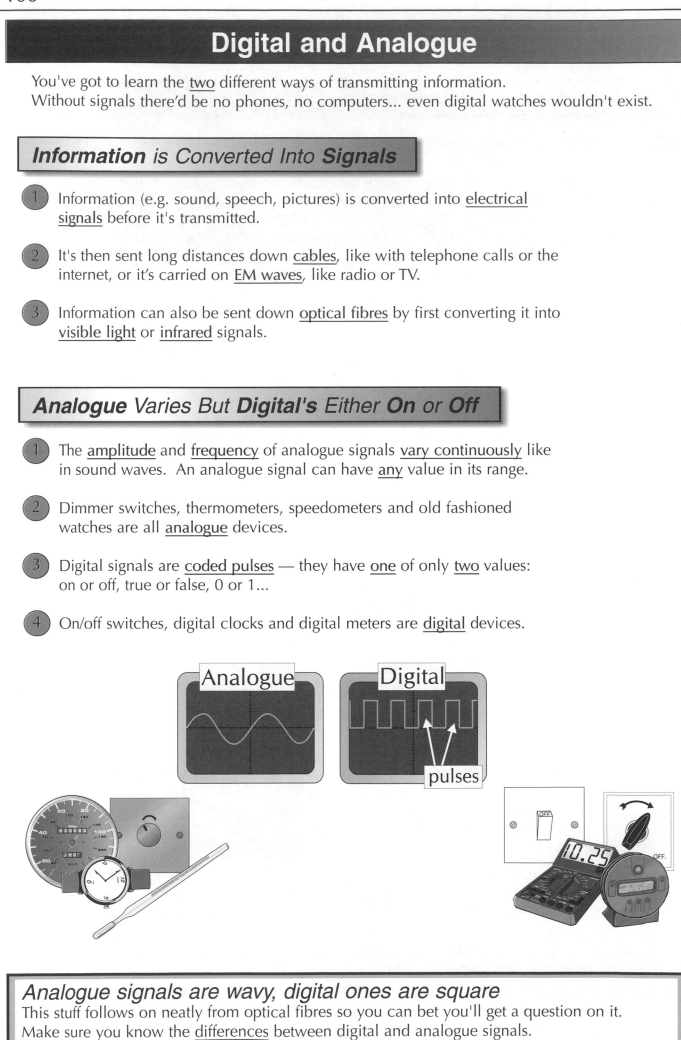

Analogue

Digital

pulses

## *Analogue signals are wavy, digital ones are square*
This stuff follows on neatly from optical fibres so you can bet you'll get a question on it.
Make sure you know the <u>differences</u> between digital and analogue signals.

# Digital and Analogue

## Signals *Have to Be* **Amplified**

Both digital and analogue signals <u>weaken</u> as they travel
so they need to be <u>amplified</u> along their route.
They also pick up <u>random disturbances</u>, called <u>noise</u>.

### Analogue *Signals* **Lose** *Quality*

Each time it's amplified, the analogue signal gets <u>less and less</u> like the original.
The different frequencies in it <u>weaken differently</u> at different times — when the
signal is amplified, the <u>differences and noise</u> are amplified too.

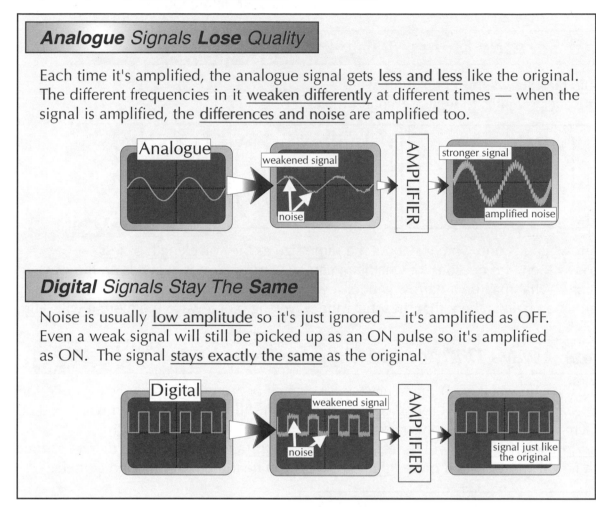

### Digital *Signals Stay The* **Same**

Noise is usually <u>low amplitude</u> so it's just ignored — it's amplified as OFF.
Even a weak signal will still be picked up as an ON pulse so it's amplified
as ON. The signal <u>stays exactly the same</u> as the original.

## Digital *Signals are* **Far Better Quality**

1) Digital signals <u>don't change</u> while they're being transmitted.
This makes them <u>higher quality</u> — the information transmitted
is the <u>same</u> as the original.

2) <u>Loads more information</u> can be sent as digital signals compared to
analogue (in a certain time). Many digital signals can be transmitted
at once by a clever way of <u>overlapping</u> them on the <u>same</u> cable or
EM wave — but you don't need to learn *how* they do it.

---

### *With digital signals it's much easier to get rid of noise*
Make sure you know <u>why</u> digital signals are better than analogue ones (for transmitting).
Learn all the details, then turn the book over and write them all down.

# Diffraction

This word sounds a lot more technical than it really is.

## *Diffraction is Just the "Spreading Out" of Waves*

All waves tend to spread out at the edges when they pass through a gap or past an object. Instead of saying that the wave "spreads out" or "bends" round a corner you should say that it diffracts around the corner. It's as easy as that. That's all diffraction means.

## *A Wave Spreads More if it Passes Through a Narrow Gap*

The ripple tank shows this effect quite well.
The same effect applies to light and sound waves too.

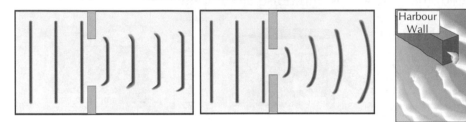

1) A "narrow" gap is one which is about the same size as the wavelength or less.
2) Obviously then, the question of whether a gap is "narrow" or not depends on the wave in question. What may be a narrow gap for a water wave will be a huge gap for a light wave.
3) It should be obvious then, that the longer the wavelength of a wave the more it will diffract.

## *Sounds Always Diffract Quite a Lot, Because λ is Quite Big*

1) A typical sound wave in air might have a wavelength of around 0.1 m, which is quite long.
2) This means they spread out round corners so you can still hear people even when you can't see them directly (the sound usually reflects off walls too which also helps).
3) Higher frequency sounds will have shorter wavelengths and so they won't diffract as much, which is why things sound more "muffled" when you hear them from round corners.

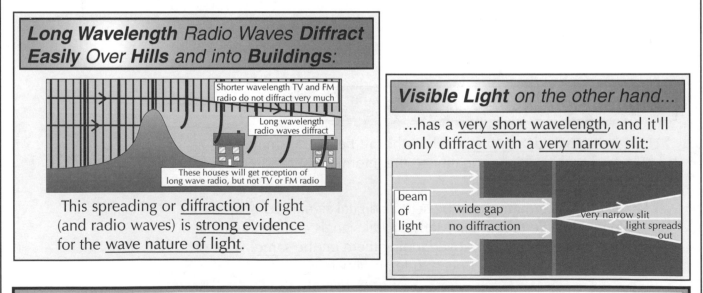

### *Long Wavelength Radio Waves Diffract Easily Over Hills and into Buildings:*

This spreading or diffraction of light (and radio waves) is strong evidence for the wave nature of light.

### *Visible Light on the other hand...*

...has a very short wavelength, and it'll only diffract with a very narrow slit:

## *The radio waves example often comes up in Exams*

People usually don't know much about diffraction, mainly because there are so few lab demos you can do to show it, and there's also very little to say about it — about one page's worth, in fact. The thing is though, if you just learn this page properly, then you'll know all you need to.

# Warm-Up and Worked Exam Question

## Warm-up Questions

1) Why are prisms used in binoculars and periscopes instead of ordinary mirrors?
2) In communications, what are the main advantages of using optical signals over using electrical signals?
3) List three examples of analogue devices and three examples of digital devices.
4) Draw an analogue signal and a digital signal, then draw them both again after they've been amplified several times.

## Worked Exam Question

1    a)    The diagram shows an optic fibre as used in an endoscope.

(i)    Complete the diagram to show how light travels along the fibre.

*(2 marks)*

(ii)    Describe what endoscopes are used for in medicine.
*For looking inside patients without having to make a big hole.*

*(2 marks)*

b)    A periscope can be used to see over walls.  The diagram shows a ray of light entering the periscope, and being reflected into an observer's eye.

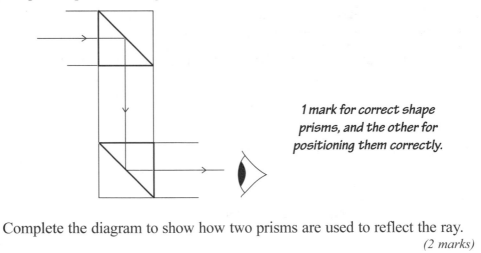

*1 mark for correct shape prisms, and the other for positioning them correctly.*

(i)    Complete the diagram to show how two prisms are used to reflect the ray.

*(2 marks)*

*Any one of these answers would get you the mark.*

(ii)    Why are prisms used instead of mirrors?
*better reflection, easier to support, more robust*

*(1 mark)*

# Exam Questions

1) Signals can be classed as either analogue or digital.

   a) What is meant by a digital signal?

   ...................................................................................................................................

   *(1 mark)*

   b) Complete the following table with words from the list,
      to show examples of analogue and digital devices.

   | Analogue | Digital |
   |----------|---------|
   |          |         |

   dimmer switch

   thermometer

   electronic scales

   speedometer

   light switch

   *(2 marks)*

   c) The diagram shows an analogue signal.

      Describe what happens to the analogue signal as it is amplified repeatedly.

   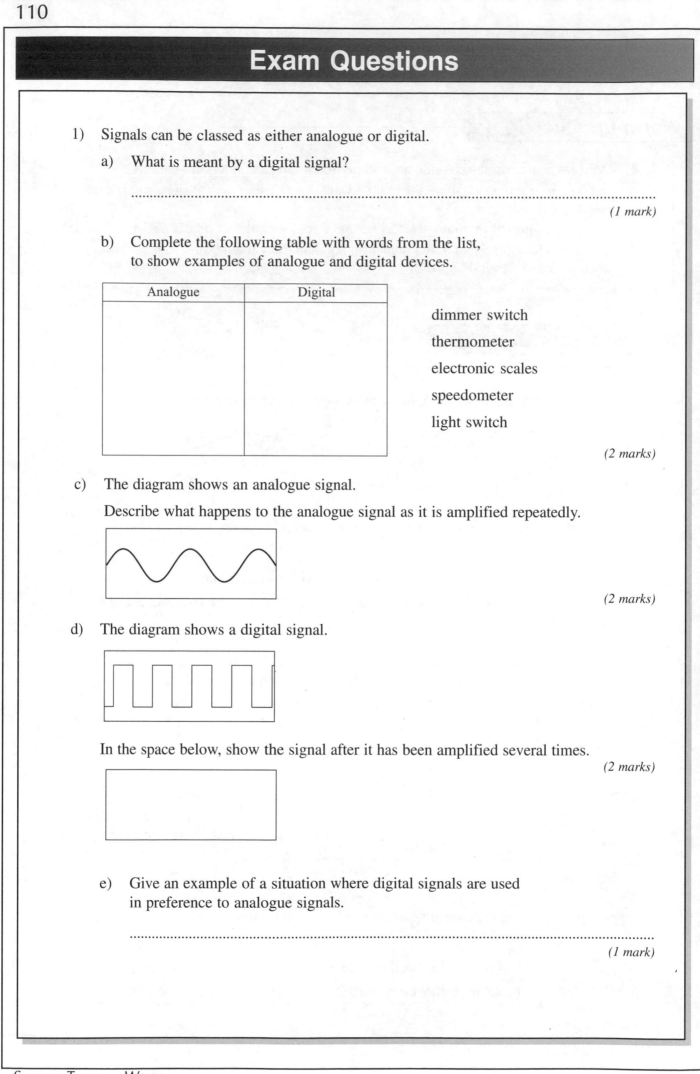

   *(2 marks)*

   d) The diagram shows a digital signal.

      In the space below, show the signal after it has been amplified several times.

      *(2 marks)*

   e) Give an example of a situation where digital signals are used
      in preference to analogue signals.

   ...................................................................................................................................

   *(1 mark)*

# Exam Questions

2) In a science lesson, the teacher shows the students a ripple tank with water in it.
She makes straight water waves move towards the gap in a straight barrier as shown:

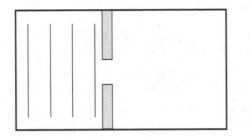

a) Complete the diagram to show what happens to the waves that have passed through the gap.

*(1 mark)*

b) What is this effect called?

..............................................................................................................................

*(1 mark)*

The teacher moves the barrier to widen the gap as shown below.

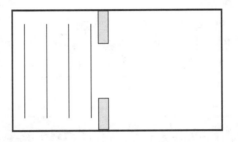

c) Sketch how the waves on the other side of the gap would look now.

*(1 mark)*

d) If the waves were made to have a higher frequency, in what way would the effect differ?

..............................................................................................................................

*(1 mark)*

This same effect is experienced by radio waves as they travel over hills.

e) Which are diffracted more — long wavelength radio waves, or shorter wavelength TV and FM radio waves?

..............................................................................................................................

*(1 mark)*

# Electromagnetic Spectrum

## The Seven Types of EM Wave Travel At the Same Speed

The properties of electromagnetic waves (EM waves) change as the frequency (or wavelength) changes. We split them into seven basic types as shown below. These EM waves form a continuous spectrum so the different regions do actually merge into each other.

| RADIO WAVES | MICRO WAVES | INFRA RED | VISIBLE LIGHT | ULTRA VIOLET | X-RAYS | GAMMA RAYS |
|---|---|---|---|---|---|---|
| $1m-10^4m$ | $10^{-2}m$ (3cm) | $10^{-5}m$ (0.01mm) | $10^{-7}m$ | $10^{-8}m$ | $10^{-10}m$ | $10^{-12}m$ |

Our eyes can only detect a very narrow range of EM waves — the ones we call (visible) light. All EM waves travel at exactly the same speed as light in a vacuum, and pretty much the same speed as light in other media like glass or water — though this is always slower than their speed in a vacuum.

## As the Wavelength Changes, so do the Properties

1) As the wavelength of EM radiation changes, its interaction with matter changes. In particular, the way any EM wave is absorbed, reflected or transmitted by any given substance depends entirely on its wavelength — that's the whole point of these three pages of course.

2) As a rule the EM waves at each end of the spectrum tend to be able to pass through material, whilst those nearer the middle are absorbed.

3) Also, the ones at the top end (high frequency, short wavelength) tend to be the most dangerous, whilst those lower down are generally harmless.

4) When any EM radiation is absorbed it can cause two effects:

   a) Heating    b) Creation of a tiny alternating current with the same frequency as the radiation.

5) You need to know all the details that follow about all the different parts of the EM spectrum.

## Radio Waves are Used Mainly for Communications

1) Radio Waves are used mainly for communication and for controlling things like model aeroplanes.

2) Both TV and FM Radio use short wavelength radio waves of about 1m wavelength.

3) To receive these wavelengths you need to be more or less in direct sight of the transmitter, because they will not bend (diffract) over hills or travel very far through buildings.

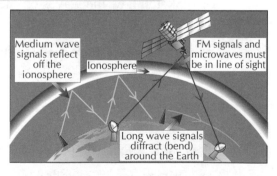

Medium wave signals reflect off the ionosphere

FM signals and microwaves must be in line of sight

Ionosphere

Long wave signals diffract (bend) around the Earth

4) Long Wave radio on the other hand has wavelengths of about 1 km and these waves will bend over the surface of the Earth and also diffract into tunnels and all sorts.

5) Medium Wave radio signals which have wavelengths of about 300 m can be received long distances from the transmitter because they are reflected from the ionosphere, which is an electrically charged layer in the Earth's upper atmosphere. However, these signals are usually very fuzzy.

## That's why hilly areas can get French radio clearer than Radio 4

There are lots of details on this page that you definitely need to know. The top diagram is an absolute must — they usually give it you with one or two missing labels to be filled in. Learn the three sections on this page then write a mini-essay for each one to see what you know.

# Electromagnetic Spectrum

## Microwaves Are Used For Cooking and Satellite Signals

1) Microwaves have two main uses: cooking food and satellite transmissions.

2) These two applications use two different frequencies of microwaves.

3) Satellite transmissions use a frequency which passes easily through the Earth's atmosphere, including clouds, which seems pretty sensible.

4) The frequency used for cooking, on the other hand, is one which is readily absorbed by water molecules. This is how a microwave oven works. The microwaves pass easily into the food and are then absorbed by the water molecules and turn into heat inside the food.

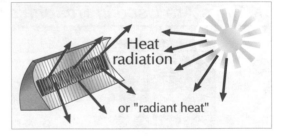

5) Microwaves can therefore be dangerous because they can be absorbed by living tissue and the heat will damage or kill the cells causing a sort of "cold burn".

## Infrared Radiation — Night-Vision and Remote Controls

1) Infrared (or IR) is otherwise known as heat radiation. This is given out by all hot objects and you feel it on your skin as radiant heat. Infrared is absorbed by all materials and causes heating.

2) Radiant heaters (i.e. those that glow red) use infrared radiation, including toasters and grills.

3) Infrared is also used for night-vision equipment. This works by detecting the heat radiation given off by all objects, even in the dark of night, and turning it into an electrical signal which is displayed on a screen as a clear picture. The hotter an object is, the brighter it appears. Police and the military use this to spot miscreants running away, like you've seen on TV.

4) Infrared is also used for all the remote controls of TVs and videos. It's ideal for sending harmless signals over short distances without interfering with radio frequencies (like the TV channels).

## It's a shame facts aren't as easily absorbed as infrared radiation

Each part of the EM spectrum is different, and you definitely need to know all the details about each type of radiation. These are just the kind of things they'll test in your Exams. Do mini-essays for microwaves and IR. Then check to see how you did. Then try again... and again...

# Electromagnetic Spectrum

## Visible light is Used To See With and In Optical Fibres

1) Visible light is pretty useful. We use it for <u>seeing</u> with for one thing.

2) It's also used in <u>Optical Fibre Digital Communications</u> which is the best use by far for an answer <u>in the Exam</u>.

3) You could say that one use of it is in <u>endoscopes</u> for seeing inside a patient's body, but then there are also <u>microscopes</u>, <u>telescopes</u>, <u>kaleidoscopes</u>, seeing in the dark (torch, lights, glow stars, etc.) and for controlling things like model aeroplanes.

## Ultraviolet Light Causes Skin Cancer

1) <u>Skin cancer</u> is caused by spending <u>too much time</u> soaking up the UV rays from the <u>Sun</u>.

2) It makes you <u>tan</u>. <u>Sunbeds</u> give out fewer of the more harmful UV rays than the Sun, but they're still <u>harmful</u>.

3) <u>Tanned or darker skin</u> protects against UV rays. It <u>stops</u> them reaching more <u>vulnerable skin tissues</u> deeper down.

4) <u>Special coatings</u> which <u>absorb UV light</u> and then <u>give out visible light</u> instead are used to coat the inside of <u>fluorescent tubes</u> and lamps.

5) UV is also useful for <u>hidden security marks</u> which are written in special ink that can only be seen with an ultraviolet light.

*Note: only UV light causes sunburn*

Fluorescent Tube

UV light produced inside tube

Coating on glass absorbs UV and emits visible light

## X-Rays Are Used in Hospitals, but are Pretty Dangerous

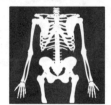

*The <u>brighter bits</u> are where <u>fewer</u> X-rays get through. This is a <u>negative image</u>. The plate starts off <u>all white</u>.*

1) <u>Radiographers</u> in <u>hospitals</u> take <u>X-ray photographs</u> of people to see whether they have any <u>broken bones</u>.

2) X-rays pass <u>easily through flesh</u> but not through <u>denser material</u> like <u>bones</u> or <u>metal</u>.

3) X-rays can cause <u>cancer</u>, so radiographers wear <u>lead aprons</u> and stand behind a <u>lead screen</u> or <u>leave the room</u> to keep their <u>exposure</u> to X-rays to a <u>minimum</u>.

## Gamma Rays Treat Cancer Without Surgery

1) Gamma rays are used to kill <u>harmful bacteria</u> to keep food <u>fresher for longer</u> and <u>sterilise medical instruments</u>.

2) In <u>high doses</u>, gamma rays, X-rays and UV rays can <u>kill normal cells</u>.

3) In <u>lower doses</u>, these three types of EM waves can cause normal cells to become <u>cancerous</u>.

4) If the dose is just right, gamma rays can be used to treat cancer <u>without surgery</u> because they <u>kill cancer cells</u>.

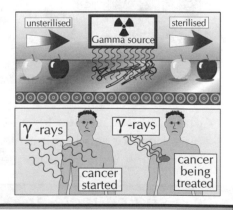

## Gamma rays can cause or treat cancer, depending on the dose

Here are the other four parts of the EM spectrum for you to learn.

Do a <u>mini-essay</u> for each section, then <u>check</u>, <u>re-learn</u>, <u>re-scribble</u>, <u>re-check</u>, etc. etc.

# Warm-Up and Worked Exam Question

## Warm-up Questions

1) What type of radiation lies between visible light and X-rays in the electromagnetic spectrum?
2) Give one use for each of the following types of radiation:
   a) microwaves   b) infrared   c) gamma rays.
3) Which type of radiation has the longer wavelength — gamma rays or radio waves?

## Worked Exam Question

This is a classic question type — fill in the electromagnetic spectrum and reel off a load of facts about electromagnetic waves. It's not too complicated, but you need to remember all the details to get all the marks. If you can't answer all of these questions, you need to go back and learn some more.

1) This diagram shows part of the electromagnetic spectrum.

| radio | microwaves | infrared | y | ultraviolet | z | gamma |
|-------|-----------|----------|---|-------------|---|-------|

a) Name the missing types of radiation, y and z.

y: *visible light*

z: *X-rays*

[1 mark]

b) Which type of radiation has the shortest wavelength?

*gamma rays*

[1 mark]

c) Which radiation has the highest frequency?

*gamma rays*

[1 mark]

d) Name one method of detecting X-rays.

*photographic film*

[1 mark]

e) Describe one danger of gamma rays.

*can kill cells / cause cancer*

[1 mark]

## Exam Questions

1    Microwaves are used for sending communications between satellite dishes.
    These signals cannot be detected by an ordinary radio.

    a)    Describe a difference between radio and microwave signals.

    .................................................................................................................................

    .................................................................................................................................

    *[1 mark]*

    b)    How do the microwaves used in communications differ from those used
    for cooking?

    .................................................................................................................................

    .................................................................................................................................

    *[1 mark]*

    c)    How does this difference make the microwaves used for cooking more dangerous
    to people than those used in communications?

    .................................................................................................................................

    .................................................................................................................................

    *[1 mark]*

## Exam Questions

2  Most types of electromagnetic radiation can be dangerous to people, as well as being useful.

a) Give one danger and one use of each of the following types of radiation.

(i)   ultraviolet

(ii)  X-rays

(iii) infrarcd

(iv)  visible light

|            | Danger | Use |
|------------|--------|-----|
| ultraviolet |        |     |
| X-rays      |        |     |
| infrared    |        |     |
| visible light |      |     |

*[8 marks]*

b) Name one property that all electromagnetic waves have in common.

.............................................................................................................................

*[1 mark]*

# The Earth

## Seismic Waves Are Caused By Earthquakes

1. We can only drill about 10 km or so into the crust of the Earth, which is not very far, so seismic waves are really the only way of investigating the inner structure.

2. When there's an earthquake somewhere the seismic waves travel out from it and we detect them all over the surface of the planet using seismographs.

3. There are two different types of seismic wave (see below). The time it takes for them to reach each seismograph is measured.

4. Seismologists also note the parts of the Earth which don't receive the waves at all.

5. From this information you can work out all sorts of stuff about the inside of the Earth as shown below:

## S-Waves and P-Waves Take Different Paths

### P-Waves are Longitudinal

P-Waves travel through both solids and liquids. They travel faster than S-waves.

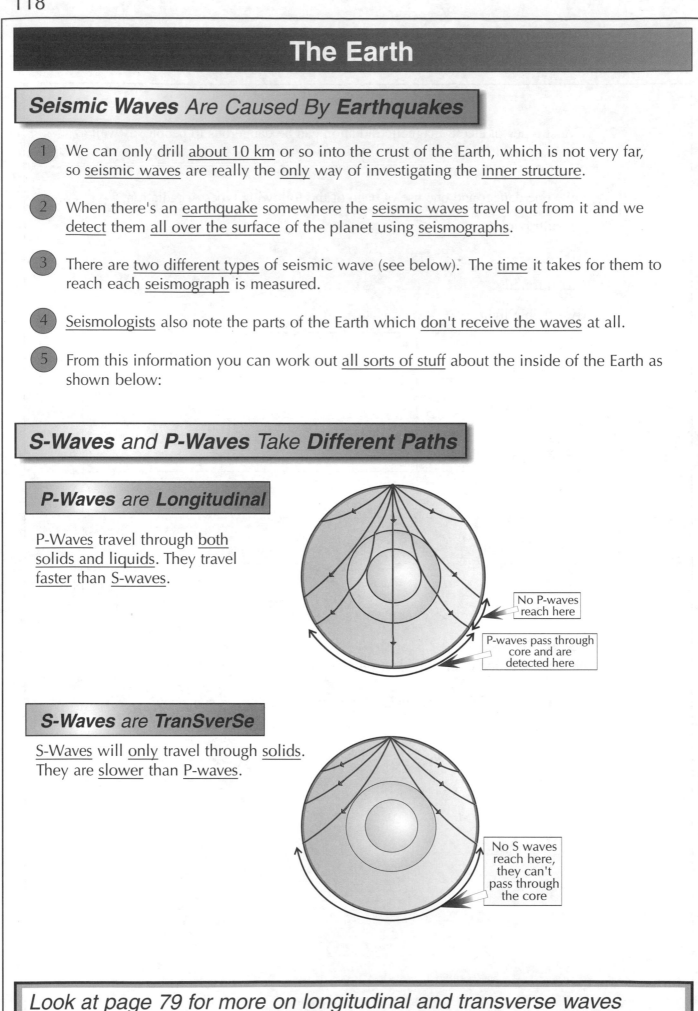

No P-waves reach here

P-waves pass through core and are detected here

### S-Waves are TranSverSe

S-Waves will only travel through solids. They are slower than P-waves.

No S waves reach here, they can't pass through the core

## Look at page 79 for more on longitudinal and transverse waves

Write a mini-essay describing what happens in terms of seismic waves when there's an earthquake. Draw the diagrams and shade the area that misses out on the S-waves.

# The Earth

## The **Seismograph** Results Tell Us What's **Down There**

(1) About <u>halfway through</u> the Earth, there's an <u>abrupt change in direction</u> of both types of wave. This indicates that there's a <u>sudden increase in density</u> at that point — the <u>core</u>.

(2) The fact that S-waves are <u>not detected</u> in the <u>shadow</u> of this core tells us that it's very <u>liquid</u>.

(3) It's also found that <u>P-waves</u> travel <u>slightly faster</u> through the <u>middle</u> of the core, which strongly suggests that there's a <u>solid inner core</u>.

(4) Note that <u>S-waves</u> do travel through the <u>mantle</u> which suggests that it's pretty <u>solid</u>, although I always thought it was made of <u>molten lava</u> which looks fairly <u>liquid</u> when it comes <u>rushing</u> out of volcanoes. Another one of life's mysteries.

## The Paths **Curve** Due to **Increasing Density** (causing **Refraction**)

(1) Both <u>S-waves</u> and <u>P-waves</u> travel <u>faster</u> in <u>more dense</u> material.

(2) The <u>curvature of their paths</u> is due to the <u>increasing density</u> of the <u>mantle</u> and <u>core</u> with depth.

(3) When the density changes <u>suddenly</u>, the waves change direction <u>abruptly</u>, as shown opposite.

(4) The paths <u>curve</u> because the density of both the mantle and the core <u>increases steadily</u> with increasing depth. The waves <u>gradually change direction</u> because their speed is <u>gradually changing</u>, due to gradual changes in the <u>density</u> of the medium. This is <u>refraction</u>, of course.

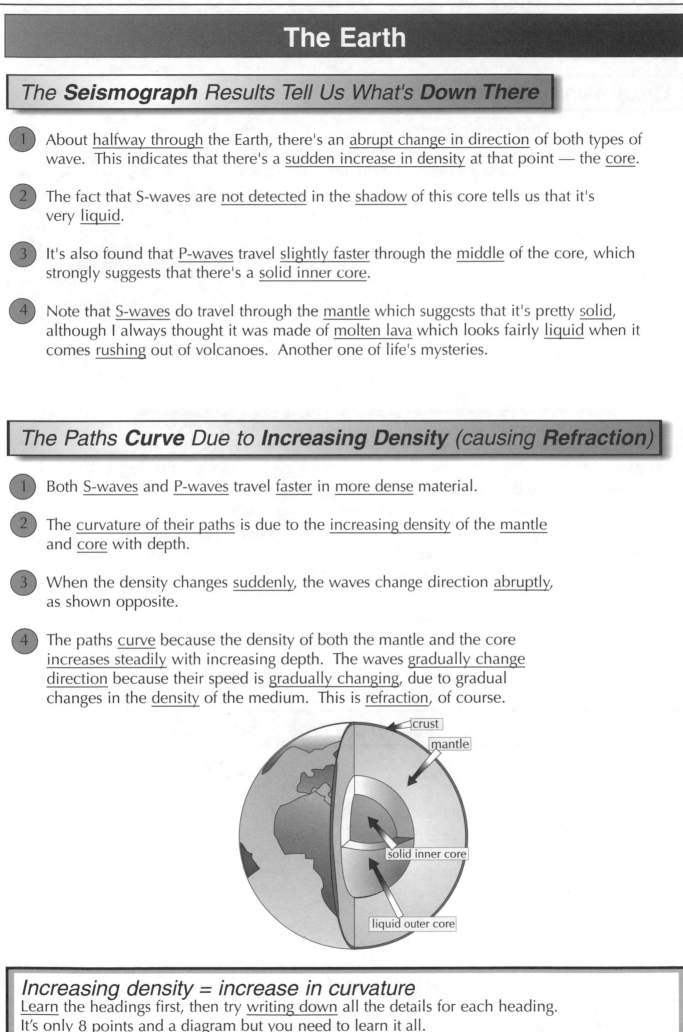

crust

mantle

solid inner core

liquid outer core

---

*Increasing density = increase in curvature*
<u>Learn</u> the headings first, then try <u>writing down</u> all the details for each heading.
It's only 8 points and a diagram but you <u>need to learn it all</u>.

# The Earth

## Crust, Mantle, Outer and Inner Core

1) The <u>crust</u> is very <u>thin</u> (about 20km).

2) The <u>mantle</u> extends almost <u>halfway</u> to the centre of the Earth.

3) It's got all the properties of a <u>solid</u> but it can flow very <u>slowly</u>.

4) The <u>core</u> is just over <u>half</u> the Earth's radius.

5) The <u>core</u> is made of <u>iron and nickel</u>. This is where the Earth's <u>magnetic field</u> originates.

6) The core is <u>solid in the middle</u> and <u>liquid at the edge</u>.

7) <u>Radioactive decay</u> creates a lot of the <u>heat</u> inside the Earth.

8) This heat causes the <u>convection currents</u> which cause the <u>plates</u> of the crust to <u>move</u>.

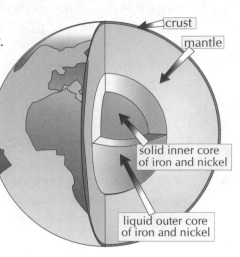

## Big Clues: Seismic Waves, Magnetism and Meteorites

1) We can tell how <u>dense</u> the Earth is by measuring <u>seismic waves</u> and the <u>Earth's motion</u>. We find that the inner core is much <u>too dense</u> to be made out of rock.

2) <u>Meteorites</u> which crash to Earth are often made of <u>iron and nickel</u>.

3) If the <u>core</u> of the Earth was made of <u>iron and nickel</u> it would explain a lot: iron and nickel are about the <u>right density</u>, and being metals, this would explain the <u>Earth's magnetic field</u> (it's like a giant electromagnet).

4) Also, by following the paths of <u>seismic waves</u> as they travel through the Earth, we can tell that there is a <u>change</u> to <u>liquid</u> about <u>halfway</u> through the Earth.

5) There must be a <u>liquid outer core of iron and nickel</u>. The seismic waves also indicate a solid inner core.

## The Earth's Surface is made up of Large Plates of Rock

1) These <u>plates</u> are like <u>big rafts</u> that float across the mantle.

2) The map shows the <u>edges</u> of these plates. As they <u>move</u>, the <u>continents</u> move too.

3) Most of the plates are moving at about <u>1 cm or 2 cm per year</u>.

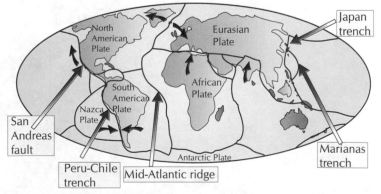

## There's more to the Earth than meets the eye

This is pretty easy — just make sure you learn <u>all</u> the details.
There's <u>nothing dafter</u> than missing easy marks. <u>Cover the page and check you know it all</u>.

# Warm-Up and Worked Exam Question

## Warm-up Questions

1) All waves are either longitudinal or transverse.
   What types of wave are S-waves and P-waves?

2) Complete the following sentences:

   The Earth is made up of_____ layers.  These layers are the _____,
   the _____, the _____ core and the _____ core.

3) What two metals are thought to make up the core of the Earth?

## Worked Exam Question

1    During an earthquake, scientists take readings of the seismic waves that travel
     through the Earth.

a)   What do we call the equipment used for measuring these waves?

     *a seismograph*
                                                                          *[1 mark]*

b)   Which type of seismic wave, P or S, will travel through both liquid and solid?

     *P-wave*
                                                                          *[1 mark]*

Readings from earthquakes have helped us to build up a picture of the structure of the
Earth.  The surface crust appears to be made of large plates of rock that move around
very slowly.

c)   Describe what makes these plates move.

     *Heat from the core causes convection currents in the mantle,*

     *which moves the tectonic plates that float on them.*
                                                                         *[2 marks]*

d)   How do the inner and outer cores differ?

     *The inner core is solid, whereas the outer core is liquid.*
                                                                          *[1 mark]*

# Exam Questions

1    The diagram shows the inside of the earth.

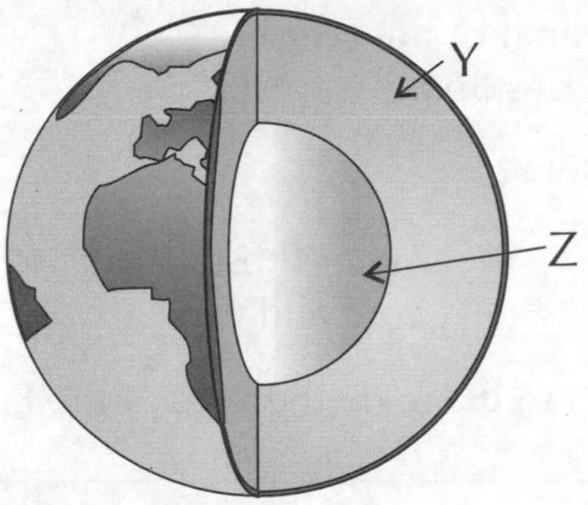

a)    Name the layers marked Y and Z.

Y: .................................................            Z: ...................................................

*[2 marks]*

b)    Two types of seismic waves produced during an earthquake are P-waves and
S-waves.  What type of wave are P-waves — longitudinal or transverse?

...........................................................................................................................

*[1 mark]*

c)    The diagram shows S-waves travelling through the earth after an earthquake.

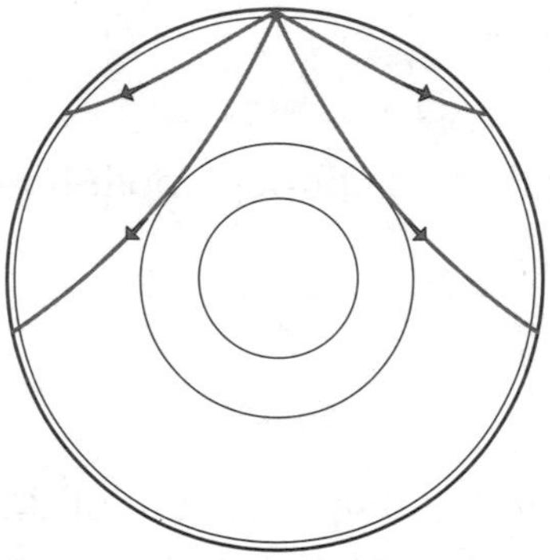

None of the S-waves travel through the Earth's core.
What does this tell us about the structure of the Earth's core?

...........................................................................................................................

*[1 mark]*

d)    Scientists believe that the Earth's inner core is solid.  Why do they believe this?

...........................................................................................................................

*[1 mark]*

e)    Explain why the paths of the waves curve as shown in the diagram above.

...........................................................................................................................

*[1 mark]*

# Plate Tectonics

The old theory was that all the features of the Earth's surface, e.g. mountains, were due to shrinkage of the crust as it cooled. In the Exam they may ask you about that, and then they'll ask you for evidence in favour of plate tectonics as a better theory.

## 1) *Jigsaw Fit* — *the supercontinent "Pangaea"*

a) There's a very obvious jigsaw fit between Africa and South America.

b) The other continents can also be fitted in without too much trouble.

c) It's widely believed that they once all formed a single land mass, now called Pangaea.

## 2) *Matching Fossils* in Africa and South America

a) Identical plant fossils of the same age have been found in rocks in South Africa, Australia, Antarctica, India and South America, which strongly suggests they were all joined once upon a time.

b) Animal fossils support the theory too. There are identical fossils of a freshwater crocodile found in both Brazil and South Africa. It certainly didn't swim across.

## 3) *Identical* Rock Sequences

a) When rock strata of similar ages are studied in various countries they show remarkable similarity.

b) This is strong evidence that these countries were joined together when the rocks formed.

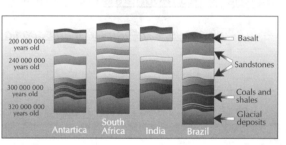

## 4) *Living Creatures*: The Earthworm

a) There are various living creatures found in both America and Africa.

b) One such creature is a particular earthworm which is found living at the tip of South America and the tip of South Africa.

c) Most likely it travelled across ever so slowly on the land mass that's now America.

## *Wegener's Theory* of Crustal Movement

This stuff was noticed hundreds of years ago, but nobody really believed that the continents could once have actually been joined. In 1915, a chap called Alfred Wegener proposed his theory of "continental drift" saying that they had definitely been joined and that they were slowly drifting apart. This wasn't accepted for two reasons:

      a) he couldn't give a convincing reason why it happened,

      b) he wasn't a qualified geologist.

Only in the 1960s with fossil evidence and the magnetic pattern (see page 52) from the mid-Atlantic ridge was the theory widely accepted.

---

## *Learn about Plate Tectonics — but don't get carried away*

Four bits of evidence support the theory that there are big plates of rock moving about. Learn all four well enough to be able to answer a question like this: *"Describe evidence which supports the theory of Plate Tectonics (4 marks)."*

# Plate Tectonics

At the <u>boundaries</u> between tectonic plates there's usually <u>trouble</u> like <u>volcanoes</u> or <u>earthquakes</u>. There are <u>three</u> different ways that plates interact:  <u>colliding</u>, <u>separating</u> or <u>sliding</u> past each other.

## *Oceanic* and *Continental* Plates Colliding:  *The Andes*

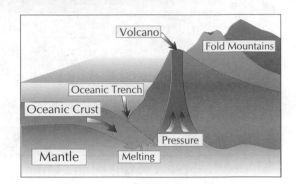

1) The <u>oceanic plate</u> is always <u>forced underneath</u> the continental plate because it's <u>denser</u>.

2) This is called a <u>subduction zone</u>.

3) As the oceanic crust is pushed down it <u>melts</u> and <u>pressure builds up</u> due to the melting rock.

4) This <u>molten rock</u> finds its way to the <u>surface</u> and <u>volcanoes</u> form.

5) There are also <u>earthquakes</u> as the two plates slowly <u>grind</u> past each other.

6) A <u>deep trench</u> forms on the ocean floor where the <u>oceanic plate</u> is being <u>forced down</u>.

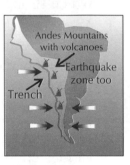

7) The <u>continental</u> crust <u>crumples</u> and <u>folds</u> forming <u>mountains</u> at the coast.

8) The classic example of all this is the <u>west coast of South America</u> where the <u>Andes mountains</u> are.  That region has <u>all the features</u>:

> *<u>Volcanoes</u>, <u>earthquakes</u>, an <u>oceanic trench</u> and <u>mountains</u>.*

## *Two Continental* Plates Collide:  *The Himalayas*

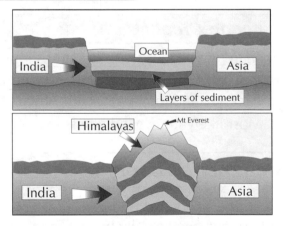

1) The <u>two continental plates</u> meet <u>head on</u>, neither one being subducted.

2) Any <u>sediment layers</u> lying between the two continent masses get <u>squeezed</u> between them.

3) These sediment layers inevitably start <u>crumpling and folding</u> and soon form into <u>big mountains</u>.

4) The <u>Himalayan mountains</u> are the classic case of this.

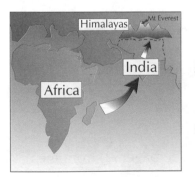

5) <u>India</u> actually <u>broke away</u> from the side of <u>Africa</u> and <u>pushed</u> into the bottom of <u>Asia</u>, and is <u>still</u> doing so, <u>pushing the Himalayas up</u> as it goes.

6) <u>Mount Everest</u> is there and is <u>getting higher</u> by a few cm every year as India continues to push up into the continent of Asia.

---

## *And Mount Everest is STILL growing*

<u>Make sure you learn all these diagrams</u> — they summarise all the information in the text. They may well ask you for examples in the Exam, so make sure you know the two different kinds of situation that the Andes and the Himalayas actually represent.  <u>Cover and scribble...</u>

# Plate Tectonics

## *Sea Floor Spreading:* **The Mid-Atlantic Ridge**

1) When tectonic plates move <u>apart</u>, <u>magma</u> <u>rises up</u> to fill the gap and produces <u>new crust</u> made of <u>basalt</u> (of course). Sometimes it comes out with <u>great force</u> producing <u>undersea volcanoes</u>.

2) The <u>Mid-Atlantic ridge</u> runs the <u>whole length</u> of the Atlantic and cuts through the middle of <u>Iceland</u>, which is why they have <u>hot underground water</u>.

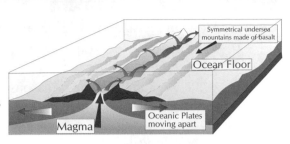

3) As the magma rises up through the gap, it forms <u>ridges</u> and <u>underwater mountains</u>.

4) These form a <u>symmetrical pattern</u> either side of the ridge, providing strong <u>evidence</u> for the theory of <u>continental drift</u>.

5) However the most <u>compelling</u> evidence comes from the <u>magnetic orientation</u> of the rocks.

6) As the <u>liquid magma</u> erupts out of the gap, the <u>iron particles</u> in it tend to <u>align themselves</u> with the <u>Earth's magnetic field</u> and as the rocks cool they <u>set</u> in position.

7) Every half million years or so the Earth's magnetic field tends to <u>swap direction</u>.

8) This means the rock on <u>either side of the ridge</u> has bands of <u>alternate magnetic polarity</u>.

9) This pattern is found to be <u>symmetrical</u> either side of the ridge.

## *Plates* **Sliding Past** *Each Other:* **San Francisco**

*(They didn't know about this when they built the city.)*

1) Sometimes the plates are just <u>sliding past each other</u>.
2) The best known example of this is the <u>San Andreas Fault</u> in California.
3) A narrow strip of the coastline is <u>sliding north</u> at about <u>7cm a year</u>.
4) Big plates of rock <u>don't glide smoothly</u> past each other.
5) They <u>catch</u> on each other and as the <u>forces build up</u> they suddenly <u>lurch</u>.
6) This <u>sudden lurching</u> only lasts <u>a few seconds</u> — but it can bring buildings down easily.
7) The city of <u>San Francisco</u> sits <u>astride</u> this fault line.
8) The city was <u>destroyed</u> by an earthquake in <u>1906</u> and hit by another quite serious one in <u>1991</u>. They could have another one <u>any time</u>.
9) In <u>earthquake zones</u> they try to build <u>earthquake-proof buildings</u> which are designed to withstand a bit of shaking.
10) Earthquakes usually cause <u>much greater devastation</u> in <u>poorer countries</u> where they may have <u>overcrowded cities</u>, <u>poorly constructed buildings</u>, and <u>inadequate rescue services</u>.
11) It's impossible to accurately predict <u>when</u> an earthquake will occur because there are so many reasons <u>why</u> they happen — and it's hard to <u>take measurements</u>.

## *San Francisco — a disaster waiting to happen*

<u>Let me remind you</u> of the benefits of the <u>mini-essay method</u>. You read the stuff and try and learn it. Then you cover the page and scribble yourself a mini-essay on each topic. Then you look back and see what stuff you missed. Then you try again, and again — until you get it all.

# Warm-Up and Worked Exam Question

## Warm-up Questions

1) How do the shapes of South America and Africa provide evidence for plate tectonics?

2) Give three reasons, other than shape, why scientists think South America and Africa were once joined together.

3) What were the two main reasons Wegener's theory of 'continental drift' was not accepted in the early part of the 20th century?

## Worked Exam Question

1 The Earth's crust is made up from sections of rock called tectonic plates.

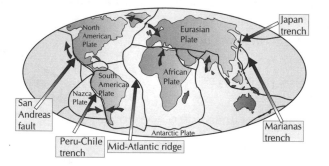

There are two types of plates — oceanic plates and continental plates.
Oceanic plates are denser than continental plates.

a) When oceanic and continental plates meet, the oceanic plate is always forced down underneath the continental plate. Explain why this happens.

*The oceanic plate is denser, so it will sink*

*beneath the continental plate.*

*(2 marks)*

b) As the oceanic plate is pushed into the mantle it melts.
What does this create around this area?

*an increase in pressure, which can lead to volcanoes or earthquakes*

*(1 mark)*

c) Name two other features that are formed where oceanic and continental plates meet.

*oceanic trenches Mountains*

*...or any other two sensible answers*
*NOT mentioned in part (b), e.g. volcano.*

*(2 marks)*

# Exam Questions

1    The city of San Francisco sits on a fault line.

    a)    Explain what is meant by a fault line.

    .................................................................................................................................................

    .................................................................................................................................................

                                                                       *(1 mark)*

    b)    At San Francisco the Pacific plate is rubbing against the North American plate.
        Explain how this causes frequent earthquakes in this area.

    .................................................................................................................................................

    .................................................................................................................................................

                                                                       *(2 marks)*

2    The Himalayas contain the highest mountains in the world.
    These mountains were formed by the Indian plate and Asian plate meeting.

    a)    What happened to the layers of sediment between these two plates as they came
        together?

    .................................................................................................................................................

    .................................................................................................................................................

                                                                         *(1 mark)*

    b)    In 1960 the height of Mount Everest, the highest mountain in the Himalayas, was
        measured at 8820 m. Explain why this height might now be different.

    .................................................................................................................................................

    .................................................................................................................................................

                                                                          *(1 mark)*

  In the middle of the Atlantic, the South American plate and the African plate are moving apart.

    c)    Explain how this movement has resulted in the formation of the Mid-Atlantic Ridge.

    .................................................................................................................................................

    .................................................................................................................................................

                                                                        *(2 marks)*

# Revision Summary for Section Three

There are plenty of easy facts to learn about waves. There are still some bits that need thinking about, but most of it is fairly easy stuff that just needs learning. This book contains all the important information they've specifically mentioned in the specifications, and this is precisely the stuff they're going to test you on in the Exams. Practise these questions over and over again until they're easy.

1) Define frequency and time period for a wave. Give three examples of waves carrying energy.
2) Sketch transverse and longitudinal waves. Define them and give examples of both types.
3) Write down the speed of sound in air. Describe the bell jar experiment. What does it demonstrate?
4) Sketch graphs of normal and damaged hearing. Write down three ways of reducing noise pollution.
5) What's the connection between amplitude and the energy carried by a wave?
6) What effect does greater amplitude have on a) sound waves, and b) light waves?
7) What's the relationship between frequency and pitch for a sound wave?
8) Sketch CRO screens showing higher and lower pitch and quiet and loud sounds.
9) What is ultrasound? Give full details of five applications of ultrasound.
10) What are the three formulas involved with waves? How do you decide which one to use?
11) Are SI units important? What are the SI units for: wavelength; frequency; velocity; time?
12) Convert these to SI units: a) 500kHz, b) 35cm, c) 4.6MHz, d) 4 cm/s, e) 2½ mins.
13) Find the speed of a wave with frequency 50 kHz and wavelength 0.3 cm.
14) Find the time period of a wave of wavelength 1.5 km and speed $3 \times 10^8$ m/s.
15) A crash of thunder is heard 6 seconds after the flash of lightning. How far away is it?
16) If the sea bed is 600 m down, how long will it take to receive a sonar echo from it?
17) Sketch the patterns when plane ripples reflect at a) a plane surface, and b) a curved surface.
18) Sketch the reflection of curved ripples at a plane surface.
19) What is the law of reflection? Give a sketch to illustrate the diffuse reflection of light.
20) Draw a neat ray diagram to show how to locate the position of the image in a plane mirror.
21) What is refraction? What causes it? How does it affect wavelength and frequency?
22) Sketch a ray of light going through a rectangular glass block, showing the angles of incidence and reflection.
23) How fast does light travel in glass? Which way does it bend as it enters glass? What if it enters at right angles?
24) What is dispersion? Sketch a labelled diagram to illustrate it.
25) Sketch the three diagrams to illustrate Total Internal Reflection and the Critical Angle.
26) Sketch three applications of total internal reflection which use 45° prisms, and explain them.
27) Give details of the two main uses of optical fibres. How do optical fibres work?
28) Describe analogue and digital signals. Why are digital signals better?
29) What is diffraction? Sketch the diffraction of: a) water waves b) sound waves c) light.
30) What aspect of EM waves determines their differing properties?
31) Sketch the EM spectrum with all its details. What happens when EM waves are absorbed?
32) Give full details of the uses of radio waves. How do the three different types get "around"?
33) Give full details of the two main uses of microwaves, and the three main uses of infrared.
34) Give a sensible example of the use of visible light. What is its main use?
35) Detail three uses of UV light, two uses of X-rays and three uses of gamma rays.
36) What harm will UV, X-rays and gamma rays do in <u>high</u> doses? What about in <u>low</u> doses?
37) What causes seismic waves? Sketch diagrams showing the paths of both types, and explain.
38) Draw a diagram of the internal structure of the Earth, with labels.
39) How big are the various parts in relation to each other?
40) What is the core made of? What are the three big clues that tell us about the Earth?
41) What was the old theory about the Earth's surface? What is the theory of Plate Tectonics?
42) Give details of the four bits of evidence which support the theory of Plate Tectonics.
43) What are the three different ways that tectonic plates interact at boundaries?
44) What happens when an oceanic plate collides with a continental plate? Draw a diagram.
45) What four features does this produce? Which part of the world is the classic case of this?
46) What happens when two continental plates collide? Draw diagrams.
47) What features does this produce? Which part of the world is the classic case of this?
48) What is the mid-Atlantic ridge? What happens there? What is the evidence for this?
49) Which country lies on top of it? Do they get earthquakes? What *do* they get?
50) Where is the San Andreas fault? What are the tectonic plates doing along this fault line?
51) Why does it cause earthquakes — and why did they build San Francisco right on top of it?

## Celestial Bodies

You need to revise the <u>order</u> of the planets, which is made easier by using this mnemonic:

| Mercury, | Venus, | Earth, | Mars, | (Asteroids) | Jupiter, | Saturn, | Uranus, | Neptune, | Pluto |
|----------|--------|--------|-------|-------------|----------|---------|---------|----------|-------|
| (My | Very | Energetic | Maiden | Aunt | Just | Swam | Under | North | Pier) |

<u>Mercury</u>, <u>Venus</u>, <u>Earth</u> and <u>Mars</u> are known as the <u>inner planets</u>.
<u>Jupiter</u>, <u>Saturn</u>, <u>Uranus</u>, <u>Neptune</u> and <u>Pluto</u> are much further away and are the <u>outer planets</u>.

## *Planets* **Reflect Sunlight** *and Orbit in* **Ellipses**

1  You can <u>see</u> some of the nearer planets with the <u>naked eye</u> at night, e.g. Mars and Venus.

2  They look just like <u>stars</u>, but they are of course <u>totally different</u>.

3  Stars are <u>huge</u> and <u>very far away</u> and <u>give out</u> lots of light.  The planets are <u>smaller</u> and <u>nearer</u> and they just <u>reflect sunlight</u> falling on them.

4  The Sun, like other stars, produces <u>heat</u> from <u>nuclear fusion reactions</u> which turn <u>hydrogen</u> into <u>helium</u>.  It gives out the <u>full spectrum</u> of <u>EM radiation</u>.

5  Planets always orbit around <u>stars</u>.  In our Solar System the planets orbit the <u>Sun</u> of course.

6  These orbits are all <u>slightly elliptical</u> (elongated circles).

7  All the planets in our Solar System orbit in the <u>same plane</u> except Pluto (as shown in the picture above).

8  The <u>further</u> the planet is from the Sun, the <u>longer</u> its orbit takes (see next page about Gravity).

## *We're just the third rock from the Sun*

Isn't the Solar System great.  All those pretty coloured planets and all that big black empty space.  You can look forward to one or two easy questions on the planets — or you might get two real horrors instead.  Be ready — <u>learn</u> all the <u>little details</u> till you know it all inside out.

# Celestial Bodies

## *Gravity* Decreases *Quickly* as you get Further Away

1. With <u>very large</u> masses like <u>stars</u> and <u>planets</u>, gravity is <u>very big</u> and acts <u>a long way out</u>.

2. The <u>closer</u> you get to a star or a planet, the <u>stronger</u> the <u>force of attraction</u>.

3. To <u>counteract</u> the stronger gravity, planets nearer the Sun move <u>faster</u> and cover their orbit <u>quicker</u>.

4. <u>Comets</u> are also held in <u>orbit</u> by gravity, as are <u>moons</u> and <u>satellites</u> and <u>space stations</u>.

5. The size of the force of gravity follows the fairly famous "<u>inverse square</u>" relationship. The main effect of that is that the force <u>decreases very quickly</u> with increasing <u>distance</u>. The <u>formula</u> is $F \propto 1/d^2$, but it's probably <u>easier</u> just to remember the basic idea <u>in words</u>:

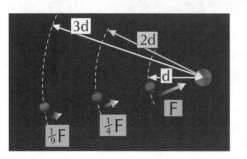

a) If you <u>double the distance</u> from a planet, the size of the <u>force</u> will <u>decrease</u> by a <u>factor of four</u> ($2^2$).

b) If you <u>treble the distance</u>, the <u>force</u> of gravity will <u>decrease</u> by a <u>factor of nine</u> ($3^2$), and so on.

c) On the other hand, if you get <u>twice as close</u> the gravity becomes <u>four times stronger</u>.

## *Planets* in the *Night Sky* Seem to *Move* across the *Constellations*

1. The <u>planets</u> look just like stars except that they <u>wander</u> across the constellations over periods of <u>days or weeks</u>, sometimes going back on themselves a bit.

2. Their position and movement depends on <u>where</u> they are in their orbit, compared to <u>us</u>.

3. This <u>peculiar movement</u> of the planets made the <u>early astronomers</u> realise that the Earth was <u>not the centre</u> of the Universe after all, but was in fact just the <u>third rock from the Sun</u>. It's <u>very strong evidence</u> for the <u>Sun-centred</u> model of the Solar System.

4. Unfortunately, the <u>Spanish Inquisition</u> saw this as heresy, and <u>Copernicus</u> had a pretty hard time of it for a while. In the end though, "<u>the truth will out</u>".

---

## *So the Earth moves much faster than, say, Neptune*

That "inverse square" relationship foxes quite a lot of people. If you struggle with formulas, just make sure you learn points a) to c) in the middle of the page. Then if you can't remember the formula in the exam, you can explain it using those examples instead.

# Celestial Bodies

## *Moons* are Heavenly Bodies Which *Orbit Planets*

1. The Earth only has <u>one</u> moon of course, but some of the <u>other planets</u> have <u>quite a few</u>.

2. We can only <u>see</u> the Moon because it <u>reflects sunlight</u>.

3. The <u>phases of the Moon</u> happen depending on <u>how much</u> of the <u>illuminated side</u> of the Moon we can <u>see</u>, as shown:

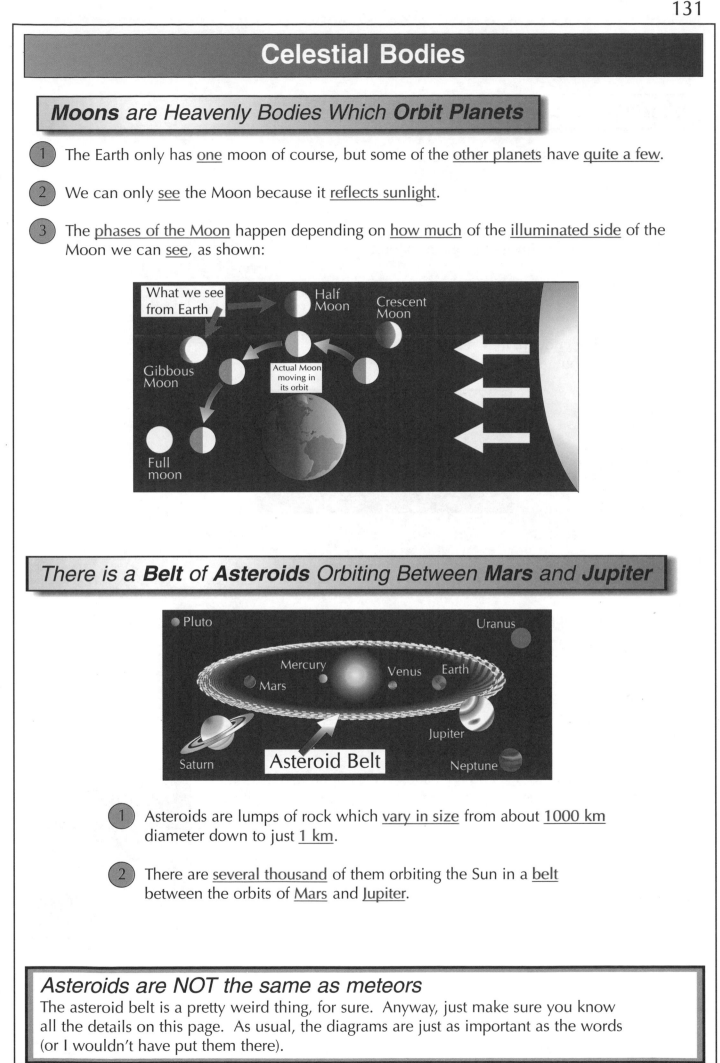

## There is a *Belt* of *Asteroids* Orbiting Between *Mars* and *Jupiter*

1. Asteroids are lumps of rock which <u>vary in size</u> from about <u>1000 km</u> diameter down to just <u>1 km</u>.

2. There are <u>several thousand</u> of them orbiting the Sun in a <u>belt</u> between the orbits of <u>Mars</u> and <u>Jupiter</u>.

## *Asteroids are NOT the same as meteors*

The asteroid belt is a pretty weird thing, for sure.  Anyway, just make sure you know all the details on this page.  As usual, the diagrams are just as important as the words (or I wouldn't have put them there).

# Celestial Bodies

## Meteors are Lumps of Rock that Enter the Earth's Atmosphere

1. Don't confuse meteors with asteroids.

2. Asteroids stay in a nice steady orbit round the Sun.

3. Meteors are bodies that enter the Earth's atmosphere. Most of them are tiny — like grains of dust.

4. When they enter the Earth's atmosphere they burn up, and we then see them as shooting stars.

5. If they're big enough, they reach the Earth's surface. This is rare, but it can be serious when they do.

## Comets Orbit the Sun, but have very Eccentric (elongated) Orbits

1. Comets only appear rarely because their orbits take them very far from the Sun and then back in close, which is when we see them. (Their orbits can vary from a few years to thousands of years.)

2. The Sun is not at the centre of the orbit but near one end, as shown.

3. Comet orbits can be in different planes from the orbits of the planets.

4. Comets are made of ice and rock and, as they approach the Sun, the ice melts leaving a bright tail of debris which can be millions of km long.

5. The comet travels much faster when it's nearer the Sun than it does when it's in the more distant part of its orbit. This is because the pull of gravity makes it speed up as it gets closer, and then slows it down as it gets further away from the Sun.

Comet

## Meteors are NOT the same as Asteroids

Two more 'celestial bodies' for you to know about. There's more to the Solar System than just planets — make sure you learn all the details about these different things. It's all in the syllabus, so they could ask you about any of it. Two mini-essays will check you know it all.

# Warm-Up and Worked Exam Question

## Warm-up Questions

1) Which planet takes the longest time to orbit the Sun?
2) What are meteors?
3) The Moon doesn't create light. Why then can it be seen?
4) Why are planets sometimes referred to as 'wanderers'?
5) Name two planets that can easily be seen at night using just your eyes.
6) Write a phrase to remember the order of the planets (including the Asteroid Belt), starting with Mercury.

## Worked Exam Question

1 Gravity gets smaller as the distance ($d$) between two masses increases. The pull of gravity decreases in proportion to the square of the distance ($d^2$).

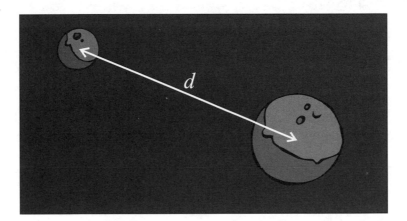

*Remember that gravity DECREASES as distance increases — that can catch a lot of people out.*

a) The force of gravity between a planet and its moon is $F$ when they are distance $d$ apart. What will the force of gravity become if they move to distance $2d$ apart?

*The distance increases by a factor of 2, so the force decreases by a factor of $2^2 = 4$. So the force of gravity is reduced to F/4.*

*[2 marks]*

b) The moon moves further away from the planet, and the force of gravity between them becomes $F/16$. How far apart are the planet and moon compared to distance $d$?

*The force of gravity decreases by a factor of 16, so the distance increases by a factor of $\sqrt{16} = 4$. So the distance is 4d.*

*[2 marks]*

# Exam Questions

1    This question is about objects which orbit the Sun.

a)    Name the four inner planets.

...................................................................................................................................
*[1 mark]*

b)    Explain why the Earth orbits the Sun.

...................................................................................................................................

...................................................................................................................................
*[2 marks]*

c)    Describe the orbit of the Earth around the Sun.

...................................................................................................................................
*[1 mark]*

d)    The asteroid belt lies between the planets Mars and Jupiter.
What are asteroids?

...................................................................................................................................

...................................................................................................................................
*[1 mark]*

e)    The comet Hale-Bopp could be seen from Britain with the naked eye in 1997.
Why can this comet no longer be seen, even on a very clear night?

...................................................................................................................................

...................................................................................................................................

...................................................................................................................................
*[2 marks]*

# Satellites

Moons are sometimes called natural satellites.

Artificial satellites are sent up by humans for four main purposes:

1) Monitoring Weather.
2) Communications, e.g. phone and TV.
3) Space research such as the Hubble Space Telescope.
4) Spying.

There are two different orbits useful for satellites:

## 1) *Geostationary Satellites* are Used For *Communications*

1) These can also be called geosynchronous satellites.
2) They are put in quite a high orbit over the equator which takes exactly 24 hours to complete.
3) This means that they stay above the same point on the Earth's surface because the Earth rotates with them — hence the name Geo(*Earth*)–stationary.
4) This makes them ideal for telephone and TV because they're always in the same place and they can transfer signals from one side of the Earth to the other in a fraction of a second.
5) There is room for about 400 geostationary satellites — any more and their orbits will interfere.

## 2) *Low Polar Orbit* Satellites are for *Weather* and *Spying*

1) In a low polar orbit, the satellite sweeps over both poles whilst the Earth rotates beneath it.
2) The time taken for each full orbit is just a few hours.
3) Each time the satellite comes round it can scan the next bit of the globe.
4) This allows the whole surface of the planet to be monitored each day.
5) Geostationary satellites are too high to take good weather or spying photos, but the satellites in polar orbits are nice and low.

## The *Hubble Space Telescope* has no *Atmosphere* in the way

1) The big advantage of having telescopes on satellites is that they can look out into space without the distortion and blurring caused by the Earth's atmosphere.
2) This allows much greater detail to be seen of distant stars and also the planets in the Solar System.

## *Learn the difference between the two types of artificial satellite*

You can actually see the low polar orbit satellites on a nice dark clear night. They look like stars except they move quite fast in a perfect straight line across the sky. You'll never see the geostationary ones though. Learn all the details about satellites.

# Searching for Life on Other Planets

There's a good chance that life exists somewhere else in the Universe.
Scientists use <u>three methods</u> to search for anything from amoebas to little green men.

## 1) *SETI* Looks for *Radio Signals* from *Other Planets*

1) We are constantly beaming <u>radio</u>, <u>TV</u> and <u>radar</u> into space that could be detected.
There might be intelligent life out there that's as clever as we are — or even more clever.
They may have built <u>transmitters</u> to send out signals like ours.

2) <u>SETI</u> stands for "Search for Extra Terrestrial Intelligence".  Scientists on the SETI project are looking
for <u>narrow bands</u> of <u>radio wavelengths</u> coming to Earth from outer space.  They're looking for
<u>meaningful signals</u> in all the 'noise' (see page 107).

3) Signals on a narrow band can <u>only</u> come from a <u>transmitter</u>.
The 'noise' comes from giant stars and gas clouds.

4) It takes <u>ages</u> to analyse all the radio waves so the SETI people get help from the public —
you can download a <u>screensaver</u> off the Internet which analyses a chunk of radio waves.

5) SETI has been going for the last <u>40 years</u> but they've <u>not found anything</u> yet.

6) Scientists are now looking for possible <u>laser</u> signals from outer space.  Watch this space...

## 2) *Robots* Collect *Photos* and *Samples*

*This <u>could</u> be a microscopic fossil of a bacteria-like organism from Mars...*

*500 nm*

1) Scientists have sent robots in spacecraft to <u>Mars</u> and <u>Europa</u>
(one of Jupiter's moons) to look for microorganisms.

2) The robots wander round the planet, sending <u>photographs</u>
back to Earth or <u>collecting samples</u> for analysis.

3) Scientists can detect living things or <u>evidence</u> of them, such as <u>fossils</u> or <u>remains</u>, in the samples.
This "fossil" is from Mars, though no one really seems sure *what* it is.

4) A couple of bacteria may not seem like much, but that's how we started out on Earth...

## 3) *Chemical Changes* and *Reflected Light* Are *Big Clues*

### *Changes* Show There's Life

1) Scientists are looking for <u>chemical changes</u> in the atmospheres of other planets.
2) Some changes are just caused by things like volcanoes but others are a <u>clue</u> that there's life there.
3) The amounts of <u>oxygen</u> and <u>carbon dioxide</u> in Earth's atmosphere have <u>changed</u> over time — it's
<u>very different</u> to what it'd be like if there was <u>no life</u> here.  Plants have made oxygen levels <u>go up</u>
but carbon dioxide levels <u>go down</u>.
4) They can look at planets' atmospheres from Earth — no spacecraft required.

### *Light* Gives Away What's On The *Surface*

A planet's <u>reflected light</u> (from the Sun) is <u>different</u> depending on
whether it's bounced off rock, trees, water or whatever.  It's a good
way to find out what's on the <u>surface</u> of a planet.  Scientists haven't
yet found anything exciting, but they are using these methods to
search for planets with <u>suitable conditions</u> for life.

## *Mini-essay time again*

You need to learn these <u>three</u> different ways that scientists are looking for life on other planets.
You definitely need to learn this stuff, even if you get given more information in the exam.
Cover the page and write notes about <u>how</u> the methods work and <u>what</u> they've found.

# Warm-Up and Worked Exam Question

## Warm-up Questions

1) Name a natural satellite of the Earth.
2) Give four uses for artificial satellites.
3) Why can the Hubble telescope take such clear pictures?
4) What has happened to the amount of oxygen in the Earth's atmosphere since green plants have grown on the surface?
5) TV signals from Earth might be picked up by aliens in the future. Explain how this could happen.

## Worked Exam Question

1    The Sky TV satellite is in a geostationary orbit beaming TV signals down to Earth. A technical book says its orbit takes 86 400 seconds.

a)    How many hours does it take for the Sky TV satellite to orbit?

*86 400 s = 86 400 ÷ 60 ÷ 60 h = 24 h*

*[1 mark]*

b)    As long as the satellite continues in its orbit, a dish correctly fitted on a house will point at the satellite. Explain why this happens.

*The Earth takes 24 hours to spin on its axis. This matches the satellite's orbit time, and the satellite is positioned over the equator, so the satellite always stays over the same point on the Earth, and the dish continues to point at it all the time.*

*[1 mark]*

c)    The Sky TV satellite is positioned over Africa. How can Sky TV engineers transmit Spanish football matches to the UK?

*The football transmission is sent up to the satellite from Spain. The satellite then beams this down to the UK*

*[2 marks]*

d)    Heavy rain sometimes disrupts the TV picture people receive in their homes. Explain why this happens.

*The rain can block the signal from the satellite to the receiving dish.*

*[1 mark]*

# Exam Questions

1  Geostationary satellites are used for communications.

a) Describe the orbit of a geostationary satellite.

    ...........................................................................................................................

    ...........................................................................................................................

    ...........................................................................................................................
    *[3 marks]*

b) Why is a satellite in a polar orbit better for spying than one in a geostationary orbit?

    ...........................................................................................................................

    ...........................................................................................................................
    *[2 marks]*

2  Scientists are searching for other life in the Universe.

a) How could robots sent from Earth help in this task?

    ...........................................................................................................................

    ...........................................................................................................................
    *[1 mark]*

b) Describe two different methods that scientists on Earth can use to search for life elsewhere.

    ...........................................................................................................................

    ...........................................................................................................................

    ...........................................................................................................................

    ...........................................................................................................................

    ...........................................................................................................................
    *[2 marks]*

# The Universe — Stars and Solar Systems

## Stars and Solar Systems form from Clouds of Dust

1. Stars form from <u>clouds of dust</u> which <u>spiral in together</u> due to <u>gravitational attraction</u>.

2. The gravity <u>compresses</u> the matter so much that <u>intense heat</u> develops and sets off <u>nuclear fusion reactions</u> and the star then begins <u>emitting light</u> and other <u>radiation</u>.

3. At the <u>same time</u> that the star is forming, <u>other lumps</u> may develop in the <u>spiralling dust clouds</u> and these eventually gather together and form <u>planets</u> which orbit <u>around the star</u>.

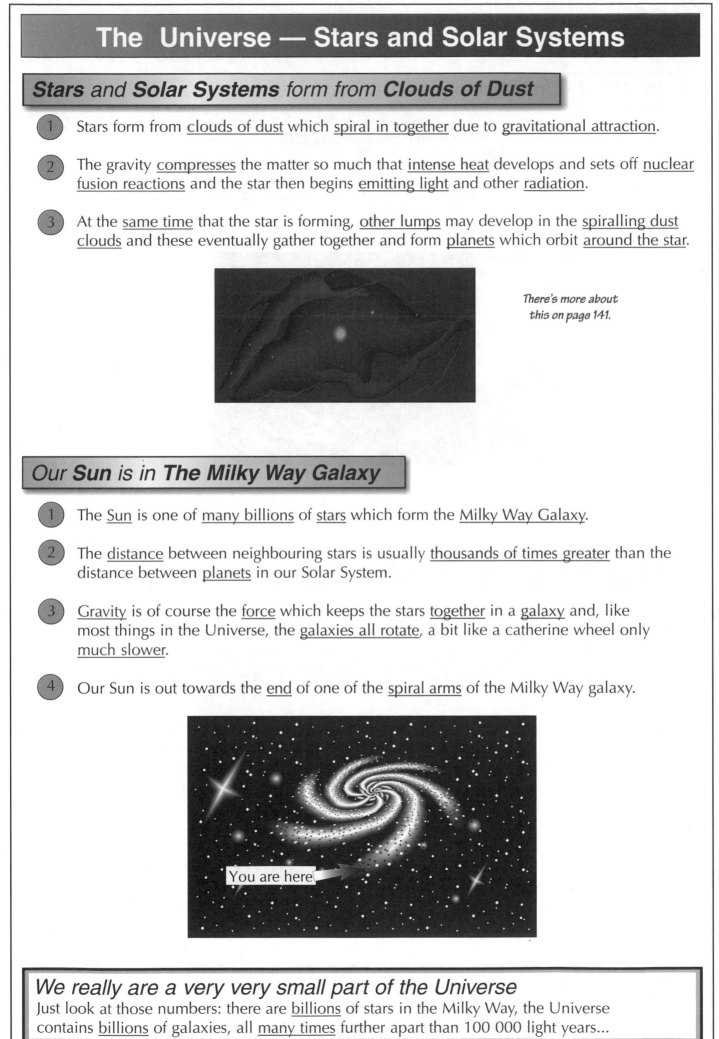

*There's more about this on page 141.*

## Our Sun is in The Milky Way Galaxy

1. The <u>Sun</u> is one of <u>many billions</u> of <u>stars</u> which form the <u>Milky Way Galaxy</u>.

2. The <u>distance</u> between neighbouring stars is usually <u>thousands of times greater</u> than the distance between <u>planets</u> in our Solar System.

3. <u>Gravity</u> is of course the <u>force</u> which keeps the stars <u>together</u> in a <u>galaxy</u> and, like most things in the Universe, the <u>galaxies all rotate</u>, a bit like a catherine wheel only <u>much slower</u>.

4. Our Sun is out towards the <u>end</u> of one of the <u>spiral arms</u> of the Milky Way galaxy.

You are here

## We really are a very very small part of the Universe

Just look at those numbers: there are <u>billions</u> of stars in the Milky Way, the Universe contains <u>billions</u> of galaxies, all <u>many times</u> further apart than 100 000 light years...

# The Universe — Galaxies and Black Holes

## The **Whole Universe** has More Than **A Billion Galaxies**

1) <u>Galaxies</u> themselves are often <u>millions of times further apart</u> than the <u>stars</u> are within a galaxy.

2) So you should begin to realise that the Universe is <u>mostly empty space</u> and is <u>really really big</u>.

You are here

## **Black Holes** Don't Let **Anything** Escape

1) The gravity on neutron stars, white dwarfs and black dwarfs is <u>so strong</u> that it <u>crushes atoms</u>. The stuff in the stars gets <u>squashed up</u> so much that they're <u>MILLIONS OF TIMES DENSER</u> than anything on Earth.

2) If <u>enough</u> matter is left behind after a supernova explosion, it's <u>so dense</u> that <u>nothing</u> can escape the powerful gravitational field. Not even electromagnetic waves. The dead star is then called a <u>black hole</u>. Black holes <u>aren't visible</u> because any light being emitted is sucked right back in there (that's why it's called 'black').

3) Astronomers can detect black holes in other ways — they can observe <u>X-rays</u> emitted by <u>hot gases</u> from other stars as they spiral into the black hole.

## *Revision is like a black hole — there's no escaping it*

Black holes are pretty interesting as Physics goes. Make sure you learn all the details about how they're formed and why they're 'black'. Write a mini-essay to check you know it all.

# The Life Cycle of Stars

Stars go through <u>many traumatic stages</u> in their lives.

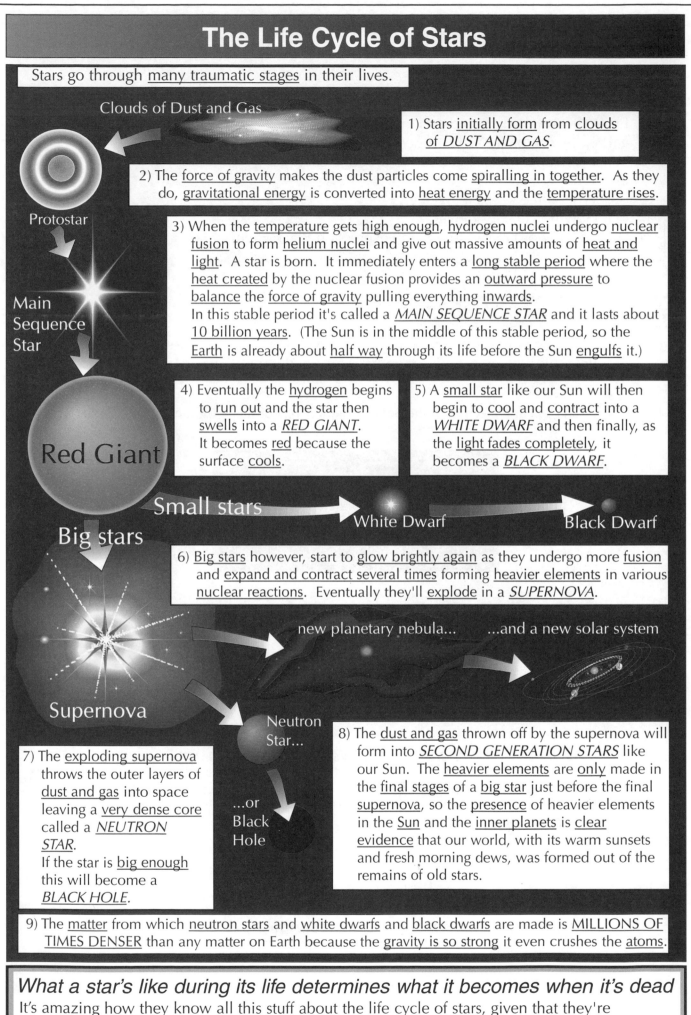

Clouds of Dust and Gas

Protostar

Main Sequence Star

Red Giant

Big stars

Supernova

Small stars

White Dwarf

Black Dwarf

new planetary nebula...

...and a new solar system

Neutron Star...

...or Black Hole

1) Stars <u>initially form</u> from <u>clouds</u> of _DUST AND GAS_.

2) The <u>force of gravity</u> makes the dust particles come <u>spiralling in together</u>. As they do, <u>gravitational energy</u> is converted into <u>heat energy</u> and the <u>temperature rises</u>.

3) When the <u>temperature</u> gets <u>high enough</u>, <u>hydrogen nuclei</u> undergo <u>nuclear fusion</u> to form <u>helium nuclei</u> and give out massive amounts of <u>heat and light</u>. A star is born. It immediately enters a <u>long stable period</u> where the <u>heat created</u> by the nuclear fusion provides an <u>outward pressure</u> to <u>balance</u> the <u>force of gravity</u> pulling everything <u>inwards</u>.
In this stable period it's called a _MAIN SEQUENCE STAR_ and it lasts about <u>10 billion years</u>. (The Sun is in the middle of this stable period, so the <u>Earth</u> is already about <u>half way</u> through its life before the Sun <u>engulfs</u> it.)

4) Eventually the <u>hydrogen</u> begins to <u>run out</u> and the star then <u>swells</u> into a _RED GIANT_. It becomes <u>red</u> because the surface <u>cools</u>.

5) A <u>small star</u> like our Sun will then begin to <u>cool</u> and <u>contract</u> into a _WHITE DWARF_ and then finally, as the <u>light fades completely</u>, it becomes a _BLACK DWARF_.

6) <u>Big stars</u> however, start to <u>glow brightly again</u> as they undergo more <u>fusion</u> and <u>expand and contract several times</u> forming <u>heavier elements</u> in various <u>nuclear reactions</u>. Eventually they'll <u>explode</u> in a _SUPERNOVA_.

7) The <u>exploding supernova</u> throws the outer layers of <u>dust and gas</u> into space leaving a <u>very dense core</u> called a _NEUTRON STAR_.
If the star is <u>big enough</u> this will become a _BLACK HOLE_.

8) The <u>dust and gas</u> thrown off by the supernova will form into _SECOND GENERATION STARS_ like our Sun. The <u>heavier elements</u> are <u>only</u> made in the <u>final stages</u> of a <u>big star</u> just before the final <u>supernova</u>, so the <u>presence</u> of heavier elements in the <u>Sun</u> and the <u>inner planets</u> is <u>clear evidence</u> that our world, with its warm sunsets and fresh morning dews, was formed out of the remains of old stars.

9) The <u>matter</u> from which <u>neutron stars</u> and <u>white dwarfs</u> and <u>black dwarfs</u> are made is <u>MILLIONS OF TIMES DENSER</u> than any matter on Earth because the <u>gravity is so strong</u> it even crushes the <u>atoms</u>.

_What a star's like during its life determines what it becomes when it's dead_
It's amazing how they know all this stuff about the life cycle of stars, given that they're all billions and billions of km away. Still, amazing or not, it's all got to be learnt.

# Warm-Up and Worked Exam Question

## Worked Exam Question

You will almost certainly get questions on the life cycles of stars — make sure
you can follow this worked example then have a go at the questions on page 143.

1    The sun at the centre of the Solar System is one of many billions of stars in the Milky

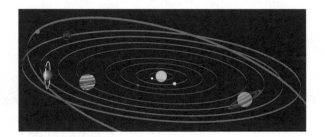

Way.  Like all stars, our Sun has gone through many stages up to the present day.
Briefly outline each of these stages.

*Clouds of dust and gas* (1 mark) *were drawn together by gravity* (1 mark).

*Gravitational energy was converted to heat energy* (1 mark),

*which caused nuclear fusion of hydrogen to form helium nuclei,*

*releasing more heat* (1 mark).

*The resulting outward pressure balanced out the gravity pulling the*

*star inwards, creating a stable star* (1 mark).

*(5 marks)*

# Exam Questions

1    Our Sun is now in its stable period as a main sequence star.
     Describe the stages that will happen to the Sun between now and it becoming a
     black dwarf.

     ..................................................................................................................................

     ..................................................................................................................................

     ..................................................................................................................................

     ..................................................................................................................................

     ..................................................................................................................................

     ..................................................................................................................................

     ..................................................................................................................................

     ..................................................................................................................................

                                                                                                    *(5 marks)*

2    A massive star will eventually turn into a supernova.

     a)   What is left behind after a supernova?

          ..............................................................................................................................
                                                                                                    *(1 mark)*

     b)   How is a supernova linked to the presence of heavier elements in our Sun and its
          inner planets?

          ..............................................................................................................................

          ..............................................................................................................................

          ..............................................................................................................................
                                                                                                    *(3 marks)*

# Red-Shift and Background Radiation

The <u>Big Bang Theory</u> of the Universe is the <u>most convincing</u> at the present time. There is also the <u>Steady State Theory</u> which is quite presentable but it <u>doesn't explain</u> some of the observed features too well.

## Red-shift and Background Radiation need Explaining

There are <u>three important bits of evidence</u> you need to know about:

### 1) Light From Other Galaxies is Red-Shifted

1) When we look at <u>light from distant galaxies</u> we find that <u>all the frequencies</u> are <u>shifted</u> towards the <u>red end</u> of the spectrum.

2) In other words, the <u>frequencies</u> are all <u>slightly lower</u> than they should be. It's the same effect as a car <u>horn</u> sounding lower-pitched when the car is travelling <u>away</u> from you. The sound <u>drops in frequency</u>.

3) This is called the <u>Doppler effect</u>.

4) <u>Measurements</u> of the red-shift suggest that <u>all the galaxies</u> are <u>moving away from us</u> very quickly — and it's the <u>same result</u> whichever direction you look in.

### 2) The Further Away a Galaxy is, The Greater The Red-Shift

1) <u>More distant</u> galaxies have <u>greater</u> red-shifts than nearer ones.

2) This means that more distant galaxies are <u>moving away faster</u> than nearer ones.

3) The inescapable <u>conclusion</u> appears to be that the whole Universe is <u>expanding</u>.

### 3) There is Uniform Microwave Radiation From All Directions

1) This <u>low frequency radiation</u> comes from <u>all directions</u> and from <u>all parts</u> of the Universe.

2) It's known as the <u>background radiation</u> (of the Big Bang). It's nothing to do with radioactive background radiation on Earth.

3) For complicated reasons this background radiation is <u>strong evidence</u> for an <u>initial Big Bang</u>, and as the Universe <u>expands and cools</u>, so this background radiation "<u>cools</u>" and <u>drops in frequency</u>.

---

*'Describe and explain 3 pieces of evidence for the Big Bang Theory'* (5 marks)
Red-shift is a tricky concept to get your head round, but it's a crucial piece of evidence for the Big Bang, and you can guarantee you'll get a question asking you to explain it.

# The Steady State Theory

## *The **Steady State Theory** of the Universe — **Not Popular***

 This is based on the idea that the Universe appears pretty much the <u>same everywhere</u> and <u>always has done</u>.

 In other words the Universe has <u>always existed</u> and <u>always will</u> in the same form that it is now.

3 This theory explains the <u>apparent expansion</u> of the Universe by suggesting that <u>matter</u> is being <u>created</u> in the spaces as the Universe expands.

4 However, as yet, there's <u>no convincing explanation</u> of <u>where</u> this new matter <u>comes from</u>.

5 There isn't much support for the Steady State Theory, especially since the discovery of <u>background radiation</u> which fits in <u>much better</u> with the idea of a Big Bang.

 But you <u>just never know</u>...

---

### *Background radiation doesn't fit with the Steady State Theory*

Not many people agree with the Steady State Theory, but you have to explain why, so learn all the numbered points and try the old 'mini-essay' method of checking you know it.

# The Big Bang Theory

## The **Big Bang Theory** — Very Popular

1) Since all the galaxies appear to be <u>moving apart</u> very rapidly, the <u>obvious conclusion</u> is that there was an <u>initial explosion</u>: the <u>Big Bang</u>.

2) All the matter in the Universe must have been <u>compressed into a very small space</u> and then it <u>exploded</u> and the <u>expansion</u> is still going on.

3) The Big Bang is believed to have happened around <u>14 billion years ago</u>.

4) The age of the Universe can be <u>estimated</u> from the <u>current rate of expansion</u>.

5) The rate at which the expansion is <u>slowing down</u> is an <u>important factor</u> in deciding the <u>future</u> of the Universe.

6) <u>Without gravity</u> the Universe would expand at the <u>same rate forever</u>.

7) However, the <u>attraction</u> between all the mass in the Universe tends to <u>slow</u> the expansion down.

### The Big Bang seems to explain most of the evidence from page 144

Seven numbered points to describe the history of the Universe. You can't really complain — 14 billion years' worth of revision in just seven points. Cover the page, write it all out, etc.

# The Future of the Universe

## The *Future* of the *Universe*:

## It Could **Expand Forever** — or Collapse into **The Big Crunch**

 1. The eventual fate of the Universe depends on <u>how fast</u> the galaxies are <u>moving apart</u> and how much <u>total mass</u> there is in it.

 2. We can <u>measure</u> how fast the galaxies are <u>separating</u> quite easily, but we'd also like to know just <u>how much mass</u> there is in the Universe in order to <u>predict its future</u>.

 3. This is proving <u>tricky</u> as most of the mass appears to be <u>invisible</u>, e.g. <u>black holes</u>, <u>particles</u>, <u>weird kinds of matter</u>, etc.

4. Depending on <u>how much mass</u> there is, there are <u>two ways</u> the Universe could go:

### 1) **The Big Crunch** — But Only if there's **Enough Mass**

- If there's <u>enough mass</u> compared to <u>how fast</u> the galaxies are currently moving, the Universe will eventually <u>stop expanding</u> and <u>begin contracting</u>.
- This would end in a <u>Big Crunch</u>.
- The Big Crunch could be followed by another Big Bang and then <u>endless cycles</u> of <u>expansion and contraction</u>.

### 2) If there's **Too Little Mass** — then it's **Expansion For Ever**

- But there may be <u>too little mass</u> in the Universe to slow the expansion down.
- In that case it could <u>expand forever</u>.
- The Universe would become <u>more and more</u> <u>spread out</u> into eternity.

---

*Where will it end — Big Crunch or endless expansion...*

These are questions scientists everywhere are still pondering. Learn the theories for both ideas, and make sure you know it all well enough to be able to explain it to a 5-year-old.

# Warm-Up and Worked Exam Question

## Warm-up Questions

1) From what directions does background radiation from the Big Bang come to us?
2) Which part of the electromagnetic spectrum does this radiation fall into?
3) Why is it at a lower frequency now than it was billions of years ago?
4) Explain what is meant by the steady state theory of the Universe.
5) How does this theory explain the expansion of the Universe?
6) What was the Big Bang?
7) How long ago is it believed to have happened?
8) What force may prevent the Universe from expanding forever?

## Worked Exam Question

1) What eventually will happen to the Universe depends on how fast the galaxies are moving apart and how much total mass it has.

a) How do scientists measure how fast the galaxies are moving?

*by looking at the amount of red shift of near and far galaxies*

*(1 mark)*

b) Why is it difficult to estimate the total mass?

*a lot of the mass is difficult to see, e.g. black holes*

*(1 mark)*

c) State the conditions which would be necessary for the following possible endings:

(i) The Universe will stop expanding and start contracting under gravity.

*The Universe would need to contain enough mass to provide the necessary gravitational attraction to reverse the separation of the galaxies.*

*(1 mark)*

(ii) The Universe will continue expanding forever.

*The Universe would contain too little mass, hence gravitational forces would be too weak to reverse the expansion.*

*(1 mark)*

# Exam Questions

1    The diagrams show the spectra produced by light from one type of element from three
different sources: on Earth, a near galaxy, a distant galaxy.

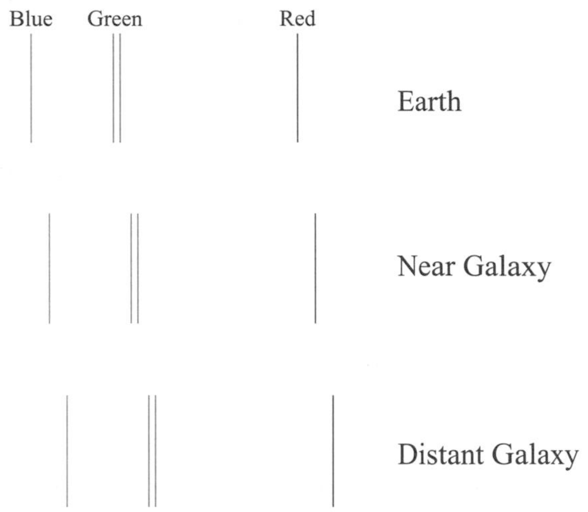

Blue    Green           Red

Earth

Near Galaxy

Distant Galaxy

a)    What is the name of the effect shown by the galaxies?

.................................................................................................................................
*(1 mark)*

b)    What does this tell us about the near galaxy?

.................................................................................................................................
*(1 mark)*

c)    Why is the spectrum for the distant galaxy different from that of the near one?

.................................................................................................................................

.................................................................................................................................
*(1 mark)*

d)    What does this suggest about the whole Universe?

.................................................................................................................................
*(1 mark)*

e)    The Universe is believed to have started from a Big Bang, approximately 14 billion
years ago.  How was this age estimated?

.................................................................................................................................

.................................................................................................................................
*(1 mark)*

# Revision Summary for Section Four

The Universe is completely mindblowing in its own right. But surely the most mindblowing thing of all is the very fact that we are actually here, sitting and contemplating the truly outrageous improbability of our own existence. If your mind isn't blown, then it hasn't sunk in yet. Think about it. 14 billion years ago there was a huge explosion, but there was no need for the whole chain of events to happen which allowed (or caused?) intelligent life to evolve and develop to the point where it became conscious of its own existence, not to mention the very disturbing unlikelihood of it all. But we have. We're here. And yet the Universe could so easily have existed without conscious life ever evolving. Or come to that, the Universe needn't exist at all. Just black nothingness. So why does it exist? And why are we here? And why do we have to do so much revision? Who knows — but stop dreaming and get on with it.

1) List the eleven parts of the Solar System starting with the Sun, and get them in the right order.

2) What do planets look like in the night sky? Name two that can be seen with the naked eye.

3) How does the Sun produce all its heat? What does the Sun give out?

4) Which planet has an unusual orbit?

5) What is it that keeps the planets in their orbits? What shape are their orbits?

6) What is the famous "inverse square" relationship all about? Sketch a diagram to explain it.

7) What are constellations? What do planets do in the constellations?

8) Who had trouble with the Spanish Inquisition? Why did he have such trouble?

9) Sketch a diagram to explain the phases of the Moon.

10) What and where are the asteroids? What and where are meteors? Is there a difference?

11) What and where are comets? What are they made of? Sketch a diagram of a comet orbit.

12) What are natural and artificial satellites? What four purposes do we have for satellites?

13) Explain fully what a geostationary satellite does, and state what they're used for.

14) Explain fully what a low polar orbit satellite does, and state what they're used for.

15) What is the Hubble Telescope and where is it? What is its big advantage?

16) What does SETI stand for? Why are they looking for narrow-band signals?

17) What two things can robots on planets send back? Which places have they sent robots to?

18) Describe 2 ways that scientists can look for life on a planet without sending a spacecraft there.

19) Has life been found on other planets?

20) What do stars and solar systems form from? What force causes it all to happen?

21) What is the Milky Way? Sketch it and show our Sun in relation to it.

22) What is the Universe made up of?

23) What's unusual about the gravity on neutron stars, white dwarfs and black dwarfs?

24) Why would a black hole form? Why's it called 'black'? How can you spot one?

25) Describe the first stages of a star's formation. Where does the initial energy come from?

26) What process eventually starts inside the star to make it produce so much heat and light?

27) What is a "main sequence" star? How long might a sun like ours last? What happens after that?

28) What are the final two stages of a small star's life?

29) What are the two final stages of a big star's life?

30) What is meant by a "second generation" star? How do we know our Sun is one?

31) What are the two main theories for the origin of the Universe? Which one is more likely?

32) What are the three important bits of evidence which need explaining by these theories?

33) Give brief details of both theories. How long ago did each suggest the Universe began?

34) What are the two possible futures for the Universe?

35) What do these possible futures depend upon?

# Different Types of Energy

## Learn all The Ten Types Of Energy

You should know all of these well enough to list them from memory, including examples:

1) Electrical Energy ........................................................ — whenever a current flows.
2) Light Energy ............................................................ — from the Sun, light bulbs etc.
3) Sound Energy .......................................................... — from loudspeakers or anything noisy.
4) Kinetic Energy, or Movement Energy ...................... — anything that's moving has it.
5) Nuclear Energy ........................................................ — released only from nuclear reactions.
6) Thermal Energy or Heat Energy ............................. — flows from hot objects to colder ones.
7) Radiant Heat Energy, or Infra Red Heat ................. — given out as EM radiation by hot objects.
8) Gravitational Potential Energy .............................. — possessed by anything which can fall.
9) Elastic Potential Energy ......................................... — stretched springs, elastic, rubber bands, etc.
10) Chemical Energy ................................................... — possessed by foods, fuels and batteries.

## Potential- and Chemical- are forms of Stored Energy

The last three above are forms of stored energy because the energy is not obviously doing anything, it's just waiting to happen, i.e. waiting to be turned into one of the other forms.

## There are Often Exam Questions on Energy Transfers

These are very important examples. You must learn them till you can repeat them all easily.

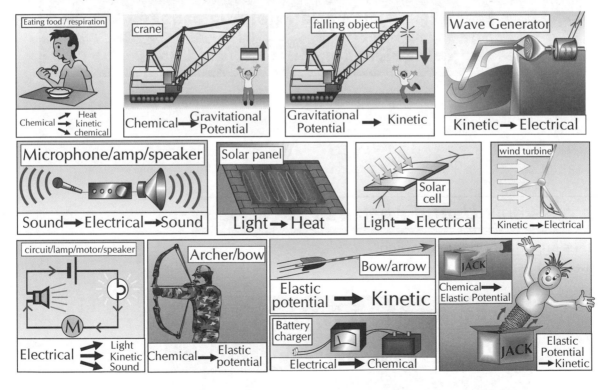

And DON'T FORGET — ALL types of ENERGY are measured in JOULES

## Learn about Energy — and just keep working at it...

You'll definitely get an Exam question on different types of energy and energy transfers, and if you learn all the stuff on this page, you should have it pretty well covered.

# Energy Conservation

## There are **Two Types** of "Energy Conservation"

Make sure you fully understand the difference between these two.

① "Energy conservation" is all about <u>using less fossil fuels</u> because of the damage it does and because they might <u>run out</u>. This is all to do with protecting the <u>environment</u>.

② The "<u>Principle of the Conservation of Energy</u>" on the other hand, is one of the <u>major cornerstones</u> of modern Physics. It's an <u>all-pervading principle</u> which governs the workings of the <u>entire physical Universe</u>. If this principle were not so, then life as we know it would simply cease to be.

## The **Principle of the Conservation of Energy** can be stated thus:

<u>ENERGY</u> CAN NEVER BE <u>CREATED NOR DESTROYED</u>
— IT'S ONLY EVER <u>CONVERTED</u>
FROM ONE FORM TO ANOTHER.

## Energy is only **Useful** when it's **Converted**

Another <u>important principle</u> which you need to <u>learn</u> is this one:

ENERGY IS <u>ONLY USEFUL</u> WHEN
IT'S <u>CONVERTED</u> FROM ONE
FORM TO ANOTHER.

## Two types of Energy Conservation — Learn the Difference

Don't get confused between these two. When people talk about 'energy loss', what they mean is energy being used in a non-useful way, e.g. leaving the window open and the heating on means that air is being heated up then leaving the house, which is useless... but it's not actually 'lost'.

# Efficiency

## Most *Energy Transfers* Involve Some *Losses*, as *Heat*

1. Useful devices are only useful because they convert energy from one form to another.

2. In doing so, some of the useful input energy is always lost or wasted, often as heat.

3. The less energy that is wasted, the more efficient the device is said to be.

4. The energy flow diagram is pretty much the same for all devices. You must learn this basic energy flow diagram:

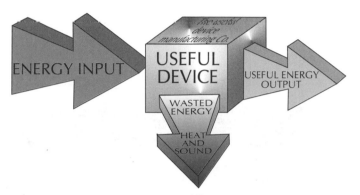

For any specific example you can give more detail about the types of energy being input and output, but remember this:

> ### NO device is 100% efficient
> ### and the WASTED ENERGY is always dissipated as HEAT and SOUND.

Electric heaters are the exception to this. They're 100% efficient because all the electricity is converted to "useful" heat. What else could it become? Ultimately, all energy ends up as heat energy. If you use an electric drill, it gives out various types of energy but they all quickly end up as heat. That's an important thing to realise. Don't forget it.

## Learn the *Formula* for *Efficiency*

A machine is a device which turns one type of energy into another.
The efficiency of any device is defined as:

$$\text{Efficiency} = \frac{\text{USEFUL Energy OUTPUT}}{\text{TOTAL Energy INPUT}}$$

Energy out / Efficiency × Energy in

You can give efficiency as a fraction, decimal or percentage. i.e. ¾ or 0.75 or 75%

---

## *Don't forget you can use the formula triangle for efficiency*

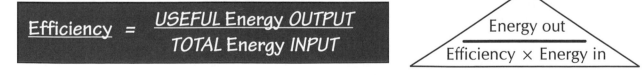

The thing about loss of energy is it's always the same — it always disappears as heat and sound, and even the sound ends up as heat pretty quickly. So when they ask "Why is the input energy more than the output energy?", the answer is always the same... Learn it.

# Efficiency

## *Efficiency is Really Straightforward*

(1) You find how much energy is <u>supplied</u> to a machine. (The Total Energy <u>input</u>.)

(2) You find how much <u>useful energy</u> the machine <u>delivers</u>. (The Useful Energy <u>OUTPUT</u>.) They either tell you this directly or they tell you how much it <u>wastes</u> as heat/sound.

(3) Either way, you get those <u>two important numbers</u> and then just <u>divide</u> the <u>smaller one</u> by the <u>bigger one</u> to get a value for <u>efficiency</u> somewhere between <u>0 and 1</u> (or <u>0 and 100%</u>). Easy.

(4) The other way they might ask it is to tell you the <u>efficiency</u> and the <u>input energy</u> and ask for the <u>energy output</u>.
The best way to tackle that is to <u>learn</u> this <u>other version</u> of the formula:

$$\underline{\textit{USEFUL ENERGY OUTPUT}} = \underline{\textit{Efficiency}} \times \textit{TOTAL Energy INPUT}$$

## *Five Important Examples on Efficiency for you to Learn*

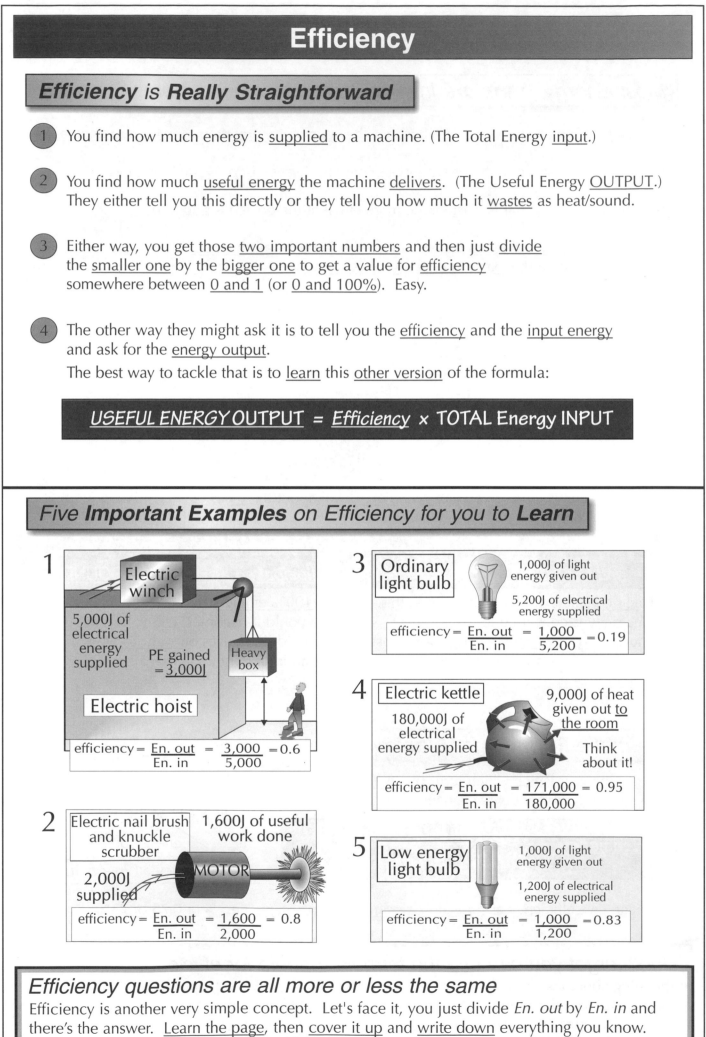

**1** Electric winch

5,000J of electrical energy supplied

PE gained = 3,000J

Heavy box

**Electric hoist**

efficiency = $\dfrac{\text{En. out}}{\text{En. in}}$ = $\dfrac{3,000}{5,000}$ = 0.6

**2** Electric nail brush and knuckle scrubber

1,600J of useful work done

MOTOR

2,000J supplied

efficiency = $\dfrac{\text{En. out}}{\text{En. in}}$ = $\dfrac{1,600}{2,000}$ = 0.8

**3** Ordinary light bulb

1,000J of light energy given out

5,200J of electrical energy supplied

efficiency = $\dfrac{\text{En. out}}{\text{En. in}}$ = $\dfrac{1,000}{5,200}$ = 0.19

**4** Electric kettle

180,000J of electrical energy supplied

9,000J of heat given out <u>to the room</u>

Think about it!

efficiency = $\dfrac{\text{En. out}}{\text{En. in}}$ = $\dfrac{171,000}{180,000}$ = 0.95

**5** Low energy light bulb

1,000J of light energy given out

1,200J of electrical energy supplied

efficiency = $\dfrac{\text{En. out}}{\text{En. in}}$ = $\dfrac{1,000}{1,200}$ = 0.83

## *Efficiency questions are all more or less the same*

Efficiency is another very simple concept. Let's face it, you just divide *En. out* by *En. in* and there's the answer. <u>Learn the page</u>, then <u>cover it up</u> and <u>write down</u> everything you know.

# Warm-Up and Worked Exam Question

## Warm-up Questions

1) For each of the energy transfer devices shown below state the useful energy output.

   a) 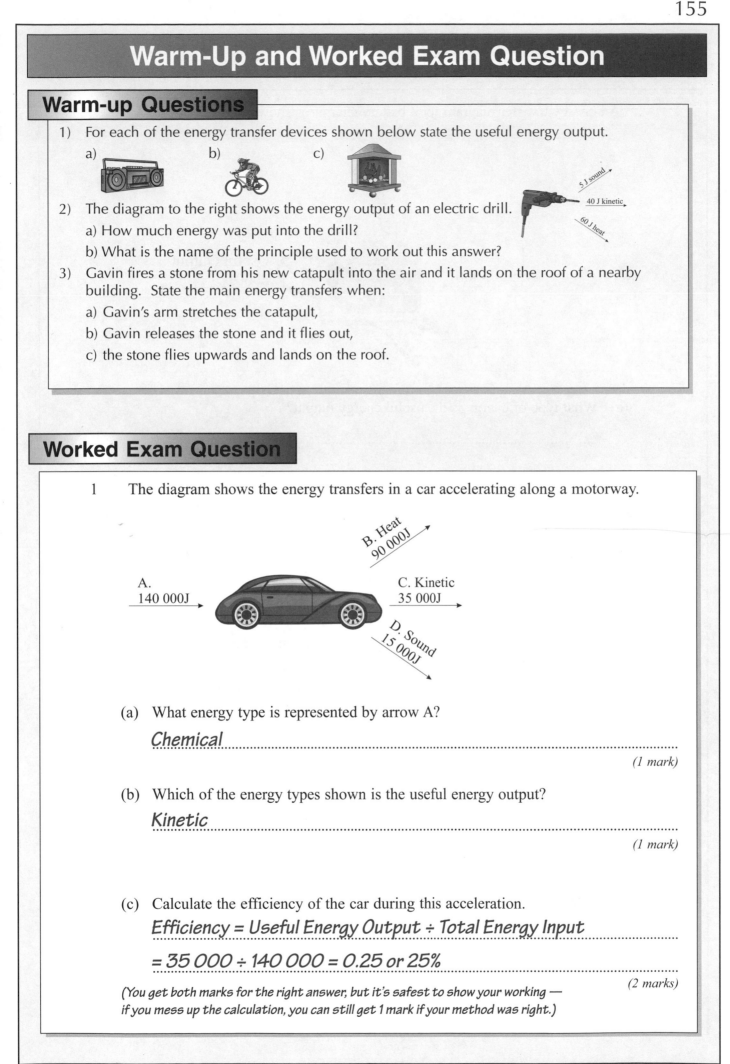   b)   c)

2) The diagram to the right shows the energy output of an electric drill.

   a) How much energy was put into the drill?

   b) What is the name of the principle used to work out this answer?

   *5 J sound*
   *40 J kinetic*
   *60 J heat*

3) Gavin fires a stone from his new catapult into the air and it lands on the roof of a nearby building. State the main energy transfers when:

   a) Gavin's arm stretches the catapult,

   b) Gavin releases the stone and it flies out,

   c) the stone flies upwards and lands on the roof.

## Worked Exam Question

1    The diagram shows the energy transfers in a car accelerating along a motorway.

*B. Heat 90 000J*

*A. 140 000J*

*C. Kinetic 35 000J*

*D. Sound 15 000J*

(a)   What energy type is represented by arrow A?

   *Chemical*

   *(1 mark)*

(b)   Which of the energy types shown is the useful energy output?

   *Kinetic*

   *(1 mark)*

(c)   Calculate the efficiency of the car during this acceleration.

   *Efficiency = Useful Energy Output ÷ Total Energy Input*

   *= 35 000 ÷ 140 000 = 0.25 or 25%*

   *(2 marks)*

   *(You get both marks for the right answer, but it's safest to show your working —*
   *if you mess up the calculation, you can still get 1 mark if your method was right.)*

156

## Exam Questions

1 An energy transfer diagram for a battery charger is shown below.

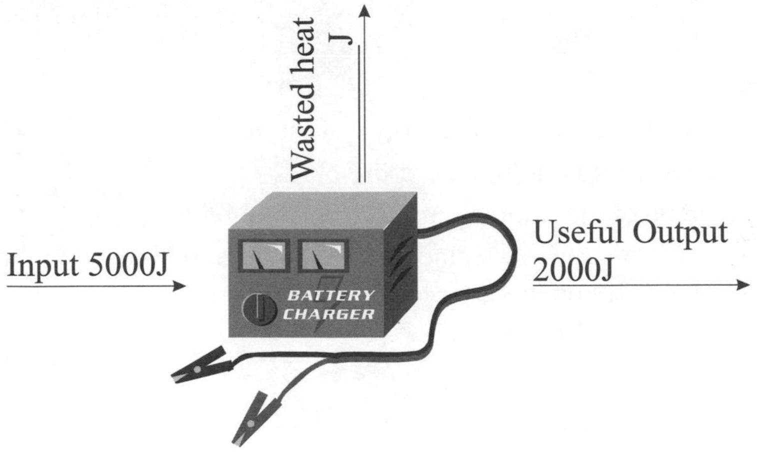

Wasted heat J

Input 5000J

BATTERY CHARGER

Useful Output 2000J

(a) What type of energy is the useful energy output?

.................................................................................................................................
*(1 mark)*

(b) How much energy is wasted as heat?

.................................................................................................................................
*(1 mark)*

(c) Calculate the efficiency of the charging process.

.................................................................................................................................
*(2 marks)*

The battery charger is used to recharge a nickel-cadmium cell.
The fully charged cell is then used in a MiniDisc player.

(d) Complete the energy transfer diagram for the MiniDisc player.

Kinetic

_____    Electrical    Sound

_____
*(2 marks)*

(e) The MiniDisc player is 12% efficient. Calculate the useful energy output for
theMiniDisc player if 2000 J of energy is supplied by the cell.

.................................................................................................................................
*(2 marks)*

SECTION FIVE — ENERGY

# Work

## Work is the Transfer of Energy

### When a force moves an object, energy is transferred and work is done.

That statement sounds complicated but it isn't. Try this:

1) Whenever something moves, something else is providing some sort of "effort" to move it.

2) The thing putting the effort in needs a supply of energy (like fuel or food or electricity etc.).

3) It then does "work" by moving the object — and one way or another it transfers the energy it receives (as fuel) into other forms.

4) Whether this energy is transferred "usefully" (e.g. by lifting a load) or is "wasted" (e.g. lost as friction), you can still say that "work is done". So "work done" and "energy transferred" are "one and the same". (And they're both in joules)

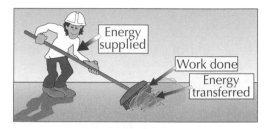

## It's Another Straightforward Formula:

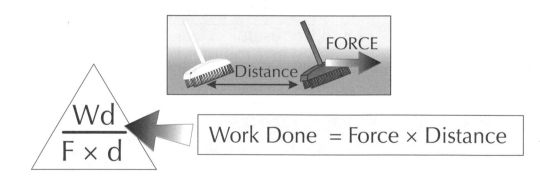

Work Done = Force × Distance

Whether the force is friction or weight or tension in a rope, it's always the same.

To find how much energy has been transferred (in joules), you just multiply the force in N by the distance moved in m. Look at this example:

Example: Mr Jones drags an old tractor tyre 5m over rough ground.
He pulls with a total force of 340N. Find the energy transferred.

Answer: Wd = F×d = 340 × 5 = 1700J.

---

## Work Done = Energy Transferred

This is a pretty easy formula — the trick is not to get it mixed up with any other formulas in this section. When you get to the end of the section, close the book and make a list of all the formulas you can remember. Keep going until you can write them all down correctly.

# Power

## Power is the "Rate of Doing Work" — i.e. how much per second

Power is not the same thing as force, nor energy. A powerful machine is not necessarily one which can exert a strong force (though it usually ends up that way).

A powerful machine is one which transfers a lot of energy in a short space of time.

This is the very easy formula for power:

$$\text{Power} = \frac{\text{Work done}}{\text{Time taken}}$$

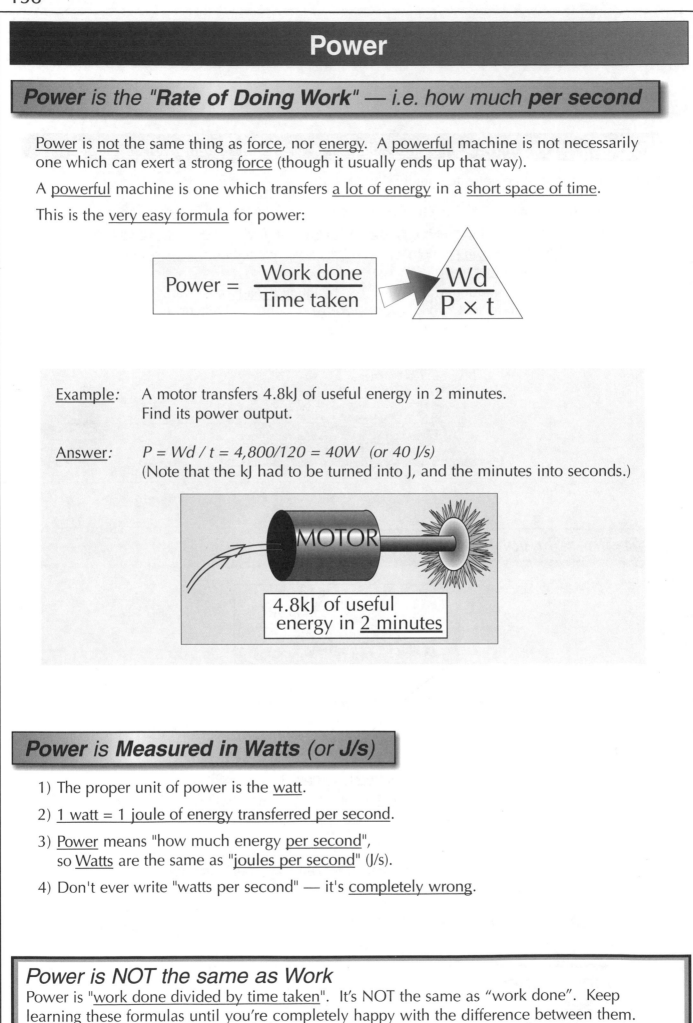

$$\frac{Wd}{P \times t}$$

Example: A motor transfers 4.8kJ of useful energy in 2 minutes.
Find its power output.

Answer: $P = Wd / t = 4,800/120 = 40W$ (or 40 J/s)
(Note that the kJ had to be turned into J, and the minutes into seconds.)

4.8kJ of useful
energy in 2 minutes

## Power is Measured in Watts (or J/s)

1) The proper unit of power is the watt.

2) 1 watt = 1 joule of energy transferred per second.

3) Power means "how much energy per second",
   so Watts are the same as "joules per second" (J/s).

4) Don't ever write "watts per second" — it's completely wrong.

## Power is NOT the same as Work

Power is "work done divided by time taken". It's NOT the same as "work done". Keep learning these formulas until you're completely happy with the difference between them. Remember the units: watts = joules per second — that'll help you remember the formulas.

# Kinetic Energy

## Kinetic Energy is Energy of Movement

Anything which is <u>moving</u> has <u>kinetic energy</u>.

There's a slightly <u>tricky formula</u> for it, so you have to concentrate a little bit <u>harder</u> for this one.

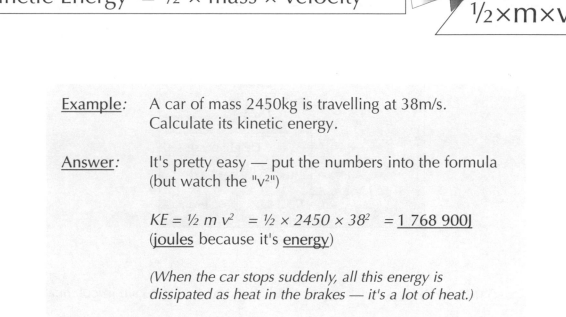

$$\text{Kinetic Energy} = \tfrac{1}{2} \times \text{mass} \times \text{velocity}^2$$

Example:   A car of mass 2450kg is travelling at 38m/s.
Calculate its kinetic energy.

Answer:   It's pretty easy — put the numbers into the formula
(but watch the "$v^2$")

$KE = \tfrac{1}{2} \, m \, v^2 = \tfrac{1}{2} \times 2450 \times 38^2 = \underline{1\ 768\ 900J}$
(<u>joules</u> because it's <u>energy</u>)

*(When the car stops suddenly, all this energy is
dissipated as heat in the brakes — it's a lot of heat.)*

## Kinetic Energy depends on Mass and Speed

Remember, the <u>kinetic energy</u> of something depends both on <u>mass</u> and <u>speed</u>.

The <u>more it weighs</u> and the <u>faster it's going</u>, the <u>bigger</u> its kinetic energy will be.

small mass, not fast
low kinetic energy

big fast
lorries Ltd

big mass, very fast
high kinetic energy

## Kinetic energy — Don't forget the "$v^2$"

This is a very important formula that crops up all over the place. Particularly in stopping
distances for cars — the $v^2$ in the formula means that doubling the speed makes the kinetic
energy 4 times bigger — which means that stopping distances increase by 4 too (see page 69).

# Potential Energy

## *Potential Energy* is Energy *Due to Height*

The proper name for this kind of "Potential Energy" is Gravitational Potential Energy (as opposed to "elastic potential energy" or "chemical potential energy").

The proper name for g is "gravitational field strength".

On Earth, g is approximately 10 N/kg.

Potential Energy = mass × g × height

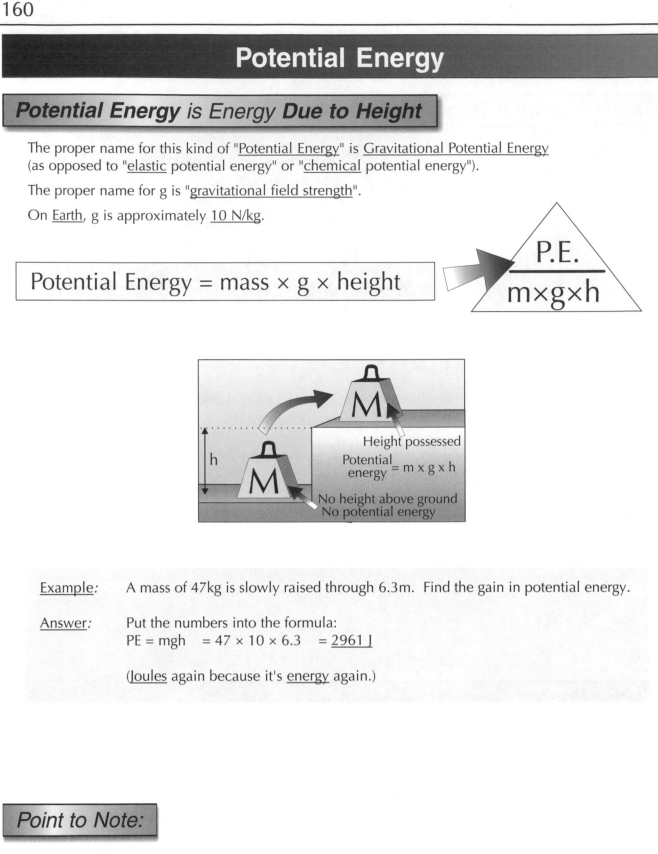

Example: A mass of 47kg is slowly raised through 6.3m. Find the gain in potential energy.

Answer: Put the numbers into the formula:
PE = mgh    = 47 × 10 × 6.3    = 2961 J

(Joules again because it's energy again.)

## Point to Note:

Strictly speaking it's the change in potential energy we're dealing with, so the formula can sometimes be written as: "Change in Potential Energy = mass × g × change in height". But that's a minor detail, because it all works out just the same.

## *Remember — Potential Energy is Stored Energy*
Don't forget to use the formula triangle — and remember there are four quantities to put in it, not the usual three. But other than that, this is a pretty easy formula to use.

# Potential and Kinetic Energy Calculations

## 1) Calculating Your Power Output

Both cases use the same formula:

$$POWER = \frac{ENERGY\ TRANSFERRED}{TIME\ TAKEN} \quad or \quad P = \frac{E}{t}$$

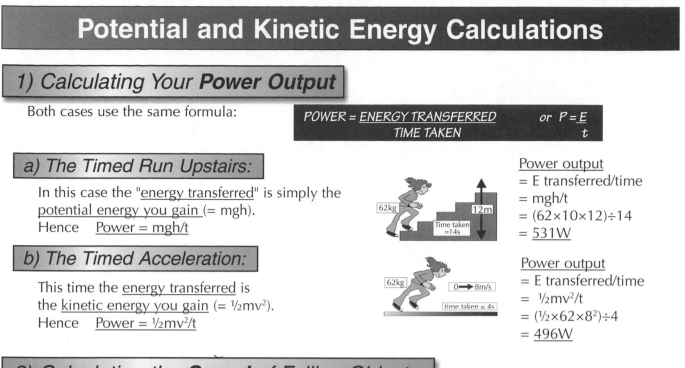

### a) The Timed Run Upstairs:

In this case the "energy transferred" is simply the potential energy you gain (= mgh).
Hence    Power = mgh/t

Power output
= E transferred/time
= mgh/t
= $(62 \times 10 \times 12) \div 14$
= 531W

### b) The Timed Acceleration:

This time the energy transferred is the kinetic energy you gain (= $\frac{1}{2}mv^2$).
Hence    Power = $\frac{1}{2}mv^2$/t

Power output
= E transferred/time
= $\frac{1}{2}mv^2$/t
= $(\frac{1}{2} \times 62 \times 8^2) \div 4$
= 496W

## 2) Calculating the Speed of Falling Objects

When something falls, its potential energy is converted into kinetic energy (Principle of Conservation of Energy — see page 152). Hence the further it falls, the faster it goes. In practice, some of the PE will be dissipated as heat due to air resistance, but in Exam questions they'll say you can ignore air resistance, in which case you'll just need to remember this simple and really quite obvious formula:

### Kinetic energy gained = Potential Energy lost

Example: A bean bag of mass 140g is dropped from a height of 1.7m.
Calculate its speed as it hits the floor.
Answer: There are four key steps to this method — and you've got to learn them:

Step 1) Find the PE lost:    = mgh = $0.14 \times 10 \times 1.7$ = 2.38J  This must also be the KE gained.
Step 2) Equate the number of Joules of KE gained to the KE formula with v in, " $\frac{1}{2}mv^2$ ":
$2.38 = \frac{1}{2}mv^2$
Step 3) Stick the numbers in:    $2.38 = \frac{1}{2} \times 0.14 \times v^2$    or    $2.38 = 0.07 \times v^2$
$2.38 \div 0.07 = v^2$          so    $v^2 = 34$
Step 4) Square root:    $v = \sqrt{34}$    = 5.83 m/s

## The Bouncing Ball — Same % Drop in Energy and Height

1) A bouncing ball is constantly swapping its energy between PE and KE, just as in the above example. As it falls it converts PE into KE. After the bounce, it rises again and converts its KE back into PE.

2) However, each time it bounces it will lose some energy in the bounce. This means it'll leave the surface a bit slower than it hits it, which means with less KE, so it won't reach the same height as the previous bounce.

3) The relation between total energy and height reached is really simple: If the ball loses say 10% of its energy each bounce, then the height reached will also be 10% lower each time. It's as simple as that.

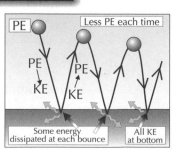

## Remember about the Principle of Conservation of Energy

This is getting into serious Physics now. You need to learn that four step method and get plenty of practice doing these kinds of question. It's the only way you're going to crack them.

# Warm-Up and Worked Exam Question

## Warm-up Questions

1) Complete the following three equations.
   a) Work Done = _force_ × Distance
   b) _P.E.G_ = Mass × g × Height
   c) Kinetic Energy = _½_ × Mass × _v²_

2) Three of these quantities are measured in joules:
   — Kinetic Energy, Work Done, Power, Potential Energy.
   a) Which is the odd one out? _Power_
   b) What unit is the odd one out measured in? _Watts_

3) For each of the examples below, state whether work is being done.
   a) Weightlifter lifting weights. _Yes_
   b) Builder standing still holding a plank. _No_
   c) Librarian lifting a book onto a shelf. _Yes_

## Worked Exam Question

1     The diagram below shows Dominic about to climb a rope during fitness training. Dominic has a mass of 65kg.

5 m

(a) Calculate how much potential energy he will have gained when he reaches the top of the rope.

    **Potential Energy = Mass × g × Height**

           **= 65 × 10 × 5**     (1 mark for formula or working)

           **= 3250J**     (1 mark for numerical answer, full marks for correct answer with units)

           *(3 marks)*

(b) Dominic takes 50 seconds to climb the rope. Calculate his power in watts.

    **Power = Work Done or Energy ÷ Time**

           **= 3250 ÷ 50**     (1 mark for formula or working)

           **= 65 W**     (1 mark)

           *(2 marks)*

# Exam Questions

1    The Brookes' car has broken down at the bottom of the hill, only <u>30 metres</u> from their home. They decide to push it home. They push with a combined force of <u>2000N</u> and it takes them 2 minutes.

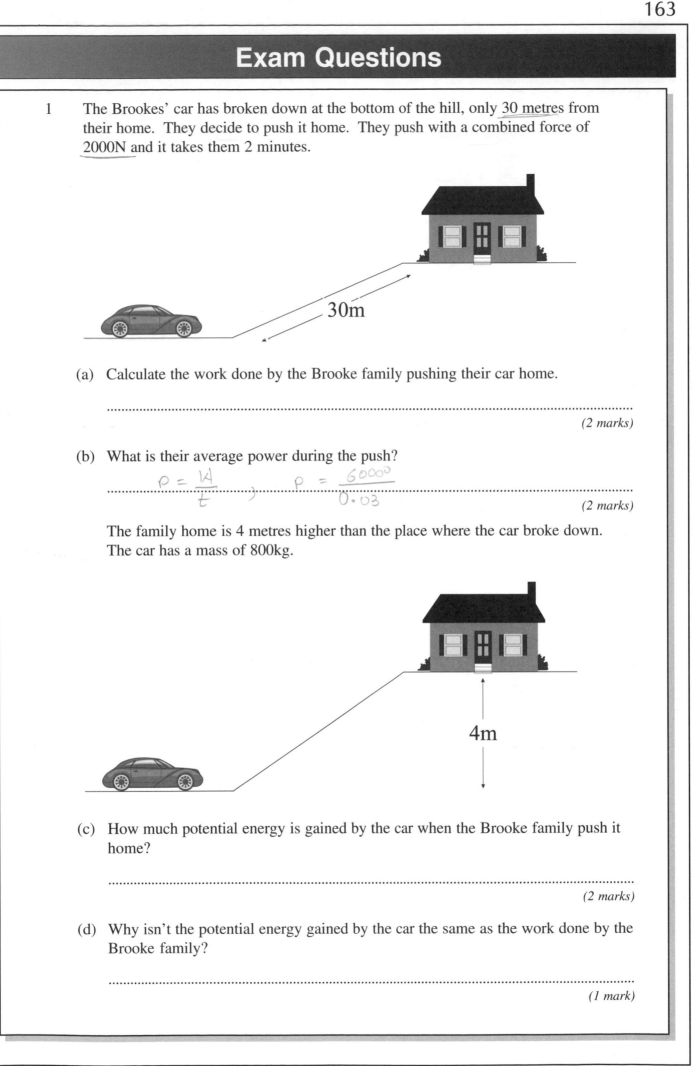

30m

(a)    Calculate the work done by the Brooke family pushing their car home.

    .............................................................................................................................
                                                                            *(2 marks)*

(b)    What is their average power during the push?

$P = \dfrac{W}{t}$  )    $P = \dfrac{60000}{0.03}$

                                                                            *(2 marks)*

The family home is 4 metres higher than the place where the car broke down. The car has a mass of 800kg.

4m

(c)    How much potential energy is gained by the car when the Brooke family push it home?

    .............................................................................................................................
                                                                            *(2 marks)*

(d)    Why isn't the potential energy gained by the car the same as the work done by the Brooke family?

    .............................................................................................................................
                                                                            *(1 mark)*

# Exam Questions

2   Marcel is working on the Eiffel Tower. He is 50 metres up, carrying his bag of tools, when he drops a spanner. The spanner has a mass of 500g.

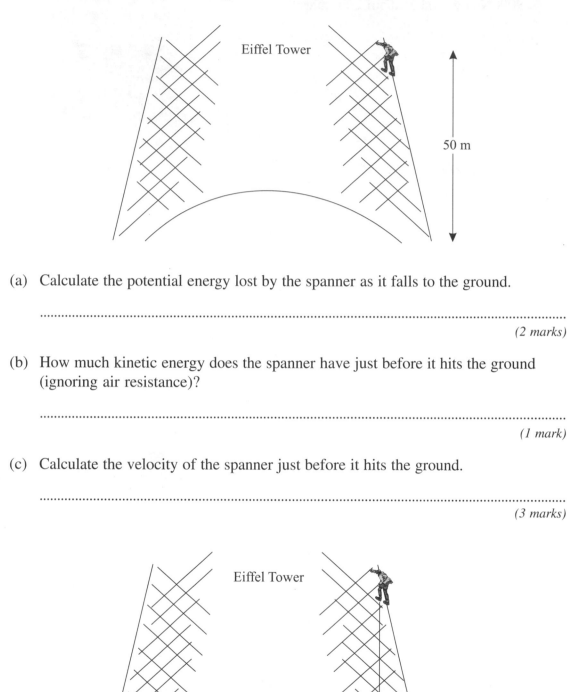

(a)  Calculate the potential energy lost by the spanner as it falls to the ground.

......................................................................................................................................

*(2 marks)*

(b)  How much kinetic energy does the spanner have just before it hits the ground (ignoring air resistance)?

......................................................................................................................................

*(1 mark)*

(c)  Calculate the velocity of the spanner just before it hits the ground.

......................................................................................................................................

*(3 marks)*

(d)  If the spanner loses 99% of its kinetic energy as heat and sound when it hits the ground, what will be the height of its first bounce?

......................................................................................................................................

*(2 marks)*

# Heat Transfer

There are three distinct methods of heat transfer: conduction, convection and radiation. To answer Exam questions, you must use those three key words in just the right places. That means you need to know exactly what they are, and all the differences between them.

## Heat Energy Causes Molecules to Move Faster

1) Heat energy causes gas and liquid molecules to move around faster, and causes particles in solids to vibrate more rapidly.
2) When particles move faster it shows up as a rise in temperature.
3) This extra kinetic energy in the particles tends to get dissipated to the surroundings.
4) In other words, the heat energy tends to flow away from a hotter object to its cooler surroundings.

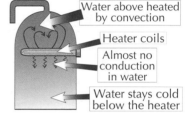
Cool Surroundings
Heat Heat Heat Heat Heat Heat Heat
Cool Surroundings

> If there's a difference in temperature between two places then *HEAT WILL FLOW* between them.

## Conduction, Convection and Radiation Compared

These differences are really important — make sure you learn them:
1) Conduction occurs mainly in solids.
2) Convection occurs mainly in gases and liquids.
3) Gases and liquids are very poor conductors — convection is usually the dominant process. Where convection can't occur, the heat transfer by conduction is very slow indeed as the diagram of the immersion heater shows. This is a classic example, so it's a good idea to learn it.
4) Radiation travels through anything see-through including a vacuum.
5) Heat Radiation is given out by anything which is warm or hot.
6) The amount of heat radiation which is absorbed or emitted depends on the colour and texture of the surface. But don't forget, convection and conduction are totally unaffected by surface colour or texture. A shiny white surface conducts just as well as a matt black one.

Water above heated by convection
Heater coils
Almost no conduction in water
Water stays cold below the heater

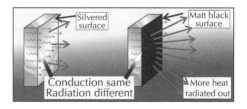
Silvered surface
Matt black surface
Conduction same Radiation different
More heat radiated out

## Convection Heaters and "Radiators" — A Classic Mistake

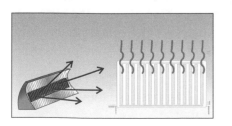

1) A "radiator" strictly should be something that glows red and gives most heat out as radiation, like a coal fire or an electric bar radiator.
2) Central heating "radiators" have the wrong name really, because they're not like that at all. They give most heat out as convection currents of warm rising air. This is what a "convection heater" does.

> *Remember: most 'radiators' don't radiate — they cause convection*
> Back to straightforward fact-learning again. You've got to make sure you get those three key processes of heat transfer all sorted out in your head so that you know exactly what they are and when they occur.

# Conduction

## *Conduction* of Heat — Occurs Mainly in *Solids*

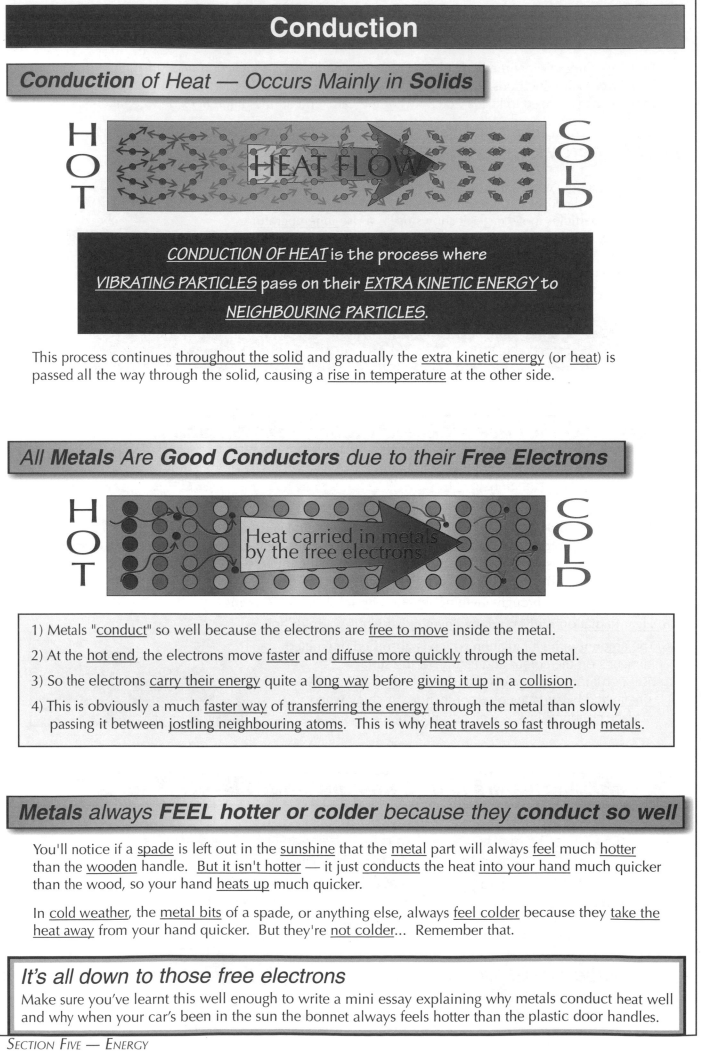

H O T

C O L D

HEAT FLOW

*CONDUCTION OF HEAT* is the process where
*VIBRATING PARTICLES* pass on their *EXTRA KINETIC ENERGY* to
*NEIGHBOURING PARTICLES*.

This process continues <u>throughout the solid</u> and gradually the <u>extra kinetic energy</u> (or <u>heat</u>) is passed all the way through the solid, causing a <u>rise in temperature</u> at the other side.

## All *Metals* Are *Good Conductors* due to their *Free Electrons*

H O T

C O L D

Heat carried in metals by the free electrons

1) Metals "<u>conduct</u>" so well because the electrons are <u>free to move</u> inside the metal.

2) At the <u>hot end</u>, the electrons move <u>faster</u> and <u>diffuse more quickly</u> through the metal.

3) So the electrons <u>carry their energy</u> quite a <u>long way</u> before <u>giving it up</u> in a <u>collision</u>.

4) This is obviously a much <u>faster way</u> of <u>transferring the energy</u> through the metal than slowly passing it between <u>jostling neighbouring atoms</u>. This is why <u>heat travels so fast</u> through <u>metals</u>.

## Metals always *FEEL hotter or colder* because they *conduct so well*

You'll notice if a <u>spade</u> is left out in the <u>sunshine</u> that the <u>metal</u> part will always <u>feel</u> much <u>hotter</u> than the <u>wooden</u> handle.  <u>But it isn't hotter</u> — it just <u>conducts</u> the heat <u>into your hand</u> much quicker than the wood, so your hand <u>heats up</u> much quicker.

In <u>cold weather</u>, the <u>metal bits</u> of a spade, or anything else, always <u>feel colder</u> because they <u>take the heat away</u> from your hand quicker.  But they're <u>not colder</u>...  Remember that.

## *It's all down to those free electrons*

Make sure you've learnt this well enough to write a mini essay explaining why metals conduct heat well and why when your car's been in the sun the bonnet always feels hotter than the plastic door handles.

# Convection

## Convection of Heat — Liquids and Gases Only

Gases and liquids are usually free to move about a lot — and that allows them to transfer heat by convection, which is a much more effective process than conduction.

Convection simply can't happen in solids because the particles can't move.

> CONVECTION occurs when the more energetic particles MOVE
> from the HOTTER REGION to the COOLER REGION — AND
> TAKE THEIR HEAT ENERGY WITH THEM.

When the more energetic (i.e. hotter) particles get somewhere cooler they then transfer their energy by the usual process of collisions which warm up the surroundings.

## A Classic Experiment

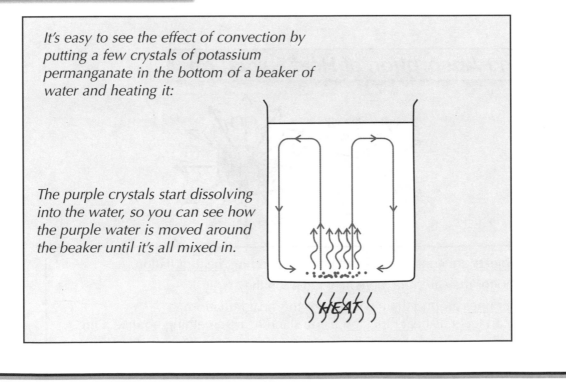

It's easy to see the effect of convection by putting a few crystals of potassium permanganate in the bottom of a beaker of water and heating it:

The purple crystals start dissolving into the water, so you can see how the purple water is moved around the beaker until it's all mixed in.

## Convection and Conduction — learn the difference

Convection is a similar word to conduction — people always mix them up. It's not just a different word though — convection is a totally different process too. Learn exactly why it isn't like conduction.

# Radiation

<u>Heat radiation</u> can also be called <u>infra-red radiation</u>, and it consists purely of electromagnetic waves of a certain frequency.  It's just below visible light in the <u>electromagnetic spectrum</u>.

## *Heat Radiation* Can Travel Through A *Vacuum*

<u>Heat radiation</u> is <u>different</u> from the <u>other two methods</u> of heat transfer in quite a few ways:

1) It travels in <u>straight lines</u> at the <u>speed of light</u>.

2) It travels through a <u>vacuum</u>.  This is the <u>only way</u> that heat can reach us from the <u>Sun</u>.

3) It can be very effectively <u>reflected away again</u> by a <u>silver surface</u>.

4) It only travels through <u>transparent media</u>, like <u>air</u>, glass and <u>water</u>.

5) Its behaviour is <u>strongly dependent</u> on <u>surface colour and texture</u>.
   This definitely <u>isn't</u> so for conduction and convection.

6) <u>No particles</u> are involved.  It's transfer of heat energy <u>purely</u> by <u>waves</u>.

## *Emission* and *Absorption* of *Heat Radiation*

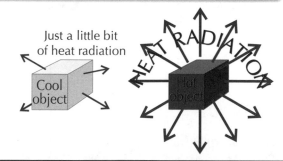

Just a little bit of heat radiation

1) <u>All objects</u> are <u>continually</u> emitting and absorbing <u>heat radiation</u>.

2) The <u>hotter</u> they are, the <u>more</u> heat radiation they <u>emit</u>.

3) <u>Cooler ones</u> around them will <u>absorb</u> this heat radiation.
   You can <u>feel</u> this <u>heat radiation</u> if you stand near something <u>hot</u> like a fire.

---

## *If radiation couldn't travel through a vacuum, we'd all be very cold*

Conduction and convection rely completely on transferring energy via molecules.
Heat radiation is totally different — it can move quite happily without involving any molecules ar all.

# Radiation

## It *Depends* An Awful Lot on *Surface Colour* and *Texture*

1) Dark matt surfaces absorb heat radiation falling on them much more strongly than bright glossy surfaces, such as gloss white or silver. They also emit heat radiation much better too.

2) Silvered surfaces reflect nearly all heat radiation falling on them.

3) In the lab, there are several experiments to demonstrate the effects of surface on emission and absorption of heat radiation. Here are two of the standard ones:

### Leslie's Cube

The matt black side emits most heat so it's that thermometer which gets hottest.

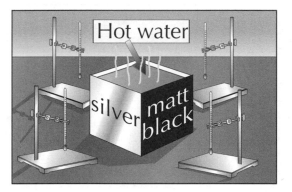

### The *Melting Wax* Trick

The matt black surface absorbs most heat so its wax melts first and the ball bearing drops.

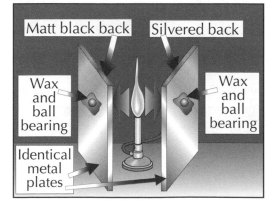

## *Revise Heat Radiation — absorb as much as you can anyway*

The main point is that heat radiation is strongly affected by the colour and texture of surfaces. The other types of heat transfer (conduction and convection) aren't affected by surface colour and texture at all.

# Applications of Heat Transfer

## Good **Conductors** and Good **Insulators**

1) All <u>metals</u> are good <u>conductors</u> e.g. iron, brass, aluminium, copper, gold, silver etc.

2) All <u>non-metals</u> are good <u>insulators</u>.

3) Gases and liquids are really <u>bad conductors</u> (but are good <u>convectors</u> don't forget).

4) The <u>best insulators</u> are ones which <u>trap pockets of air</u>. If the air <u>can't move</u>, it <u>can't</u> transfer heat by <u>convection</u> and so the heat has to <u>conduct</u> very slowly through the <u>pockets of air</u>, as well as the material in between. This really <u>slows it down</u>.
This is how <u>clothes</u>, <u>blankets</u>, <u>loft insulation</u>, <u>cavity wall insulation</u>, <u>polystyrene cups</u>, <u>woollen mittens</u>, etc. work.

## **Insulation** should also take account of **Heat Radiation**

1) <u>Silvered finishes</u> are highly effective <u>insulation</u> against heat transfer by <u>radiation</u>.

2) This can work <u>both ways</u>, either keeping heat radiation <u>out</u> or keeping heat radiation <u>in</u>.

| Keeping Heat radiation out: | Keeping Heat in: |
|---|---|
| Spacesuits | Shiny metal kettles |
| Thermos flasks | Survival blankets |
| | Thermos flasks (again) |

3) <u>Matt black</u> is rarely used for its thermal properties of <u>absorbing</u> and <u>emitting</u> heat radiation.

4) It's only <u>useful</u> where you want to <u>get rid of heat</u>, e.g. the <u>cooling fins</u> or <u>radiator</u> on an engine.

## The **Thermos Flask** — The Ultimate in **Insulation**

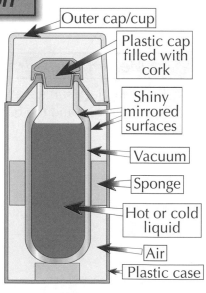

1) The glass bottle is <u>double-walled</u> with a <u>thin vacuum</u> between the two walls. This stops <u>all conduction and convection</u> through the <u>sides</u>.

2) The walls either side of the vacuum are <u>silvered</u> to keep heat loss by <u>radiation</u> to a <u>minimum</u>.

3) The bottle is supported using <u>insulating foam</u>. This minimises heat <u>conduction</u> to or from the <u>outer</u> glass bottle.

4) The <u>stopper</u> is made of <u>plastic</u> and filled with <u>cork or foam</u> to reduce any <u>heat conduction</u> through it.

In <u>Exam questions</u> you must <u>always</u> say which form of heat transfer is involved at any point, either <u>conduction</u>, <u>convection</u> or <u>radiation</u>.
*"The vacuum stops heat getting out"* will get you <u>no marks at all</u>.

*Outer cap/cup*
*Plastic cap filled with cork*
*Shiny mirrored surfaces*
*Vacuum*
*Sponge*
*Hot or cold liquid*
*Air*
*Plastic case*

## The Thermos Flask nearly always comes up in Exam questions

There's a lot more to insulation than you first realise. That's because there are <u>three ways</u> that heat can be transferred, and so effective heat insulation has to deal with <u>all three</u>, of course.
The venerable Thermos Flask is the classic example of all-in-one full-blown insulation. <u>Learn it</u>.

# Insulating Buildings

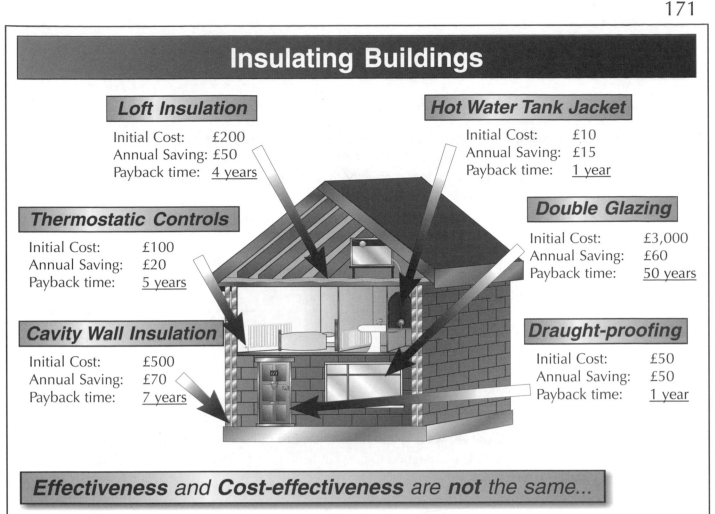

**Loft Insulation**
Initial Cost: £200
Annual Saving: £50
Payback time: 4 years

**Hot Water Tank Jacket**
Initial Cost: £10
Annual Saving: £15
Payback time: 1 year

**Thermostatic Controls**
Initial Cost: £100
Annual Saving: £20
Payback time: 5 years

**Double Glazing**
Initial Cost: £3,000
Annual Saving: £60
Payback time: 50 years

**Cavity Wall Insulation**
Initial Cost: £500
Annual Saving: £70
Payback time: 7 years

**Draught-proofing**
Initial Cost: £50
Annual Saving: £50
Payback time: 1 year

## Effectiveness and Cost-effectiveness are not the same...

1) The figures above are all roughly average, but of course it'll vary from house to house.
2) The cheaper methods of insulation tend to be a lot more cost-effective than the pricier ones.
3) The ones that save the most money each year could be considered the most "effective". i.e. cavity wall insulation. How cost-effective it is depends on what time-scale you're looking at.
4) If you subtract the annual saving from the initial cost repeatedly then eventually the one with the biggest annual saving must always come out as the winner, if you think about it.
5) But you might sell the house (or die) before that happens. If instead you look at it over say, a five year period then the cheap and cheerful draught-proofing wins.
6) But double glazing is always by far the least cost-effective, which is mildly comical, considering.

## Know Which Types of Heat Transfer are Involved:

1) Cavity Wall Insulation — foam squirted into the gap between the bricks reduces convection and radiation across the gap.
2) Loft insulation — a thick layer of fibreglass wool laid out across the whole loft floor reduces conduction and radiation into the roof space from the ceiling.
3) Draught proofing — strips of foam and plastic around doors and windows stop draughts of cold air blowing in, i.e. they reduce heat loss due to convection.
4) Double Glazing — two layers of glass with an air gap reduce conduction and radiation.
5) Thermostatic Radiator valves — these simply prevent the house being over-warmed.
6) Hot water tank jacket — lagging such as fibreglass wool reduces conduction and radiation from the hot water tank.
7) Thick Curtains — big bits of cloth you pull across the window to stop people looking in at you, but also to reduce heat loss by conduction and radiation.

## They don't seem to have these problems in Spain

Remember, the most effective insulation measure is the one which keeps the most heat in, (biggest annual saving). If your house had no roof, then a roof would be the most effective measure. But cost-effectiveness depends very much on the time-scale involved.

# Warm-Up and Worked Exam Question

## Warm-up Questions

1) Anna visits Spain and notices that many of the houses are painted white on the outside. Explain why this is likely to help the houses remain cool.

2) Choose the correct name for each of the methods of heat transfer described below.

   a) Heat energy travelling through space as electromagnetic waves.

   b) Heat travelling through a solid due to energetic atoms colliding with their neighbours and passing on energy.

   c) Heat moving upwards through a gas due to the expansion of warm areas of gas.

3) Double glazing can be an effective way of reducing energy loss from a house. Why is it unlikely to be a cost-effective way of saving energy?

## Worked Exam Question

1   The diagram below shows a vacuum flask used to transport liquid nitrogen at very low temperatures.  It is designed to prevent heat energy from entering the liquid and causing it to evaporate.

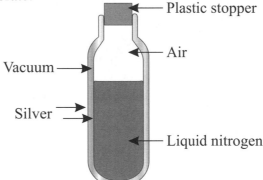

(a)   Which two methods of heat transfer is the vacuum designed to prevent?

   *Conduction and convection.*                                    Award 1 mark for each

   *(2 marks)*

The gap between the top of the liquid nitrogen and the plastic stopper is filled with air.

(b)   Describe carefully how heat is transferred through air by convection.

   *Warm air expands/becomes less dense.*       1 mark

   *So it rises, carrying heat energy with it.*       1 mark

   *(2 marks)*

(c)   Explain why heat energy cannot pass through the air gap to the liquid nitrogen by convection.

   *Convection can only carry heat energy upwards.*

   *(1 mark)*

# Exam Questions

1    The diagram shows an old copper hot water tank in a house.
The surface of the tank has become dirty, making it a dark, dull colour.

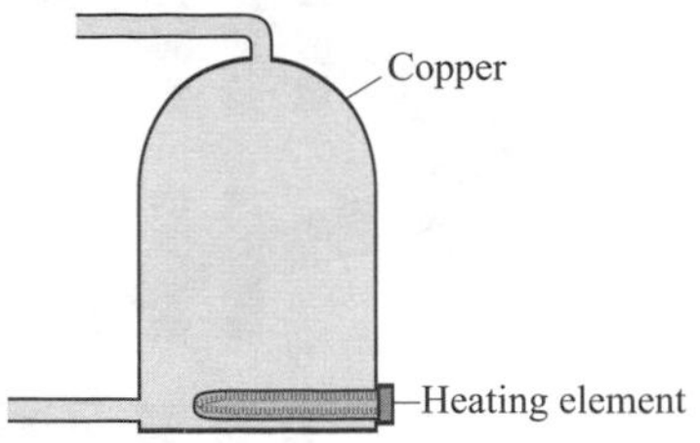

(a)   When the water is hot, heat energy travels quickly through the copper.
Explain clearly how the heat energy passes through the copper.

      .............................................................................................................................................
                                                                                          *(2 marks)*

(b)   What effect, if any, does the dirtiness of the water tank have on the rate of heat
loss from the tank?

      .............................................................................................................................................
                                                                                          *(2 marks)*

When the air near the tank warms up the heat energy is quickly transferred away.

(c)   How does the heat energy travel through the air?

      .............................................................................................................................................
                                                                                          *(1 mark)*

The new owner of the house buys a jacket for the hot water tank.  The jacket costs
£12 and is expected to save £15 per year.  The diagram shows a close-up of the
structure of the jacket.

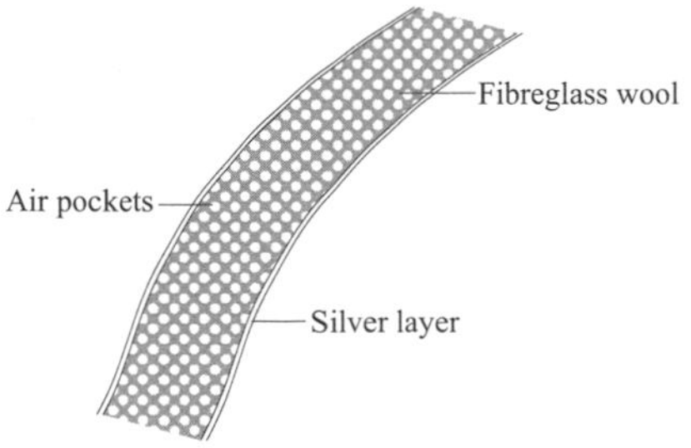

(d)   Explain how the air in the hot water jacket now reduces heat loss from the tank.

      .............................................................................................................................................
                                                                                          *(2 marks)*

A salesman suggests to the owner that it would be a good idea to invest in
cavity wall insulation, as this could save £65 per year.

(e)   Suggest reasons why cavity wall insulation may not be a good investment.

      .............................................................................................................................................
                                                                                          *(2 marks)*

# Exam Questions

2    The inside of a fridge is kept cool by the continual transfer of heat energy from inside the fridge to the outside.
A cooling element at the back of the fridge is designed to quickly release this heat into the air.

An example of such a cooling element is shown here:

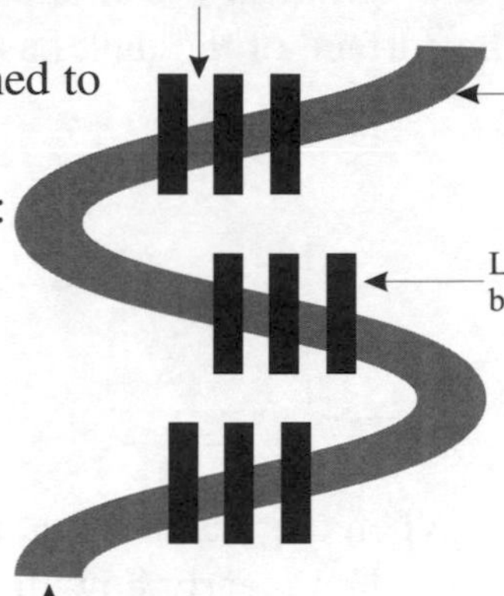

Air gaps

Black coolant pipe

Large black fins

Coolant from fridge

(a)    Identify two features of the element that help to increase heat loss by radiation.

......................................................................................................................................

*(2 marks)*

(b)    Suggest a suitable material to make the element from.  Explain your choice.

......................................................................................................................................

*(2 marks)*

Fridges are surrounded by thick layers of insulation to keep heat energy out.
A cross-section of one design for such insulation is shown here:

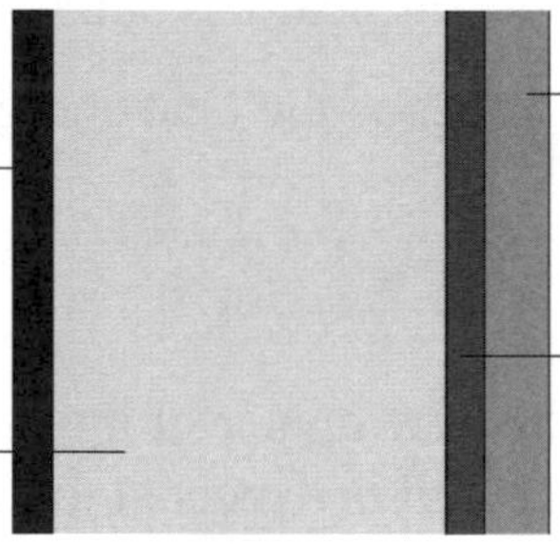

Metal outer case

Thick plastic insulation

Plastic inner case

Shiny foil (silver)

(c)    Describe what is meant by heat radiation.

......................................................................................................................................

*(1 mark)*

(d)    What feature of the insulation is designed to reduce the amount of heat radiation entering the fridge?

......................................................................................................................................

*(1 mark)*

(e)    The plastic shown is an insulator.  Despite this, heat energy is still conducted through the solid plastic in the fridge.  Give an explanation, in terms of the plastic particles, of how this happens.

......................................................................................................................................

*(2 marks)*

(f)    Suggest a modification to the insulation which would reduce the amount of heat energy conducted through the plastic.

......................................................................................................................................

*(1 mark)*

# Energy Resources

There are <u>twelve</u> different types of <u>energy resource</u>.
They fit into <u>two broad types</u>: <u>renewable</u> and <u>non-renewable</u>.

## *Non-renewable* Energy Resources Will **Run Out One Day**

The <u>non-renewables</u> are the <u>three FOSSIL FUELS</u> and <u>NUCLEAR</u>:

1) <u>Coal</u>

2) <u>Oil</u>

3) <u>Natural gas</u>

4) <u>Nuclear fuels</u> (<u>uranium</u> and <u>plutonium</u>)

> a) They will <u>all run out</u> one day.
>
> b) They all do <u>damage</u> to the environment.
>
> c) But they provide <u>most of our energy</u>.

## *Renewable* Energy Resources Will **Never Run Out**

The <u>renewables</u> are:

1) <u>Wind</u>          5) <u>Solar</u>

2) <u>Waves</u>          6) <u>Geothermal</u>

3) <u>Tides</u>          7) <u>Food</u>

4) <u>Hydroelectric</u>          8) <u>Biomass (wood)</u>

> a) These will <u>never run out</u>.
>
> b) They <u>do not damage the environment</u> (except visually).
>
> c) The trouble is they're often <u>expensive</u> and many of them are <u>unreliable</u> because they depend on the <u>weather</u>.

## *Learn which resources are renewable and which aren't*

This is the basic stuff — make absolutely sure you know all of this before you try learning anything else about energy resources. Cover the page and write out the two lists of resources.

# Energy Chains

## The Sun is the Ultimate Source for Nine of The Energy Resources

(The exceptions are tides, nuclear and geothermal — see below)

You need to know the <u>energy transfer chains</u> for all *nine* of them starting from the <u>Sun</u>.

There are however only *five* basic <u>energy chains</u>:

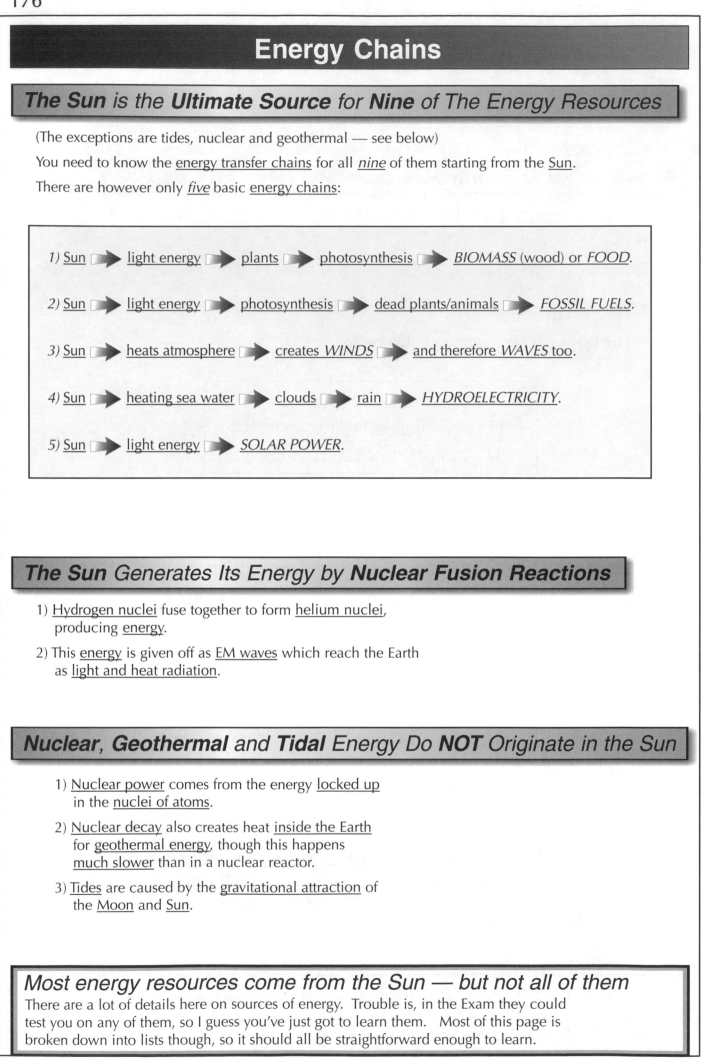

1) <u>Sun</u> ➡ <u>light energy</u> ➡ <u>plants</u> ➡ <u>photosynthesis</u> ➡ *BIOMASS* (wood) or *FOOD*.

2) <u>Sun</u> ➡ <u>light energy</u> ➡ <u>photosynthesis</u> ➡ <u>dead plants/animals</u> ➡ *FOSSIL FUELS*.

3) <u>Sun</u> ➡ <u>heats atmosphere</u> ➡ <u>creates *WINDS*</u> ➡ and therefore *WAVES* too.

4) <u>Sun</u> ➡ <u>heating sea water</u> ➡ <u>clouds</u> ➡ <u>rain</u> ➡ *HYDROELECTRICITY*.

5) <u>Sun</u> ➡ <u>light energy</u> ➡ *SOLAR POWER*.

## The Sun Generates Its Energy by Nuclear Fusion Reactions

1) <u>Hydrogen nuclei</u> fuse together to form <u>helium nuclei</u>, producing <u>energy</u>.

2) This <u>energy</u> is given off as <u>EM waves</u> which reach the Earth as <u>light and heat radiation</u>.

## Nuclear, Geothermal and Tidal Energy Do NOT Originate in the Sun

1) <u>Nuclear power</u> comes from the energy <u>locked up</u> in the <u>nuclei of atoms</u>.

2) <u>Nuclear decay</u> also creates heat <u>inside the Earth</u> for <u>geothermal energy</u>, though this happens <u>much slower</u> than in a nuclear reactor.

3) <u>Tides</u> are caused by the <u>gravitational attraction</u> of the <u>Moon</u> and <u>Sun</u>.

## Most energy resources come from the Sun — but not all of them

There are a lot of details here on sources of energy. Trouble is, in the Exam they could test you on any of them, so I guess you've just got to learn them. Most of this page is broken down into lists though, so it should all be straightforward enough to learn.

# Power Stations

<u>Most</u> of the electricity we use is <u>generated</u> from the four <u>NON-RENEWABLE</u> sources of energy (<u>coal</u>, <u>oil</u>, <u>gas</u> and <u>nuclear</u>) in <u>big power stations</u>, which are all <u>pretty much the same</u> apart from the <u>boiler</u>.

<u>Learn</u> the <u>basic features</u> of the typical power station shown here, and also the <u>nuclear reactor</u> below.

## Nuclear Reactors are Just Elaborate Boilers

1) A <u>nuclear power station</u> is mostly the same as the one shown above, where <u>heat is produced</u> in a <u>boiler</u> to make <u>steam</u> to drive <u>turbines</u> etc.  The difference is in the <u>boiler</u>, as shown here:

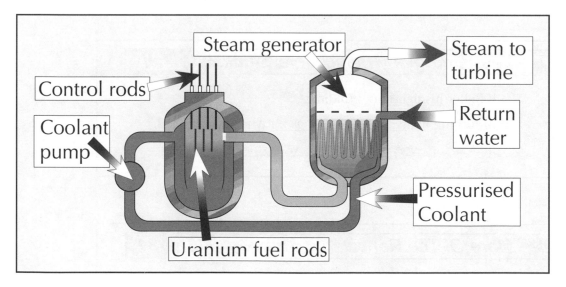

2) Nuclear reactors take the <u>longest</u> time of all the non-renewables to start up. <u>Natural gas</u> takes the shortest time.

---

## *Learn about the non-renewables — before it's too late*

Make sure you realise that we generate most of our electricity from the four non-renewables, and that the power stations are all pretty much the same, as shown by the diagram at the top of the page. Also make sure you know all the problems about them and why we should use less of them.

# Environmental Problems

## Environmental Problems With The Use Of Non-Renewables

1) All three fossil fuels, (coal, oil and gas) release $CO_2$. For the same amount of energy produced, coal releases the most $CO_2$, followed by oil then gas. All this $CO_2$ adds to the Greenhouse Effect, causing global warming. There's no feasible way to stop it being released either.

2) Burning coal and oil releases sulphur dioxide which causes acid rain.
   This is reduced by taking the sulphur out before it's burned or by cleaning up the emissions.

3) Coal mining makes a mess of the landscape, especially "open-cast mining".

4) Oil spillages cause serious environmental problems. We try to avoid it, but it'll always happen.

5) Nuclear power is clean but the nuclear waste is very dangerous and difficult to dispose of.

6) Nuclear fuel (i.e. uranium) is cheap but the overall cost of nuclear power is high due to the cost of the power plant and final de-commissioning.

7) Nuclear power always carries the risk of a major catastrophe like the Chernobyl disaster.

## The Non-Renewables Need to be Conserved

1) When the fossil fuels eventually run out we will have to use other forms of energy.

2) More importantly however, fossil fuels are also a very useful source of chemicals, (especially crude oil) which will be hard to replace when they are all gone.

3) To stop the fossil fuels running out so quickly there are two things we can do:

---

### 1) Use Less Energy by Being More Efficient With it:

(i) Better insulation of buildings,

(ii) Turning lights and other things off when not needed,

(iii) Making everyone drive cars with smaller engines.

---

### 2) Use More Of The Renewable Sources Of Energy

...as detailed in the next few pages.

---

## Make sure you learn all the sorry details

This is pretty depressing stuff, but I guess at least there's a point to learning it all. Then maybe one day you can be one of the scientists and Greenpeace activists trying to come up with viable solutions.

# Warm-Up and Worked Exam Question

## Warm-up Questions

1) What is meant by renewable and non-renewable energy sources?
   Give four examples of each type of energy resource.

2) What are the main problems with renewable energy sources?

3) Write out energy transfer chains to show how the following have the Sun as their ultimate source:   a) biomass;   b) waves;   c) hydroelectricity.

4) Describe the basic features of a typical power station.

5) Give four environmental problems with the use of non-renewables.

## Worked Exam Question

1     Electricity can be generated in several different ways.

   (a)   Describe how the energy in coal is used to drive generators in coal-fired power stations.

   *The coal is burnt to heat water in the boiler. The water is converted to steam, which is used to drive the turbine. The turbine drives the generator.*

   *Award 1 mark for each point.*

   (3 marks)

*You need to be prepared to attack different kinds of energy question. Some will be closed (requiring a specific answer) whereas some will be open-ended (requiring you to explore, discuss and compare).*

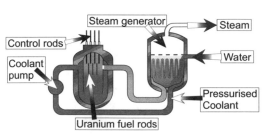

   (b)   The diagram shows a nuclear reactor at a nuclear power station. Describe how this method of electricity generation is different from using coal.

   *There is no burning of fuel. The heat is generated in the reactor and absorbed by the coolant, which heats up. Coolant pipes pass through the boiler, where the water is heated up to generate steam.*

   *Award 1 mark for each of the three main ideas.*

   (3 marks)

   (c)   In the UK very little electricity is generated using renewable sources. Suggest why this is and explain why renewable sources of energy should be used more.

   *They are not used as much because: i) they are expensive compared to fossil fuels; ii) some are unreliable because they depend on the weather.*  *Award 1 mark for each valid point up to a maximum of 2.*

   *We should use them more because: i) fossil fuels will eventually run out (using renewables can delay this); ii) fossil fuels are a very useful source of chemicals, which will be hard to replace when they are gone; iii) fossil fuels pollute the environment (e.g. increasing carbon dioxide levels and causing acid rain) whereas renewables don't (except perhaps visually)*  *Award 1 mark for each valid point up to a maximum of 2.*

   (4 marks)

180

## Exam Questions

1   Medina is an island in the north Atlantic, which has a cool climate.  It does not have any reserves of fossil fuels.  The residents are concerned that pipelines and storage terminals would spoil their environment.  They have also decided not to use nuclear power.

(a)  Give three other energy sources the islanders could use.

.......................................................................................................................

.......................................................................................................................

.......................................................................................................................

*(3 marks)*

(b)  The islanders should be encouraged to conserve energy.
Suggest two ways in which they could do this.

.......................................................................................................................

.......................................................................................................................

*(2 marks)*

(c)  Nuclear power stations produce very little pollution when they are working properly.  Despite this, many people are opposed to them.  Explain the main arguments against using nuclear power.

.......................................................................................................................

.......................................................................................................................

.......................................................................................................................

*(3 marks)*

2   Much has been said in the media about the use of alternative sources of energy.  However, most of the electricity in the UK is generated from fossil fuels or nuclear fuel.

(a)  Why are these fuels used much more than other sources of energy?

.......................................................................................................................

*(2 marks)*

(b)  Describe the environmental problems associated with the use of fossil fuels.

.......................................................................................................................

.......................................................................................................................

.......................................................................................................................

.......................................................................................................................

*(4 marks)*

*SECTION FIVE — ENERGY*

# Wind Power and Hydroelectricity

## Wind Power — Wind Turbines

1) This involves putting <u>lots of windmills</u> (wind turbines) up in <u>exposed places</u> like on <u>moors</u> or round <u>coasts</u>.

2) Each wind turbine has its own <u>generator</u> inside it so the electricity is generated <u>directly</u> from the <u>wind</u> turning the <u>blades</u>, which <u>turn the generator</u>.

3) There's <u>no pollution</u>.

4) But they do <u>spoil the view</u>. You need about <u>5000 wind turbines</u> to replace <u>one coal-fired power station</u> and 5000 of them cover <u>a lot</u> of ground — which wouldn't look very nice at all.

5) There's also the problem of <u>no power when the wind stops</u>, and it's <u>impossible</u> to <u>increase supply</u> when there's <u>extra demand</u>.

6) The <u>initial costs are quite high</u>, but there are <u>no fuel costs</u> and <u>minimal running costs</u>.

## Hydroelectricity — Flooding a Valley and Building a Dam

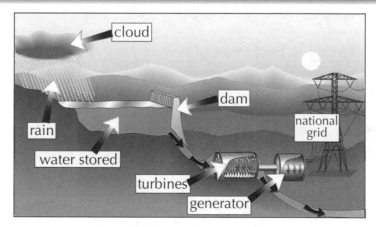

1) <u>Hydroelectric power</u> usually requires the <u>flooding</u> of a <u>valley</u> by building a <u>big dam</u>.

2) <u>Rainwater</u> is caught and allowed out <u>through turbines</u>. There is <u>no pollution</u>.

3) There is quite a <u>big impact</u> on the <u>environment</u> due to the flooding of the valley and possible <u>loss of habitat</u> for some species. The reservoirs can also look very <u>unsightly</u> when they <u>dry up</u>. Location in <u>remote valleys</u> (in <u>Scotland</u> and <u>Wales</u>) tends to avoid these problems on the whole.

4) A <u>big advantage</u> is <u>immediate response</u> to increased demand and there's no problem with <u>reliability</u> except in times of <u>drought</u> — but Britain is unlikely to have a real drought problem.

5) <u>Initial costs are high</u> but there's <u>no fuel</u> and <u>minimal running costs</u>.

---

### The main problem is the visual impact

Lots of important details here on these green sources of energy — it's a pity they make such a mess of the landscape. Two <u>mini-essays</u> should make sure you know all this stuff.

# Pumped Storage

## Pumped Storage Gives Extra Supply Just When it's Needed

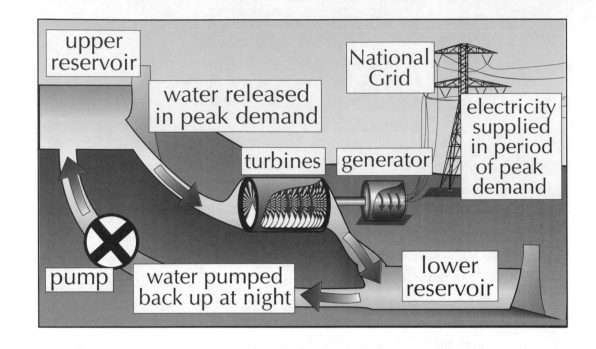

upper reservoir

water released in peak demand

National Grid

electricity supplied in period of peak demand

turbines

generator

pump

water pumped back up at night

lower reservoir

1) Most large power stations have <u>huge boilers</u> which have to be kept running <u>all night</u> even though demand is <u>very low</u>. This means there's a <u>surplus</u> of electricity at night.

2) It's surprisingly <u>difficult</u> to find a way of <u>storing</u> this spare energy for <u>later use</u>.

3) <u>Pumped storage</u> is one of the <u>best solutions</u> to the problem.

4) In pumped storage, 'spare' <u>night-time electricity</u> is used to pump water up to a <u>higher reservoir</u>.

5) This can then be <u>released quickly</u> during periods of <u>peak demand</u> such as at <u>tea time</u> each evening, to supplement the <u>steady delivery</u> from the big power stations.

6) Remember, <u>pumped storage</u> uses the same <u>idea</u> as Hydroelectric Power but it <u>isn't</u> a way of <u>generating</u> power — but simply a way of <u>storing energy</u> which has <u>already</u> been generated.

## Pumped storage is a neat way of storing energy

You need to learn both the diagram and all the notes underneath. You should be able to close the book then draw the diagram and write a mini-essay explaining why it's such a great idea.

# Wave Power

Don't confuse <u>wave power</u> with <u>tidal power</u>. They are <u>completely different</u>.

## *Wave Power* — *Lots of* *Wave Converters*

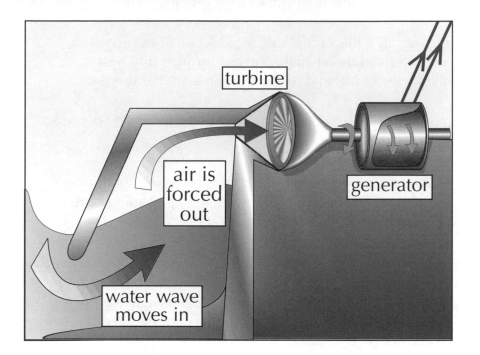

1) You need lots of small <u>wave generators</u> located <u>around the coast</u>.

2) As waves come in to the shore they provide an <u>up and down motion</u> which can be used to drive a <u>generator</u>.

3) There is <u>no pollution</u>. The main problems are <u>spoiling the view</u> and being a <u>hazard to boats</u>.

4) They are <u>fairly unreliable</u>, since waves tend to die out when the <u>wind drops</u>.

5) <u>Initial costs are high</u> but there's <u>no fuel</u> and <u>minimal running costs</u>. Wave power is never likely to provide energy on a <u>large scale</u> but it can be <u>very useful</u> on <u>small islands</u>.

## *Wave power is a nice idea — shame it's a bit unreliable*

Once again, it's "learn the diagram and the numbered points" time. Write a mini-essay explaining the advantages and disadvantages of wave power, as well as draw the diagram.

# Tidal Power

## Tidal Barrages — Using The ~~Sun~~ and Moon's Gravity

1) Tidal barrages are big dams built across river estuaries with turbines in them.

2) As the tide comes in it fills up the estuary to a height of several metres. This water can then be allowed out through turbines at a controlled speed. It also drives the turbines on the way in.

3) There is no pollution. The source of the energy is the gravity of the Sun and the Moon.

4) The main problems are preventing free access by boats, spoiling the view and altering the habitat of the wildlife, e.g. wading birds, sea creatures and creatures that live in the sand.

5) Tides are pretty reliable in the sense that they happen twice a day without fail, and always to the predicted height. The only drawback is that the height of the tide is variable — so lower (neap) tides will provide significantly less energy than the bigger "spring" tides. But tidal barrages are excellent for storing energy ready for periods of peak demand.

6) Initial costs are moderately high but there's no fuel and minimal running costs. Even though it can only be used in a few of the most suitable estuaries tidal power has the potential for generating a significant amount of energy.

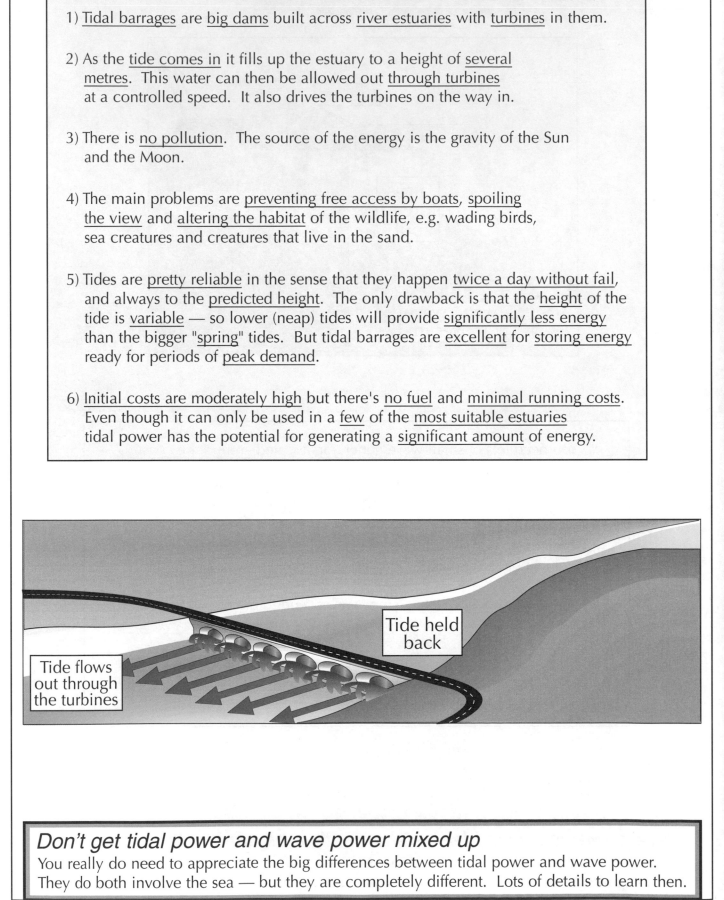

Tide held back

Tide flows out through the turbines

## Don't get tidal power and wave power mixed up

You really do need to appreciate the big differences between tidal power and wave power. They do both involve the sea — but they are completely different. Lots of details to learn then.

# Warm-Up and Worked Exam Questions

## Warm-up Questions

1) Describe briefly how wind power is used to generate electricity.
2) Draw an annotated diagram to show how hydroelectricity works.
3) What is meant by the term 'pumped storage'?
4) What is the difference between wave power and tidal power?

## Worked Exam Questions

1) a) Complete the following table by answering **Yes** or **No**.

| | Wind Power | Hydroelectric Power | Pumped Storage | Wave Power | Tidal Power |
|---|---|---|---|---|---|
| Can output be quickly increased in times of high demand? | *no* | *yes* | *yes* | *no* | *yes* |
| Is output dependent on unpredictable changes in weather? | *yes* | *no* | *no* | *yes* | *no* |

*[5 marks]*

b) Fill in the blanks in the following passage:

*There are other acceptable answers as well as "teatime", e.g. "start of working day".*

"A *pumped storage* system can provide power quickly in periods of *high* demand such as at *teatime*. To achieve this, a gate is opened in the upper *reservoir*, releasing water to flow down through the *turbines*."

*[3 marks]*

2) a) State three general advantages that wind, wave, tidal and hydroelectric power stations all have over fossil fuel and nuclear power stations.

Advantage 1: *they all cause no pollution*

Advantage 2: *they all require no fuel*

*Other possible answers include: "They all rely on power sources that will never run out."*

Advantage 3: *they all have minimal running costs*

*[3 marks]*

b) State one shared disadvantage of wind, wave, tidal and hydroelectric power stations.

*They can all spoil the view (cause visual pollution).*

*Alternative answer: they all have fairly high initial costs.*

*[1 mark]*

c) Give one specific disadvantage of <u>each</u> of wave, hydroelectric, tidal and wind power stations. Do not repeat the answer you have already given to (b).

Wind Power: *high initial costs*

*Again, there are loads of other possible answers here.*

Wave Power: *hazard to boats*

Tidal Power: *destruction of habitats of creatures that live in river estuaries*

Hydroelectric Power: *destruction of habitats when valley flooded*

*[4 marks]*

## Exam Questions

1) The following are all types of power installation built in Britain:

   *wave power, wind power, pumped storage, tidal power, hydroelectric power*

   a) Other than wind power, which one of the above produces more power when the wind is blowing more strongly?

   ......................................................................................................................................

   *[1 mark]*

   b) Which one of the above provides an amount of power that varies in a predictable cycle?

   ......................................................................................................................................

   *[1 mark]*

   c) What is the name of the device found in all of the above power installations that works by letting air or water push round blades which in turn drive a generator?

   ......................................................................................................................................

   *[1 mark]*

   d) Which of the above sources of power makes use of the gravitational attraction between the Earth, Moon and Sun?

   ......................................................................................................................................

   *[1 mark]*

2) a) Fill in the blanks in the sentences below to suggest the most suitable type of renewable energy power station to be built in each of these four locations in the UK.

   A ............................. power station would be most suitable to provide power for the small island off the west coast of Scotland.

   A ............................. power station could be built in the steep-sided mountain valley in a part of Wales with high rainfall.

   A ............................. power station could be built in the estuary where a major river runs out to sea.

   A ............................. power station could be built in the area of high, exposed, almost treeless moorland in Yorkshire.

   *[4 marks]*

   b) Which of the locations from part a) would also be suitable for a pumped storage installation?

   ......................................................................................................................................

   *[1 mark]*

# Geothermal Energy

## Geothermal Energy — Heat from Underground

1) This is <u>only possible</u> in <u>certain places</u> where <u>hot rocks</u> lie quite near to the <u>surface</u>. The source of much of the heat is the <u>slow decay</u> of various <u>radioactive elements</u> (including <u>uranium</u>) deep inside the Earth.

2) <u>Water is pumped</u> in pipes down to <u>hot rocks</u> and <u>returns as steam</u> to drive a <u>generator</u>.

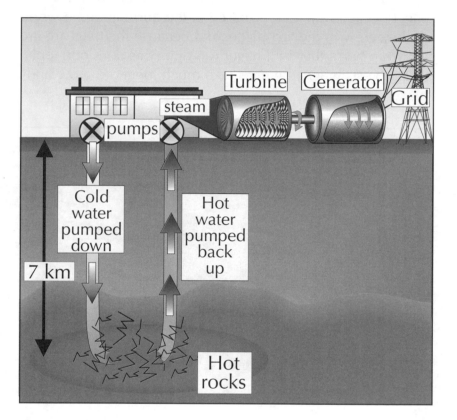

3) This is actually a very good source of <u>free energy</u> with no real environmental problems.

4) The <u>main drawback</u> is the <u>cost of drilling</u> down <u>several kilometres</u> to the hot rocks.

5) Unfortunately there are <u>very few places</u> where this seems to be an <u>economic option</u> (for now).

---

### *It might be expensive — but it'll last forever*

Geothermal energy could well be the big source of power for the next millennium or two. Just drill down 10 or 20 km and there you go — limitless free energy. Write a <u>mini-essay</u> explaining it all.

# Wood Burning

## *Wood Burning* — *Environmentally Fine*

1) This can be done <u>commercially</u> on a <u>large scale</u>.

2) It involves the cultivation of <u>fast-growing trees</u> which are then <u>harvested</u>, <u>chopped up</u> and <u>burned</u> in a power station <u>furnace</u> to produce <u>electricity</u>.

3) Unlike <u>fossil fuels</u>, wood burning does <u>not</u> cause a problem with the <u>Greenhouse Effect</u> because any $CO_2$ released in the burning of the wood was <u>removed</u> when they <u>grew in the first place</u>, and because the trees are grown <u>as quickly as they are burnt</u> they will <u>never run out</u>. This does <u>not apply</u> to the burning of <u>rainforests</u> where the trees take <u>much longer</u> to grow and are difficult to replace.

4) The <u>main drawback</u> is the <u>use of land</u> for <u>growing trees</u>, but if these woods can be made into <u>recreational areas</u> then that may be a <u>positive benefit</u> and certainly the woodlands should look quite <u>attractive</u>, as opposed to 5000 wind turbines covering miles and miles of countryside.

5) As a method of electricity generation, wood burning may seem mighty <u>old-fashioned</u>, but if enough trees are grown this is a <u>reliable and plentiful source of energy</u>, with fewer environmental drawbacks than many other energy sources.

6) Initial costs <u>aren't too high</u>, but there's some cost in <u>harvesting and processing</u> the wood.

## *Burning wood SOUNDS bad for the environment, but it does add up*

Cover the book and write a mini-essay explaining the pros and cons of this method.
There seem to be more pros than cons with this one — which could mean that at some point we'll end up with a few more forests about the place. Which would be nice.

# Solar Power

## Solar Energy — Solar *Cells*, Solar *Panels* and Solar *Furnaces*

<u>LEARN</u> the <u>three different ways</u> that solar energy can be harnessed:

1) <u>Solar cells</u> generate <u>electric currents directly</u> from sunlight. They are <u>expensive initially</u>, but can generate a lot of electricity even when it's clouded over — they don't need to have direct sunlight. Solar cells are the best source of energy for <u>calculators</u> and <u>watches</u> which don't use much electricity. <u>Satellites</u> and some other remote places don't have a choice — they have to use solar power.

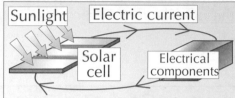

2) <u>Solar panels</u> are much less sophisticated. They simply contain <u>water pipes</u> under a <u>black surface</u>. <u>Heat radiation</u> from the Sun is <u>absorbed</u> by the <u>black surface</u> to <u>heat the water</u> in the pipes. The heat energy is used <u>directly</u>, rather than to create electricity.

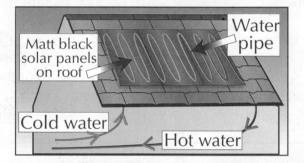

3) A <u>solar furnace</u> is a large array of <u>curved mirrors</u> which are all <u>focused</u> onto one spot to produce <u>very high temperatures</u> so water can be turned to <u>steam</u> to drive a <u>turbine</u>.

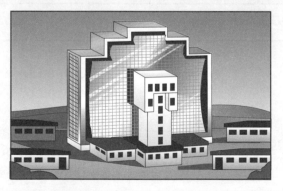

In all cases there is <u>no pollution</u>. In sunny countries solar power is a <u>very reliable source</u> of energy — but only in the <u>daytime</u>. Solar power will still provide <u>some energy</u> even in <u>cloudy countries</u> like Britain. <u>Initial costs</u> are <u>high</u> but after that the energy is <u>free</u> and <u>running costs almost nil</u> (apart from the <u>solar furnaces</u> which are more complicated).

## *Not one but three ways of harnessing the Sun's power*

Watch out for it — there are <u>three</u> different ways of using solar power directly.
<u>Learn</u> all three, along with their limitations, e.g. in Britain where we get a lot of cloud.

# Comparison of Energy Resources

## Comparison of Renewables and Non-Renewables

1) They're quite likely to give you an Exam question asking you to "evaluate" or "discuss" the relative merits of generating power by renewable and non-renewable resources.

2) The way to get the marks is to simply write down the pros and cons of each method.

3) Full details are given on the previous few pages. However there are some clear generalisations you should definitely learn to help you answer such questions. Make sure you can list these easily from memory:

## Non-Renewable Resources (Coal, Oil, Gas and Nuclear):

| Advantages: | Disadvantages: |
|---|---|
| 1) Very high output. | 1) Very polluting. |
| 2) Reliable output, entirely independent of the weather. | 2) Mining or drilling, then transportation of fuels damages the environment. |
| 3) Don't take up much land or spoil too much landscape. | 3) They're running out quickly. |

## Renewable Resources (Wind, Waves, Solar, etc.):

| Advantages: | Disadvantages: |
|---|---|
| 1) No pollution. | 1) Can require large areas of land or water. |
| 2) They will never run out. | 2) They don't always deliver when needed — if the weather isn't right, for example. |
| 3) Little damage to the environment (except visually). | 3) Often don't provide much energy. |
| 4) No fuel costs, although the initial costs are often high. | |

## Learn those tables
Make sure you learn all that summary information comparing renewables and non-renewables.

# Warm-Up and Worked Exam Question

## Warm-up Questions

1) Describe briefly how electricity is generated from geothermal energy.

2) What is the main disadvantage of using geothermal energy?

3) Why does wood burning not cause a problem with the Greenhouse Effect if the wood is from managed forests?

4) Why is burning rainforests bad for the environment?

5) Describe the three ways that solar energy is currently harnessed.

## Worked Exam Question

1    List the main advantages and disadvantages of using renewable energy.

*The main advantages of renewable energy resources are:*

*they won't run out,*

*they won't harm the environment,*

*the source of energy is free,*

*they are generally cheap to run.*

*You would get a mark for each correct point, up to a maximum of 7. I've actually listed 9 points here — but you would only need to list 7 to get full marks.*

*The main disadvantages are:*

*most require large areas of either land or water to be effective,*

*they often have a visual impact on the natural beauty of the*

*countryside or the coastline,*

*the cost of transporting the electricity to areas of need can be expensive,*

*the equipment required to convert the energy is often expensive to set up*

*for relatively little return,*

*most renewables do not provide a constant supply of energy because*

*they are affected by the weather, the time of day, month or year.*

*(7 marks)*

192

## Exam Questions

1   a)   Geothermal energy is used in Iceland to provide energy without badly damaging the environment.

   (i)   What is the original source of geothermal energy?

   .......................................................................................................................

   *(1 mark)*

   (ii)   Geothermal energy is not widely used across the world.
   Briefly explain why.

   .......................................................................................................................

   .......................................................................................................................

   *(1 mark)*

   b)   A family living in the north of Scotland is considering having a solar cell put onto their roof to provide part of their household energy requirements.

   Other than cost, give one advantage and one disadvantage of using solar power in Scotland.

   Advantage

   .......................................................................................................................

   .......................................................................................................................

   Disadvantage

   .......................................................................................................................

   .......................................................................................................................

   *(2 marks)*

# Exam Questions

2  a) Name two sources of energy which seriously affect the environment.
For each energy source, explain how it affects the environment.

Energy Source 1

........................................................................................................

Environmental Effects

..................................................................................................

........................................................................................................

Energy Source 2

..................................................................................................

Environmental Effects

................................................................................................

........................................................................................................

*(4 marks)*

(b) Renewable energy resources are generally considered more 'environmentally
friendly' than non-renewables.  However, there are many people who object to
their use.

Describe two objections to using renewable energy resources.

........................................................................................................

........................................................................................................

*(2 marks)*

(c) Explain why burning specially grown wood from a well-managed forest to
produce electricity is less harmful to the environment than burning fossil fuels.

........................................................................................................

........................................................................................................

........................................................................................................

*(2 marks)*

(d)  Name one renewable energy resource which has very little environmental impact.

........................................................................................................

*(1 mark)*

# Revision Summary for Section Five

There are five distinct parts to Section Five. First there's the two sections on power, work done, efficiency, etc., which involve a lot of formulas and calculations. Then there's a section on heat transfer, which is trickier to fully get to grips with than most people realise, and finally there's another two sections on generating power, which is basically easy but there are lots of details to learn. Make sure you realise the different approach needed for all three bits and amend your revision accordingly. Burgundy questions are for AQA syllabus only.

1)   List the ten different types of energy, and give twelve different examples of energy transfers.
2)   Write down the Principle of the Conservation of Energy. When is energy actually useful?
3)   Sketch the basic energy flow diagram for a typical "useful device".
4)    What forms does the wasted energy always take?
5)   What's the formula for efficiency? What are the three numerical forms suitable for efficiency?
6)   Give three worked examples on efficiency.
7)   What's the connection between "work done" and "energy transferred"?
8)   What's the formula for work done? A dog drags a branch 12m over the front lawn, pulling with a force of 535N. How much energy is transferred?
9)   What's the formula for power? What are the units of power?
10)  An electric motor uses 540kJ of electrical energy in 4½ minutes. What is its power consumption? If it has an efficiency of 85%, what's its power output?
11)  Write down the formulas for KE and PE. Find the KE of a 78kg man moving at 23 m/s.
12)  Calculate the power output of a 78kg man who runs up a 20m high staircase in 16.5 seconds.
13)  Calculate the speed of a 78kg mass as it hits the floor after falling from a height of 20m.
14)  If the mass bounces back up to a height of 18m calculate the % loss of KE at the bounce.
15)  What causes heat to flow from one place to another? What do particles do as they heat up?
16)  Explain briefly the differences between conduction, convection and radiation.
17)  Give a strict definition of conduction of heat and say what kinds of materials are good conductors.
18)  Give a strict definition of convection and give one use of it.
19)  List five properties of heat radiation. What kinds of object emit and absorb heat radiation?
20)  What surfaces absorb heat radiation best? What surfaces emit it best?
21)  Describe two experiments to demonstrate the effect of different surfaces on radiant heat.
22)  Describe insulation measures which reduce:  a) conduction  b) convection  c) radiation.
23)  Draw a fully labelled diagram of a Thermos Flask, and explain exactly what each bit is for.
24)  List seven ways of insulating houses and say which are the most effective and which are the most cost-effective measures. How do you decide on cost-effectiveness?
25)  List the main four non-renewable sources of energy and say why they are non-renewable.
26)  List eight kinds of renewable energy.
27)  Draw five energy chains which start with the Sun as the source of energy.
28)  Nine out of the twelve main energy resources originate in the Sun — which are they?
29)  Which three energy resources do not originate in the Sun?
30)  Which kind of resources do we get most of our energy from? Sketch the workings of a typical power station.
31)  List seven environmental hazards with non-renewables and four ways that we can use less.
32)  Give full details of how we can use wind power, including the advantages and disadvantages.
33)  Give full details of how a hydroelectric scheme works. What is pumped storage all about?
34)  Sketch a wave generator and explain the pros and cons of this as a source of energy.
35)  Explain how tidal power can be harnessed. What are the pros and cons of this idea?
36)  Explain where geothermal energy comes from. Describe how we can make use of it.
37)  Explain the principles of wood burning for generating electricity. Give the pros and cons.
38)  Write down where solar cells are used. What are the disadvantages of solar power?
39)  List the advantages and disadvantages of using renewable and non-renewable sources of energy.

# The Structure of the Atom

See the Chemistry Book for a few more details on this.

The <u>nucleus</u> contains <u>protons</u> and <u>neutrons</u>.
Most of the <u>mass</u> of the atom is contained in the <u>nucleus</u>,
but it takes up <u>virtually no space</u> — it's <u>tiny</u>.

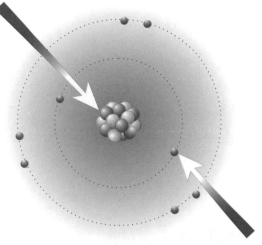

The <u>electrons</u> fly around the <u>outside</u>.
They're <u>negatively charged</u> and really really <u>small</u>.
They <u>occupy a lot of space</u> and this gives the atom
its <u>overall size</u>, even though it's <u>mostly empty space</u>.

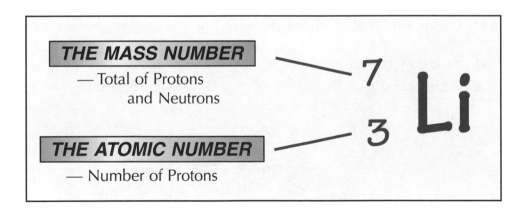

**THE MASS NUMBER**
— Total of Protons
and Neutrons

$$_{3}^{7}\text{Li}$$

**THE ATOMIC NUMBER**
— Number of Protons

Make sure you <u>learn this table</u>:

| PARTICLE | RELATIVE MASS | RELATIVE CHARGE |
|----------|---------------|-----------------|
| Proton | 1 | +1 |
| Neutron | 1 | 0 |
| Electron | $\frac{1}{2000}$ | −1 |

## *All that empty space — amazing*

Atoms are mostly empty space, which is pretty strange. Electrons have almost no mass, no size, and only a tiny little negative charge — and yet it's only their movement about the nucleus that makes atoms what they are. Strange stuff indeed — and all likely to come up in the exam. So <u>learn it</u>.

# The Structure of the Atom

## Isotopes are Different Forms of the Same Element

1) Isotopes are atoms with the same number of protons but a different number of neutrons.

2) Hence they have the same atomic number, but different mass numbers.

3) Carbon-12 and carbon-14 are good examples:

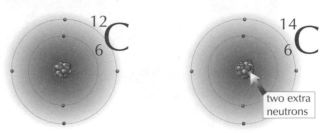

two extra neutrons

4) Most elements have different isotopes but there's usually only one or two stable ones.

5) The other isotopes tend to be radioactive, which means they decay into other elements and give out radiation. This is where all radioactivity comes from — unstable radioactive isotopes undergoing nuclear decay and spitting out high energy particles.

## Rutherford's Scattering put an end to the Plum Pudding Theory

1) In 1804 John Dalton said matter was made up of tiny solid spheres which he called atoms.

2) Later they discovered electrons could be removed from atoms. They then saw atoms as spheres of positive charge with tiny negative electrons stuck in them like plums in a plum pudding.

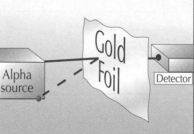

3) Then Ernest Rutherford and his colleagues tried firing alpha particles at thin gold foil. Most of them just went straight through, but the odd one came straight back at them, which was rather surprising.

They realised this meant that most of the mass of the atom was concentrated at the centre in a tiny nucleus, with a positive charge.

This means that most of an atom is just made up of empty space, which is also pretty surprising when you think about it.

---

### It was a pretty amazing discovery at the time

It might not seem too relevant learning how they discovered something that we now take for granted, but it's in the syllabus so you could get a question on it. Learn all the numbered points.

# Types of Radiation

Don't get <u>mixed up</u> between <u>nuclear radiation</u> which is <u>dangerous</u> — and <u>electromagnetic radiation</u> which <u>generally isn't</u>.  Gamma radiation is included in both, of course.

## Nuclear Radiation: Alpha, Beta and Gamma  (α, β and γ)

You need to remember <u>three things</u> about <u>each type of radiation</u>:

1) What it <u>actually is</u>.

2) How well it <u>penetrates</u> materials.

3) How strongly it <u>ionises</u> that material (i.e. by colliding with atoms and <u>knocking electrons off</u>).
There's a pattern: The <u>further</u> the radiation can <u>penetrate</u> before hitting an atom and getting stopped, the <u>less damage</u> it will do along the way and so the <u>less ionising</u> it is.

## Alpha Particles are Helium Nuclei   $^4_2He$

1) Alpha particles are relatively <u>big</u> and <u>heavy</u> and <u>slow moving</u>.

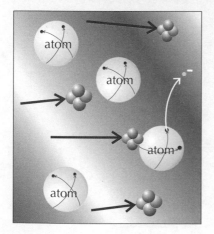

2) They therefore <u>don't penetrate far</u> into materials and are <u>stopped quickly</u>.

3) Because of their size they are <u>strongly ionising</u>, which just means they <u>collide with a lot of atoms</u> and <u>knock electrons off</u> them before they slow down, which creates lots of ions — hence the term "<u>ionising</u>".

## Learn the three types of radiation — it's easy as abc

Alpha, beta and gamma.  Those are just the first three letters of the Greek alphabet: α, β, γ — just like a, b, c.  They might sound like complex names, but they were just easy labels at the time. Anyway, <u>learn all the facts</u> about them — and <u>write them out from memory</u>.

# Types of Radiation

## Beta Particles are Electrons $_{-1}^{0}e$

1) These are in between alpha and gamma in terms of their properties.

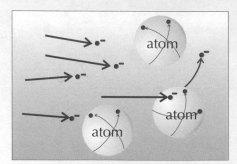

2) They move quite fast and they are quite small (they're electrons).

3) They penetrate moderately before colliding and are moderately ionising too.

4) For every β-particle emitted, a neutron turns to a proton in the nucleus.

## Gamma Rays are Very Short Wavelength EM Waves

1) They are the opposite of alpha particles in a way.

2) They penetrate a long way into materials without being stopped.

3) This means they are weakly ionising because they tend to pass through rather than colliding with atoms. Eventually they hit something and do damage.

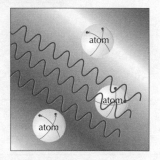

## Remember What Blocks the Three Types of Radiation...

Alpha particles are blocked by paper.
Beta particles are blocked by thin aluminium.
Gamma rays are blocked by thick lead.
Of course anything equivalent will also block them, e.g. skin will stop alpha, but not the others; a thin sheet of any metal will stop beta; and very thick concrete will stop gamma just like lead does.

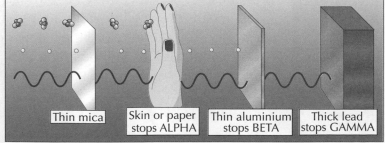

Thin mica | Skin or paper stops ALPHA | Thin aluminium stops BETA | Thick lead stops GAMMA

## Remember — alpha penetrates least, gamma penetrates most
This often comes up in Exam questions, so make sure you know what it takes to block each of the three types of radiation.

# Types of Radiation

## *Radioactivity* is a *Totally Random Process*

Unstable nuclei will decay and in the process give out radiation. This process is entirely random. This means that if you have 1000 unstable nuclei, you can't say when any one of them is going to decay, and neither can you do anything at all to make a decay happen.

Each nucleus will just decay quite spontaneously in its own good time. It's completely unaffected by physical conditions like temperature or by any sort of chemical bonding, etc.

When the nucleus does decay it will spit out one or more of the three types of radiation, alpha, beta or gamma, and in the process the nucleus will often change into a new element.

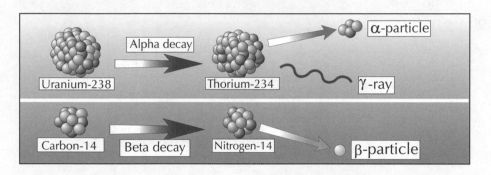

As well as this natural radioactive decay, stable nuclei can be made unstable by firing neutrons at them. When stray neutrons hit a stable nucleus they will often be absorbed into it and this generally turns it into an unstable isotope of the same element. Stray neutrons occur for example when uranium nuclei undergo fission and split in two, as shown:

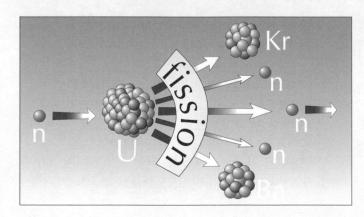

## *You can't control radioactive decay*
Learn all the facts, then close the book and write a mini-essay explaining everything you can remember. Then check back over the page — if you've missed any points, go back and try it again.

# Types of Radiation

## Background Radiation Comes from Many Sources

Natural background radiation comes from:

1) Radioactivity of naturally occurring <u>unstable isotopes</u> which are <u>all around us</u> — in the <u>air</u>, in <u>food</u>, in <u>building materials</u> and in the <u>rocks</u> under our feet.

2) Radiation from <u>space</u>, which is known as <u>cosmic rays</u>. These come mostly from the <u>Sun</u>.

3) Radiation due to <u>human activity</u>, e.g. <u>fallout</u> from <u>nuclear explosions</u> or <u>dumped nuclear waste</u>. But this represents a <u>tiny</u> proportion of the total background radiation.

The <u>relative proportions</u> of <u>radiation</u> we're exposed to:

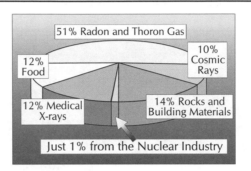

51% Radon and Thoron Gas
10% Cosmic Rays
12% Food
14% Rocks and Building Materials
12% Medical X-rays
Just 1% from the Nuclear Industry

## The Level of Background Radiation Changes Depending on Where You Are

1) At <u>high altitudes</u> (e.g. in <u>jet planes</u>) it <u>increases</u> because of more exposure to <u>cosmic rays</u>.

2) <u>Underground in mines</u>, etc. it increases because of the <u>rocks</u> all around.

3) Certain <u>underground rocks</u> can cause higher levels at the <u>surface</u>, especially if they release <u>radioactive radon gas</u>, which tends to get <u>trapped inside people's houses</u>. This varies widely across the UK depending on the <u>rock type</u>, as shown:

Coloured bits indicate more radiation from rocks

## Background Radiation — it's everywhere

Radiation is quite mysterious, but just like anything else, the <u>more you learn about it</u>, the <u>less</u> of a mystery it becomes. This page has plenty of simple, straightforward facts about radiation. Two tiny little <u>mini-essays</u> practised two or three times and all this knowledge will be yours — forever.

# Warm-Up and Worked Exam Question

The warm-up questions run quickly over the basic facts you'll need in the exam. The exam questions come later — but unless you've learnt the facts first you'll find the exams tougher than old boots.

## Warm-up Questions

1) What is the name for a negatively charged particle that orbits a nucleus?
A – a proton, B – an element, C – an electron, D – a neutron, E – a molecule, F – an ion

2) When an electron is knocked off an atom, what is the atom then called?
A – a proton, B – an element, C – an electron, D – a neutron, E – a molecule, F – an ion

3) Why do certain parts of the UK have a higher level of natural background radiation than others?

4) What type of radiation source (alpha, beta or gamma) must be stored in a lead-lined container?

## Worked Exam Question

I'm afraid this helpful blue writing won't be there in the exam so if I were you I'd make the most of it and make sure you fully understand it now.

1) a) Define the terms "mass number" and "atomic number" and state the atomic number and mass number of $^{14}_{6}C$.

mass number: *the number of protons and neutrons in the nucleus*

atomic number: *the number of protons in the nucleus*

mass number of $^{14}_{6}C$: *14*          atomic number of $^{14}_{6}C$: *6*

*(3 marks)*

b) Describe two changes which take place inside a carbon-14 atom when it decays.

*the atom emits an electron (beta particle)*

*and one neutron is converted into a proton*

*(2 marks)*

c) What usually happens when a nucleus of a fairly stable isotope of uranium is hit by a neutron, which it then absorbs?

*The absorption of an extra neutron makes the isotope unstable,*

*and this then decays.*

*(2 marks)*

d) What is meant by the statement
"The decay of unstable radioactive nuclei is totally random"?

*It means you cannot predict when any one nucleus is going*

*to decay. Nor can you do anything to make the decay*

*happen or stop it happening.*

*(2 marks)*

# Exam Questions

1    In a famous Physics experiment Ernest Rutherford and his team fired alpha particles at very thin sheets of gold foil.  He observed the following results:

A    Most of the alpha particles went straight through the foil.

B    A very few (about 1 in 20,000) alpha particles were deflected a lot (90° or more).

C    Some alpha particles were deflected a little, by one or two degrees.

a)    Explain each of these results.

.........................................................................................................................................

.........................................................................................................................................

.........................................................................................................................................

.........................................................................................................................................

*(3 marks)*

b)    Rutherford's layer of foil was very thin indeed — only a few hundred atoms thick.

Would he still have been able to carry out the experiment if he had made the foil any thicker, e.g. as thick as a sheet of paper?  Briefly explain your answer.

.........................................................................................................................................

.........................................................................................................................................

.........................................................................................................................................

*(2 marks)*

c)    A strontium-90 radiation source is separated from a radiation detector by a panel of aluminium and a sheet of paper.  No radiation is detected.  When the aluminium is taken away, radiation is detected.

Is this alpha, beta or gamma radiation?

.........................................................................................................................................

*(1 mark)*

# Exam Questions

2    This question is about radioactive isotopes.

a)    If a sample of uranium is left for a time, the proportion of unstable isotopes within it will gradually decrease.  Briefly explain why.

......................................................................................................................................

*(1 mark)*

b)    How can you determine whether an unstable isotope is present in a sample?

......................................................................................................................................

*(1 mark)*

c)    This diagram shows how uranium-238 decays, emitting alpha and gamma radiation.

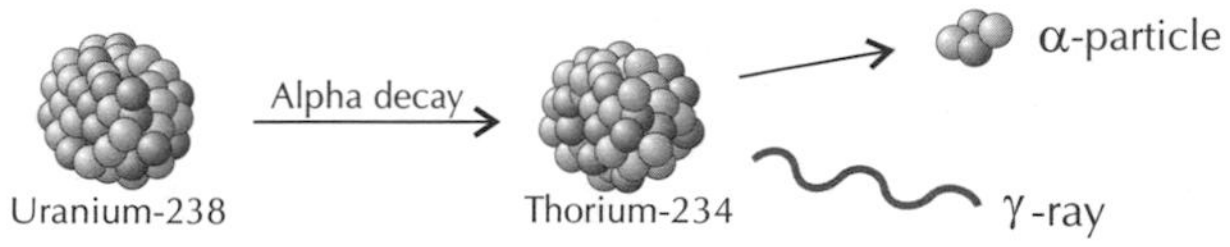

Which of the these two types of nuclear radiation is the more strongly ionising, and what quality of this radiation makes it so?

......................................................................................................................................

......................................................................................................................................

*(2 marks)*

d)    A student writes the following in answer to a Physics question:

*"The place with the lowest level of background radiation is down a mine,*
*because there you are shielded from the cosmic rays that come from space."*

Explain what is wrong with this statement.

......................................................................................................................................

......................................................................................................................................

*(2 marks)*

# Uses of Radioactive Materials

This is a <u>nice easy bit</u> of <u>straightforward learning</u>. On these two pages are the <u>seven main uses</u> for radioactive isotopes. Make sure you <u>learn all the details</u>. In particular, make sure you know why each application uses a <u>particular radioisotope</u> according to its <u>half-life</u> and the <u>type of radiation</u> it gives out.

## 1) Tracers in Medicine — always Short Half-life γ-emitters

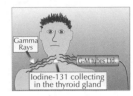

Iodine-131 collecting in the thyroid gland

1) Certain <u>radioactive isotopes</u> can be <u>injected</u> into people (or they can just <u>swallow</u> them) and their progress <u>around the body</u> can be followed using an external <u>detector</u>. A computer converts the reading to a <u>picture</u> showing where the <u>strongest reading</u> is coming from.

2) A well-known example is the use of <u>iodine-131</u> which is absorbed by the <u>thyroid gland</u>, just like normal iodine-127, but it gives out <u>radiation</u> which can be <u>detected</u> to indicate whether or not the thyroid gland is <u>taking in the iodine</u> as it should.

3) <u>All isotopes</u> which are taken <u>into the body</u> must be <u>gamma or beta</u> (never alpha), so that the radiation <u>passes out of the body</u>. They should only last a matter of <u>hours</u> (no longer than a few days, at least) so that the radioactivity inside the patient <u>quickly disappears</u> (i.e. they should have a <u>short half-life</u>).

## 2) Tracers in Industry — For Finding Leaks

This is much the same technique as the medical tracers.

1) Radioisotopes can be used to <u>detect leaks in pipes</u>.

2) You just <u>squirt it in</u>, and then go along the <u>outside</u> of the pipe with a <u>detector</u> to find areas of <u>extra high</u> radioactivity, which indicates the stuff is <u>leaking out</u>. This is really useful for <u>concealed</u> or <u>underground</u> pipes, to save you <u>digging up half the road</u> trying to find the leak.

3) The isotope used <u>must</u> be a <u>gamma emitter</u>, so that the radiation can be <u>detected</u> even through <u>metal or earth</u> which may be <u>surrounding</u> the pipe. Alpha and beta rays wouldn't be much use because they are <u>easily blocked</u> by any surrounding material.

4) It should also have a <u>short half-life</u> so as not to cause a <u>hazard</u> if it collects somewhere.

## 3) Sterilisation of Food and Surgical Instruments Using γ-Rays

1) <u>Food</u> can be exposed to a <u>high dose</u> of <u>gamma rays</u> which will <u>kill</u> all <u>microbes</u>, thus keeping the food <u>fresh for longer</u>.

2) <u>Medical instruments</u> can be <u>sterilised</u> in just the same way, rather than by <u>boiling them</u>.

unsterilised | Gamma source | sterilised

3) The great <u>advantage</u> of <u>irradiation</u> over boiling is that it doesn't involve <u>high temperatures</u> so things like <u>tinned food</u> or <u>plastic instruments</u> can be totally <u>sterilised</u> without <u>heat damage</u>.

4) The food is <u>not</u> radioactive afterwards, so it's <u>safe</u> to eat.

5) The isotope used for this needs to be a <u>very strong</u> emitter of <u>gamma rays</u> with a <u>reasonably long half-life</u> (at least several months) so that it doesn't need <u>replacing</u> too often.

## 4) Radiotherapy — the Treatment of Cancer Using γ-Rays

1) Since high doses of gamma rays will <u>kill all living cells</u>, they can be used to <u>treat cancers</u>.

2) The gamma rays have to be <u>directed carefully</u> and at just the right <u>dosage</u> so as to kill the <u>cancer cells</u> without damaging too many <u>normal cells</u>.

3) However, a <u>fair bit of damage</u> is <u>inevitably</u> done to <u>normal cells</u>, which makes the patient feel <u>very ill</u>. But if the cancer is <u>successfully killed off</u> in the end, then it's worth it.

# Uses of Radioactive Materials

## 5) *Thickness Control* in *Industry* and *Manufacturing*

This is a classic application and is <u>often</u> in Exams.

1) You have a <u>radioactive source</u> and you direct it <u>through whatever's being made</u>, usually a continuous sheet of <u>paper</u> or <u>cardboard</u> or <u>metal</u>, etc.

2) The <u>detector</u> is on the <u>other side</u> and is connected to a <u>control unit</u>.

3) When the amount of radiation detected <u>goes down</u>, it means the stuff is coming out <u>too thick</u> and so the control unit <u>pinches the rollers up</u> a bit to make it <u>thinner</u> again.

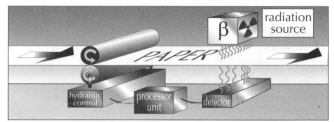

4) If the reading <u>goes up</u>, it means it's <u>too thin</u>, so the control unit <u>opens the rollers out</u> a bit. The most <u>important thing</u>, as usual, is the <u>choice of isotope</u>.

5) Firstly, it must have a <u>long half-life</u> (several <u>years</u> at least), otherwise the strength would <u>gradually decline</u> and the control unit would keep <u>pinching up the rollers</u> trying to <u>compensate</u>.

6) Secondly, the source must be a <u>beta source</u> for <u>paper and cardboard</u>, or a <u>gamma source</u> for <u>metal sheets</u>. This is because the stuff being made must <u>partly</u> block the radiation. If it <u>all</u> goes through (or <u>none</u> of it does), then the reading <u>won't change</u> at all as the thickness changes. Alpha particles are no use for this since they would <u>all be stopped</u>.

## 6) *Radioactive Dating* of *Rocks* and Archaeological *Specimens*

1) The discovery of radioactivity and the idea of <u>half-life</u> gave scientists their <u>first opportunity</u> to <u>accurately</u> work out the <u>ages</u> of <u>rocks</u> and <u>fossils</u> and <u>archaeological specimens</u>.

2) By measuring the <u>amount</u> of a <u>radioactive isotope</u> left in a sample, and knowing its <u>half-life</u>, you can work out <u>how long</u> the thing has been around. (See pages 211 and 213.)

3) <u>Igneous</u> rocks contain radioactive uranium which has a very <u>long half-life</u>. It eventually decays to become <u>stable</u> isotopes of <u>lead</u> — so the big clue to a rock sample's age is the <u>relative proportions</u> of uranium and lead isotopes.

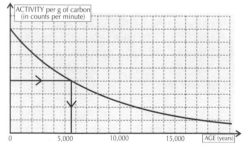

4) Igneous rocks also contain the radioisotope <u>potassium–40</u>. Its decay produces stable <u>argon gas</u> and sometimes this gets trapped in the rock. Then it's the same process again — finding the <u>relative proportions</u> of potassium–40 and argon to work out the age.

## 7) *Generating Power* from *Nuclear Fuel* *(Uranium)*

1) <u>Radioactive decay</u> always <u>gives out energy</u> in the form of <u>heat</u>.

2) The radioactive decay <u>inside the Earth</u> is responsible for much of the <u>heat</u> down there.

3) By <u>purifying uranium</u>, we can set up a <u>chain reaction</u> where each decay causes another one. In this way we can <u>increase the rate of reaction</u> to generate <u>lots of heat</u> and then use it to produce <u>electricity</u>. This is what a nuclear power station does (see page 177).

---

## *Seven main uses of radioactive materials to learn*

First <u>learn</u> the seven headings till you can write them down <u>from memory</u>. Then start <u>learning</u> all the details that go with each one of them. As usual, the best way to check what you know is to do a <u>mini-essay</u> for each section. Then check back and see what details you <u>missed</u>.

# Detection of Radiation

## The *Geiger-Müller Tube* and *Counter*

1) This is the most <u>familiar type</u> of <u>radiation detector</u>.
2) This is also the type used for <u>experiments in the lab</u>,
   as the counter allows you to record the number of <u>counts per minute</u>.
3) When <u>alpha</u>, <u>beta</u> or <u>gamma</u> radiation enters the <u>G-M tube</u>, it <u>ionises</u> the gas
   inside and triggers an <u>electrical discharge</u> (a spark) which makes a <u>clicking</u>
   <u>sound</u> and also sends a <u>small signal</u> to the electronic <u>counter</u>.

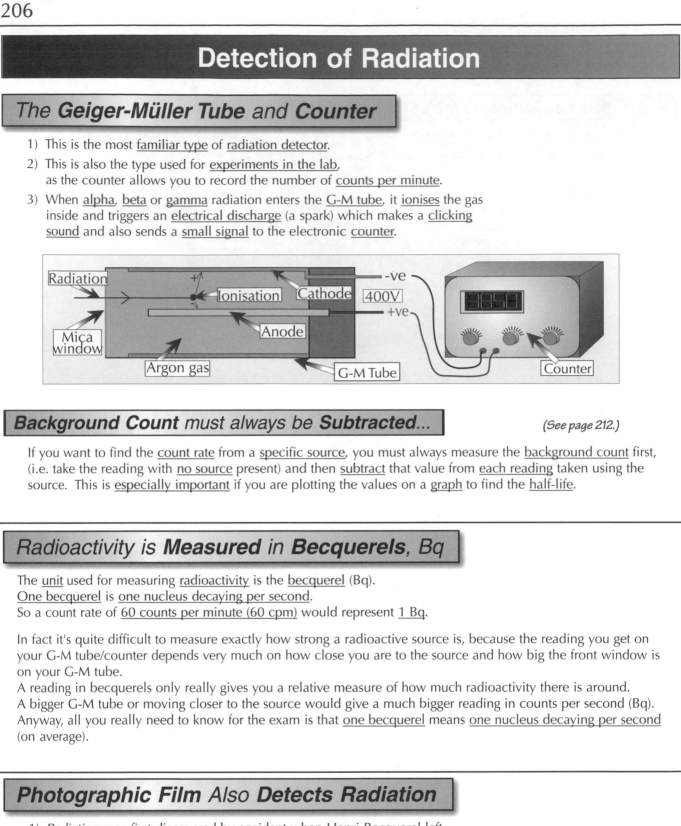

## *Background Count* must always be *Subtracted*...

*(See page 212.)*

If you want to find the <u>count rate</u> from a <u>specific source</u>, you must always measure the <u>background count</u> first,
(i.e. take the reading with <u>no source</u> present) and then <u>subtract</u> that value from <u>each reading</u> taken using the
source. This is <u>especially important</u> if you are plotting the values on a <u>graph</u> to find the <u>half-life</u>.

## *Radioactivity* is *Measured* in *Becquerels*, Bq

The <u>unit</u> used for measuring <u>radioactivity</u> is the <u>becquerel</u> (Bq).
<u>One becquerel</u> is <u>one nucleus decaying per second</u>.
So a count rate of <u>60 counts per minute (60 cpm)</u> would represent <u>1 Bq</u>.

In fact it's quite difficult to measure exactly how strong a radioactive source is, because the reading you get on
your G-M tube/counter depends very much on how close you are to the source and how big the front window is
on your G-M tube.
A reading in becquerels only really gives you a relative measure of how much radioactivity there is around.
A bigger G-M tube or moving closer to the source would give a much bigger reading in counts per second (Bq).
Anyway, all you really need to know for the exam is that <u>one becquerel</u> means <u>one nucleus decaying per second</u>
(on average).

## *Photographic Film* Also *Detects Radiation*

1) Radiation was first <u>discovered by accident</u> when <u>Henri Becquerel</u> left
   some <u>uranium</u> on some <u>photographic plates</u> which became "<u>fogged</u>" by it.
2) These days <u>photographic film</u> is a useful way of detecting radiation.
3) Workers in the <u>nuclear industry</u> or those using <u>X-ray equipment</u> such as
   <u>dentists</u> and <u>radiographers</u> wear <u>little badges</u> which have a bit of
   <u>photographic film</u> in them.
4) The film is checked <u>every now and then</u> to see if it's got fogged <u>too quickly</u>,
   which would mean the person was getting <u>too much exposure</u> to radiation.

## Learn all those details

Make sure you remember those two ways of measuring radiation: the G-M tube and
photographic film, and remember what a becquerel is. This page is ideal for the mini-essay
method of revising, just to make sure you've taken all the important points on board.

# Radiation Hazards and Safety

## Radiation Harms Living Cells

1) <u>Alpha</u>, <u>beta</u> and <u>gamma</u> radiation can easily <u>enter living cells</u> and <u>collide with molecules</u>.

2) These collisions cause <u>ionisation</u>, which <u>damages or destroys</u> the <u>molecules</u>.

3) <u>Lower doses</u> tend to cause <u>minor damage</u> without <u>killing</u> the cell.

4) This can give rise to <u>mutant cells</u> which <u>divide uncontrollably</u>. This is <u>cancer</u>.

5) <u>Higher doses</u> tend to <u>kill cells completely</u>, which causes <u>radiation sickness</u> if a lot of body cells <u>all get attacked at once</u>.

6) The <u>extent</u> of the harmful effects depends on <u>two things</u>:

   a) <u>How much exposure</u> you have to the radiation.

   b) The <u>energy and penetration</u> of the radiation emitted, since <u>some types</u> are <u>more hazardous</u> than others.

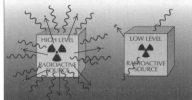

## Outside the Body, β– and γ–Sources are the Most Dangerous

This is because <u>beta and gamma</u> can get <u>inside</u> to the delicate <u>organs</u>, whereas alpha is much less dangerous because it <u>can't penetrate the skin</u>.

## Inside the Body, an α–Source is the Most Dangerous

Inside the body alpha sources do all their damage in a very localised area. However, beta and gamma sources are <u>less dangerous</u> inside the body because they mostly <u>pass straight out</u> without doing much damage.

## α, β and γ are Ionising Radiation

Ionisation is when an atom either <u>loses</u> or <u>gains</u> an <u>electron</u>.

## You Need to Learn About These Safety Precautions

In the Exam they might ask you to <u>list some specific precautions</u> that should be taken when <u>handling radioactive materials</u>. If you want those <u>easy marks</u> you'd better learn all these:

### In the School Laboratory:

1) <u>Never</u> allow <u>skin contact</u> with a source. Always handle with <u>tongs</u>.

2) Keep the source at <u>arm's length</u> to keep it <u>as far</u> from the body <u>as possible</u>.

3) Keep the source <u>pointing away</u> from the body and <u>avoid looking directly at it</u>.

4) <u>Always</u> keep the source in a <u>lead box</u> and put it back in <u>as soon</u> as the experiment is <u>over</u>.

### Extra Precautions for Industrial Nuclear Workers:

1) Wearing <u>full protective suits</u> to prevent <u>tiny radioactive particles</u> from being <u>inhaled</u> or lodging <u>on the skin</u> or <u>under fingernails</u>, etc.

2) Use of <u>lead-lined suits</u> and <u>lead/concrete barriers</u> and <u>thick lead screens</u> to prevent exposure to γ-rays from highly contaminated areas. (α and β are stopped <u>much more easily</u>.)

3) Use of <u>remote controlled robot arms</u> in highly radioactive areas.

---

## Radiation's dangerous stuff — these safety precautions are crucial

Quite a few details here. You need to know it <u>all</u> and there's only one way to find out whether you do or not. A <u>mini-essay</u> on the damage radiation can do, and another on the safety precautions you should take to protect yourself and others. With all the details in, of course.

# Warm-Up and Worked Exam Question

## Warm-up Questions

1) State whether each of the following uses of radiation should use a source of alpha, beta or gamma radiation:
   a) sterilisation of medical instruments
   b) controlling the thickness of paper
   c) killing cancer cells deep within the body

2) Match each of the following terms with its correct description.
   becquerel                    atom gaining or losing electrons
   Geiger counter               one radioactive nucleus decaying per second
   ionisation                   device used to detect radiation

3) State two precautions that should be taken by teachers when handling radioactive sources in a school.

## Worked Exam Question

1) The table below shows three possible radioactive isotopes which are being considered for use as a tracer. The tracer is to be dissolved in water in order to find leaks in underground water pipes.

| Isotope | Half-Life | Type of Emission | Description |
|---------|-----------|------------------|-------------|
| A | 17 days | beta | soluble solid |
| B | 8 minutes | gamma | soluble liquid |
| C | 28 hours | gamma | soluble solid |

a) Which of the isotopes is most suitable for the application?
Explain your answer.

isotope: ......*C*..............................................

explanation:
......*A gamma source is needed to pass through the ground,*..

*so it must be B or C, and the half-life of B would be too short.*..............

*[3 marks]*

The isotope chosen is contaminated by a fourth radioactive isotope.
This isotope is an alpha emitter with a half-life of about 4 weeks.

b) Explain why this is unlikely to be a significant hazard for the workers involved in the detection of leaks.

*Alpha particles are unable to pass through skin/clothing.*..................

*[1 mark]*

c) The new isotope contaminates the water supply and is drunk by consumers.
Does this represent a health risk for the consumers? Explain your answer.

*Yes. Alpha particles will damage a very localised area in the body if*

*taken internally, resulting in a cancer risk.*..............................

Remember — each type of radiation is dangerous in a different way — and it depends whether it's outside or inside the body, as this question shows. This stuff could easily come up in your exam.

*[2 marks]*

# Exam Questions

1) The diagram shows a device used to control the thickness of paper in a factory.
A radiation detector is used to detect how much radiation from the source penetrates the paper. The output of the detector can then be used to change the position of the rollers to alter the thickness of the paper.

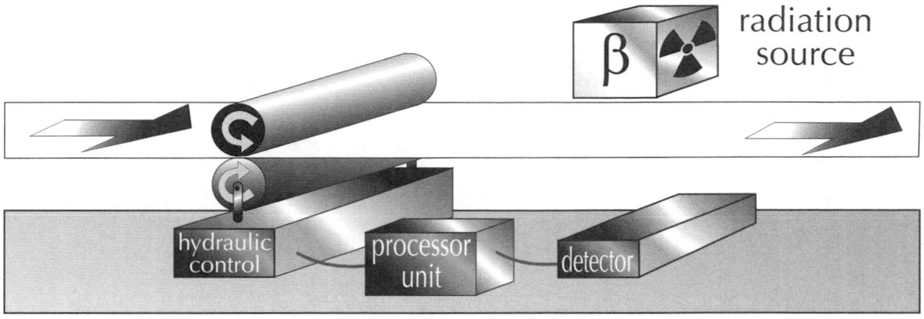

a) The designer decides to use a source of beta particles in the device.
Explain why this is a good choice of source.

.................................................................................................................................

*[2 marks]*

b) What would happen to the amount of radiation reaching the detector if the paper became thicker?

.................................................................................................................................

*[1 mark]*

A beta source of half-life 8 days is used in the device. A test run carried out over several days shows that the device gradually produces thinner and thinner paper, despite being left on the same setting.

c) Why does this happen?

.................................................................................................................................

.................................................................................................................................

*[2 marks]*

d) What should be done to solve the problem?

.................................................................................................................................

.................................................................................................................................

*[1 mark]*

It is suggested that the device could be modified to roll metal sheets to the correct thickness.

e) Explain why a source of beta particles will not work for this application.

.................................................................................................................................

*[1 mark]*

f) What type of source would be more suitable?

.................................................................................................................................

*[1 mark]*

# Balancing Nuclear Equations

## *Nuclear Equations* — *Balancing Both Sides*

Nuclear equations are all about making sure the <u>mass numbers</u> and <u>atomic numbers balance up on both sides</u>. The hardest part is <u>remembering</u> the <u>atomic- and mass- numbers</u> for α, β and γ radiations, and neutrons too. Make sure you can do all of these easily:

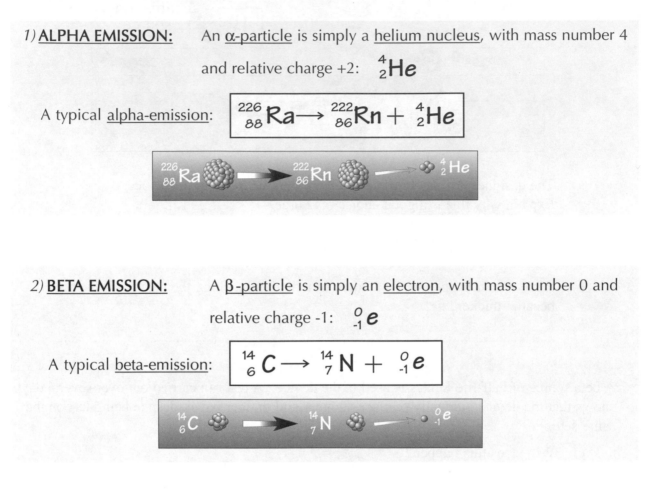

*1)* <u>ALPHA EMISSION:</u>    An <u>α-particle</u> is simply a <u>helium nucleus</u>, with mass number 4 and relative charge +2: $^{4}_{2}He$

A typical <u>alpha-emission</u>:    $^{226}_{88}Ra \longrightarrow {}^{222}_{86}Rn + {}^{4}_{2}He$

*2)* <u>BETA EMISSION:</u>    A <u>β-particle</u> is simply an <u>electron</u>, with mass number 0 and relative charge -1: $^{0}_{-1}e$

A typical <u>beta-emission</u>:    $^{14}_{6}C \longrightarrow {}^{14}_{7}N + {}^{0}_{-1}e$

*3)* <u>GAMMA EMISSION:</u>    A <u>γ-ray</u> is a <u>photon</u> with mass number 0 and no charge: $^{0}_{0}\gamma$

After an <u>alpha or beta emission</u> the nucleus sometimes has <u>extra energy to get rid of</u>. It does this by emitting a <u>gamma ray</u>. Gamma emission <u>never changes</u> the <u>atomic or mass numbers</u> of the nucleus.

A typical combined <u>α- and γ-emission</u>:    $^{238}_{92}U \longrightarrow {}^{234}_{90}Th + {}^{4}_{2}He + {}^{0}_{0}\gamma$

## *It's all about balancing, that's all*

Learn these three cases and you'll be fine. It helps to learn the diagrams too — so you've got a visual of what's actually happening.

# Half-Life Calculations

## The *Radioactivity* of a Sample Always *Decreases* Over Time

1) This is <u>pretty obvious</u> when you think about it. Each time a <u>decay</u> happens and an alpha or beta particle is given out, it means one more <u>radioactive nucleus</u> has <u>disappeared</u>.

2) Obviously, as the <u>unstable nuclei</u> all steadily disappear, the <u>activity as a whole</u> will also <u>decrease</u>. So the <u>older</u> a sample becomes, the <u>less radiation</u> it will emit.

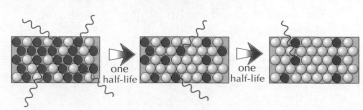

3) <u>How quickly</u> the activity <u>drops off</u> varies a lot. For <u>some</u> it takes <u>just a few hours</u> before nearly all the unstable nuclei have <u>decayed</u>, whilst others last for <u>millions of years</u>.

4) The problem with trying to <u>measure</u> this is that <u>the activity never reaches zero</u>, which is why we have to use the idea of <u>half-life</u> to measure how quickly the activity <u>drops off</u>.

5) Learn this <u>important definition</u> of <u>half-life</u>:

> *HALF-LIFE is the <u>TIME TAKEN</u> for <u>HALF</u> of the radioactive atoms initially present to DECAY*

Another definition of half-life is:
*"The time taken for the activity (or count rate) to fall by half"*.
Use either.

6) A <u>short half-life</u> means the <u>activity falls quickly</u>, because <u>lots</u> of the nuclei decay <u>quickly</u>.

7) A <u>long half-life</u> means the activity <u>falls more slowly</u> because <u>most</u> of the nuclei don't decay <u>for a long time</u> — they just sit there, <u>basically unstable</u>, but kind of <u>biding their time</u>.

## Do *Half-Life* Questions *Step By Step*

Half-life is maybe a little confusing, but Exam calculations are <u>straightforward</u> so long as you do them slowly, <u>step by step</u>. Like this one:

<u>A very simple example</u>: The activity of a radioisotope is 640 cpm (counts per minute). Two hours later it has fallen to 40 cpm. Find the half-life of the sample.
<u>ANSWER</u>: You must go through it in <u>short simple steps</u> like this:

| <u>Initial count:</u> | | after <u>one</u> half-life: | | after <u>two</u> half-lives: | | after <u>three</u> half-lives: | | after <u>four</u> half-lives: |
|---|---|---|---|---|---|---|---|---|
| 640 | ($\div 2$) $\rightarrow$ | 320 | ($\div 2$) $\rightarrow$ | 160 | ($\div 2$) $\rightarrow$ | 80 | ($\div 2$) $\rightarrow$ | 40 |

Notice the careful <u>step-by-step method</u>, which tells us it takes <u>four half-lives</u> for the activity to fall from 640 to 40. Hence <u>two hours</u> represents four half-lives, so the <u>half-life is 30 minutes</u>.

## *Learn the definition of Half-Life*

These half-life calculations are really pretty simple. Keep practising them and you'll notice that they're all pretty much the same anyway.

# Half-Life Calculations

## *Measuring* The *Half-Life* of a Sample Using a Graph

1) This can <u>only be done</u> by taking <u>several readings</u> of <u>count rate</u> using a <u>G-M tube and counter</u>.

2) The results can then be <u>plotted</u> as a <u>graph</u>, which will <u>always</u> be shaped like the one below.

3) The <u>half-life</u> is found from the graph, by finding the <u>time interval</u> on the <u>bottom axis</u> corresponding to a <u>halving</u> of the <u>activity</u> on the <u>vertical axis</u>.

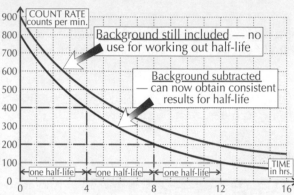

<u>One trick</u> you need to know is about the <u>background radiation</u>, which also enters the G-M tube and gives <u>false readings</u>. Measure the background count <u>first</u> and then <u>subtract it</u> from <u>every reading</u> you get, before plotting the results on the <u>graph</u>. Realistically, the <u>only difficult bit</u> is actually <u>remembering</u> about that for your <u>Exam</u>, should they ask you about it. They could also test that idea in a <u>calculation question</u>.

## *Carbon-14* Calculations — or *Radiocarbon Dating*

<u>Carbon-14</u> makes up about 1/10 000 000 (one <u>ten-millionth</u>) of the carbon in the <u>air</u>. This level stays <u>fairly constant</u> in the <u>atmosphere</u>. The same proportion of C-14 is also found in <u>living things</u>. However, when they <u>die</u>, the C-14 is <u>trapped inside</u> the wood or wool or whatever, and it <u>gradually decays</u> with a <u>half-life</u> of <u>5,600 years</u>.

By simply measuring the <u>proportion</u> of C-14 found in some old <u>axe handle</u>, <u>burial shroud</u>, etc. you can easily calculate <u>how long ago</u> the item was <u>living material</u> using the known <u>half-life</u>.

<u>Example:</u> The carbon in an an axe handle was found to contain 1 part in 40 000 000 carbon-14. How old is the axe?
<u>ANSWER:</u> The C-14 was originally <u>1 part in 10 000 000</u>. After <u>one half-life</u> it would be down to <u>1 part in 20 000 000</u>. After <u>two half-lives</u> it would be down to <u>1 part in 40 000 000</u>. Hence the axe handle is <u>two C-14 half-lives</u> old, i.e. 2 × 5,600 = <u>11,200 years old</u>. Note the same old <u>stepwise method</u>, going down one half-life at a time.

### *Carbon Dating only works on things that were once alive*

You really do need to practise all this stuff. Don't lose heart because it looks complicated. It really isn't — and once you've done a few questions you'll see what I mean.

# Half-Life Calculations

## Relative Proportions Calculations — Easy, as long as you Learn It

Uranium isotopes have a very long half-life and decay via a series of short-lived nuclei to produce stable isotopes of lead.

The relative proportions of uranium and lead isotopes in a sample of igneous rock can therefore be used to date the rock, using the known half-life of the uranium.

It's as simple as this:

| Initially | After one half-life | After two half-lives | After three half-lives |
|---|---|---|---|
| 100% uranium | 50% uranium | 25% uranium | 12.5% uranium |
| 0% lead | 50% lead | 75% lead | 87.5% lead |

Ratio of uranium to lead:                                    (half-life of uranium-238 = 4.5 billion years)

| Initially | After one half-life | After two half-lives | After three half-lives |
|---|---|---|---|
| 1:0 | 1:1 | 1:3 | 1:7 |

## The same thing applies to Potassium-40

Similarly, the proportions of potassium-40 and its stable decay product argon-40 can also be used to date igneous rocks, so long as the argon gas hasn't been able to escape.

The relative proportions will be exactly the same as for the uranium and lead example above. Learn these ratios:

| Initially | After one half-life | After two half-lives | After three half-lives |
|---|---|---|---|
| 100% : 0% | 50% : 50% | 25% : 75% | 12.5% : 87.5% |
| 1:0 | 1:1 | 1:3 | 1:7 |

## Get these ratios learnt

Half-life information is pretty useful for working out how old things are. But the main thing for now is to get on and learn how to do this, so that you can do it in the exam.

# Warm-Up and Worked Exam Question

## Warm-up Questions

1) Carbon-14 and uranium-238 are radioactive isotopes which can be used to determine the approximate age of substances.
   Decide which isotope would be better used for the following:
   a) to determine the age of a sample of granite,
   b) to determine the age of a sample of wood from a shipwreck,
   c) to determine the age of an Egyptian mummy.

2) State which types of radiation each of the following statements refer to:
   a) This is a fast moving electron.
   b) This radiation has a positive charge.
   c) Emission of this type of radiation reduces the mass number of the nucleus.
   d) This is a high energy photon.
   e) This radiation has no charge.
   f) Emission of this type of radiation increases the atomic number of the nucleus.

## Worked Exam Question

1) $^{220}_{86}\text{Rn}$ is an isotope of the gas radon.

This isotope decays by alpha emission to produce an isotope of polonium (Po).

a) Write the nuclear equation for this decay.

$$^{220}_{86}\text{Rn} \rightarrow {}^{216}_{84}\text{Po} + {}^{4}_{2}\text{He}$$

*[2 marks]*

The half-life for this decay is 56 seconds.

b) A sample of radon gas has an initial activity of 2000 counts per second, after correcting for background radiation.

How long will it be before the count rate is 250 counts per second?

*This is three half-lives — 2000 to 1000, then 1000 to 500,*
*then 500 to 250* (1 mark) *So time = 3 x 56 = 168 seconds* (1 mark)

*[2 marks]*

c) Draw on these axes a graph of the decay from 2000 to 125 counts per second.

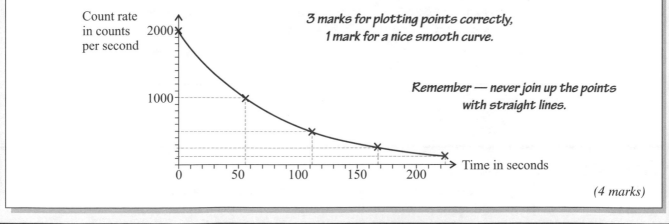

*3 marks for plotting points correctly,*
*1 mark for a nice smooth curve.*

*Remember — never join up the points*
*with straight lines.*

*(4 marks)*

# Exam Questions

1)  A university student uses a Geiger counter to measure the average radiation level in a laboratory before she carries out an experiment. She finds this to be 90 counts per minute.

   a)  What is the name given to this radiation?

   ................................................................................................................................

   *[1 mark]*

The student then brings a radioactive sample into the laboratory and measures the count rate from the sample, as shown in the table below.

| Time in Minutes | Count Rate in Counts per Minute | Corrected Count Rate in Counts per Minute |
|---|---|---|
| 0 | 490 | |
| 5 | 373 | |
| 10 | 290 | |
| 15 | 232 | |
| 20 | 190 | |

   b)  Work out the corrected count rate values for the final column of the table.

   ................................................................................................................................

   ................................................................................................................................

   *[2 marks]*

   c)  Plot your corrected values on the graph below.

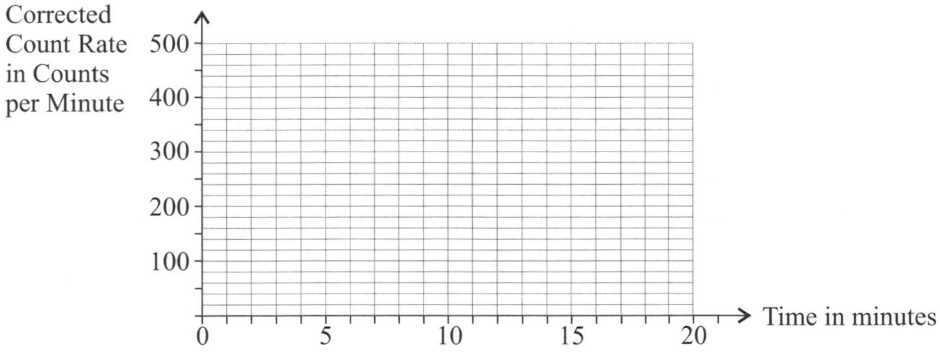

   *[2 marks]*

   d)  Use the graph to work out the half-life of the sample.
       Explain clearly how you arrived at your answer.

   ................................................................................................................................

   ................................................................................................................................

   *[2 marks]*

   e)  How long after the start of the experiment will the sample's count rate reduce to 25 counts per minute?

   ................................................................................................................................

   *[2 marks]*

# Revision Summary for Section Six

Practise these questions over and over again till you can answer them ALL effortlessly.

1) Sketch an atom.  Give three details about the nucleus and the electrons.

2) Draw up a table detailing the mass and charge of the three basic subatomic particles.

3) Explain what the mass number and atomic number of an atom represent.

4) Explain what isotopes are.  Give an example.  Are most isotopes stable or unstable?

5) What was the Plum Pudding Model?  Who put an end to that idea?

6) Describe Rutherford's Scattering Experiment with a diagram and say what happened.

7) What was the inevitable conclusion to be drawn from this experiment?

8) What is the main difference between EM radiation and nuclear radiation?

9) Describe in detail the nature and properties of the three types of radiation:  $\alpha$, $\beta$, and $\gamma$.

10) How do the three types compare in penetrating power and ionising power?

11) List several things which will block each of the three types.

12) Radioactive decay is a totally random process.  Explain what this means.

13) Will anything cause a nucleus to undergo radioactive decay?  What about nuclear fission?

14) Sketch a fairly accurate pie chart to show the six main sources of radiation we're exposed to.

15) List three places where the level of background radiation is increased and explain why.

16) Describe in detail how radioactive isotopes are used in each of the following:
    a) tracers in medicine   b) tracers in industry   c) sterilisation   d) treating cancer  e) thickness control
    f) dating of rock samples   g) generating power.

17) Draw a labelled diagram of a Geiger-Müller tube and explain what it's for and how it works.

18) What unit is radioactivity measured in?  How many of this unit are equal to 120 cpm?

19) What are the two common methods of detecting radioactivity?  Which is the simplest?

20) How is photographic film used in badges to monitor radiation?

21) Exactly what kind of damage does radiation do inside body cells?

22) What damage do low doses cause?  What effects do higher doses have?

23) Which kinds of source are most dangerous:  a) inside the body     b) outside the body?

24) List four safety precautions for handling radioactive materials in the school lab,
    and three more for nuclear workers.

25) Write down the nuclear equation for the alpha decay of   a) $^{238}_{92}\text{U}$   b) $^{230}_{90}\text{Th}$  and c) $^{226}_{88}\text{Ra}$.

26) Write down the nuclear equation for the beta/gamma decay of a) $^{230}_{90}\text{Th}$   b) $^{234}_{91}\text{Pa}$ and c) $^{14}_{6}\text{C}$ .

27) Sketch a diagram to show how the activity of a sample keeps halving.

28) Give a proper definition of half-life.  How long and how short can half-lives be?

29) Sketch a typical graph of activity against time.  Show how the half-life can be found.

30) What's the single most important thing to remember when doing half-life calculations?

31) An old bit of cloth was found to have 1 atom of C-14 to 80 000 000 atoms of C-12.
    Using the information on page 212, calculate the age of the bit of cloth.

32) A rock contains uranium-238 atoms and stable lead atoms in the ratio 1:3.
    If the half-life of uranium-238 is $4.5 \times 10^9$ years, how old is the rock?

# Analogue and Digital Signals

## Analogue Signals are Different from Digital Ones

1) An <u>analogue</u> quantity can take <u>any value</u> between set limits,
   e.g. the length of an adult's foot could be anything between, say, 150 mm and 300 mm.
2) A <u>digital</u> quantity has a value from a <u>set number of levels</u>,
   e.g. an adult's shoe size can only have set levels, e.g. size 4 or size 9½ but not 5¾.

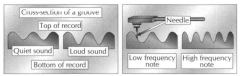

## Records are Analogue, CDs are Digital

You don't need all these details if you're doing the <u>AQA syllabus</u>.

1) **RECORDS — ANALOGUE STORAGE** Records are flat discs of plastic with a <u>narrow groove</u> starting at the edge and spiralling in to the centre. Information is stored as <u>bumps</u> in the groove — the <u>louder</u> the sound the <u>bigger</u> the bump, and the <u>higher</u> the <u>frequency</u> the <u>closer</u> together the bumps. A <u>needle</u> tracks the groove and responds to the bumps, creating an <u>electrical signal</u> that passes to an <u>amplifier</u> and <u>loudspeaker</u>. A scratch on the track causes <u>disruption</u> to the music.

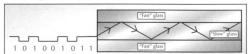

2) **COMPACT DISC (CD) — DIGITAL STORAGE**  CDs are <u>reflective discs</u> that have digital data stored as <u>pits</u> or <u>bumps</u> (sometimes called "lands") in a continuous groove spiralling from the centre of the disc.

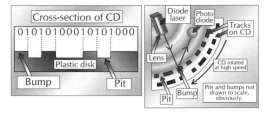

A <u>laser beam</u> (produced by a <u>diode laser</u>) is moved across the surface of the disk — and the length of the reflected beam is read by a <u>photo-diode</u> detector. Each <u>transition</u> from <u>pit to bump</u> (or <u>bump to pit</u>) is read as a <u>1</u>, and <u>no transition</u> is read as a <u>0</u>. Like records, the music is recorded <u>in sequence</u>, but more information is stored as <u>error correction codes</u> — which enables the CD player to 'guess' what information was <u>destroyed</u> by a scratch.

3) **OPTICAL FIBRES — USED TO SEND DIGITAL INFORMATION** (See page 105.)
   Once information has been <u>digitally coded</u> (as 0s and 1s), it can be sent down optical fibres. The information is carried as a <u>series of pulses</u> of <u>infrared electromagnetic radiation</u>. Although copper cables can also carry digital signals, optical fibres have the following advantages:  greater information <u>capacity</u>, <u>lighter</u>, signals <u>more secure</u>, less subject to <u>noise</u> and <u>travel further</u> before needing amplification or regeneration.

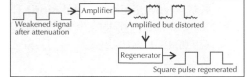

## Digital Storage and Transmission is More Convenient...

1) <u>Digital storage</u> media only require <u>two states</u> for storage — <u>0 and 1</u>. The <u>transmitter</u> need <u>only</u> produce <u>two different values</u>. A <u>receiver</u> likewise <u>only</u> needs to recognise these <u>two values</u>.

2) During transmission a <u>signal</u> will get <u>weaker</u> the further it has to travel — this is <u>attenuation</u>. <u>Repeater</u> stations at regular intervals <u>amplify the signal</u> but they can introduce <u>distortion</u> and <u>noise</u>. Using a <u>regenerator</u> can <u>restore</u> the <u>wave shape</u> of a digital signal.

3) Signals are subject to <u>random noise</u> — we hear it as <u>hiss</u>. Noise can <u>add to an analogue signal</u> but is very <u>hard to remove</u>, because the system can't distinguish the noise from the original signal. A <u>digital signal</u> can be '<u>cleaned</u>' of noise. As long as the noise is not enough for the system to mistake a 0 for a 1, or vice versa, the 0s and 1s will be read correctly.

# Communication Systems

## Marconi worked out how to send Radio Waves over the Atlantic

Methods of communication progressed from shouting, through lighting signal fires, writing letters and using carrier pigeons, to sending telegraphs (or telegrams) — and eventually telephone calls via wires. Then in 1901, Guglielmo Marconi demonstrated how to send signals using radio waves across the Atlantic Ocean. These 'wireless' transmissions lead to the development of radio and TV. More recently such information has been sent very quickly around the world via satellite communications.

## Communication Systems have Eight Parts...

Communication systems can be broken down into eight blocks, each having specific functions:

1) **Encoder** — converts information into a suitable form for transmission.
2) **Modulator** — can be used to encode information onto a steady high-frequency carrier wave. Amplitude modulation (AM), frequency modulation (FM) or phase modulation (PM) are commonly used.
3) **Amplifier** — amplifies the signal before transmission, after reception and before passing to a transducer (like a loudspeaker — see below). This has to occur with the minimum of distortion to the signal to reduce loss of information.
4) **Transmitter** — produces the signal that actually travels through the air, along a wire or through an optical fibre. In the case of radio waves, it involves an aerial.
5) **Receiver** — accepts the signal sent by the transmitter. It can be an aerial, or it can be at the other end of a wire or optical fibre.
6) **Decoder** — takes the received coded signal and extracts the information from it. If the signal contained a modulated carrier wave then a demodulator is used to obtain the original signal from it.
7) **Transducer** — a device that transforms energy from one form to another, often used at the beginning and at the end of an information system.
   *E.g. a microphone converts a sound wave into an electrical signal ready for encoding. Once the signal has been encoded, sent, received and decoded it can be converted back to sound using a loudspeaker.*
8) **Storage system** — e.g. CD, hard drive, magnetic tape, etc. — if the information needs to be recorded.

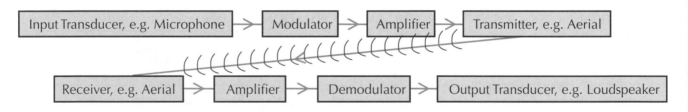

## AM and FM are Ways of Transmitting Radio Signals

A radio transmitter sends out a continuous high-frequency radio carrier wave.
The signal (e.g. music) is imposed or encoded on the carrier wave in one of two ways:

1) **AM — Amplitude Modulation** The sound wave from the music 'modulates' or changes the carrier wave by changing its amplitude. AM signals have a longer range than FM signals.

2) **FM — Frequency Modulation** The sound wave from the music 'modulates' the carrier wave by changing its frequency. FM radio gives a higher quality sound than AM as the signals are less susceptible to noise and interference.

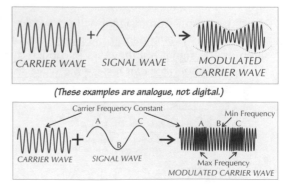

# Transmitting Radio Waves

This page covers the nuts and bolts of how radio waves are transmitted.

## Learn how *AM Radio* gets from *Studio* to *Kitchen*...

You could get a question like: "Describe what happens between the radio DJ speaking in the studio and you hearing their voice coming out of your kitchen AM radio." And this will be your answer...

1) The <u>sound</u> wave from the DJ's voice is <u>converted</u> into an <u>electrical</u> signal of the same <u>frequency</u> by a <u>microphone</u>.

2) This electrical signal <u>modulates</u> the <u>amplitude</u> (see page 218) of a <u>carrier wave</u> produced by the transmitter. (The carrier wave frequency is specific to that radio station, e.g. 430 kHz for Radio 5.)

3) The <u>transmitter</u> aerial sends the modulated <u>carrier wave</u> out in <u>all directions</u>. This is a <u>high-frequency</u> radio wave that travels at the <u>speed of light</u>.

4) <u>Your radio</u> has an <u>aerial</u> that <u>receives</u> the modulated <u>carrier wave</u> and it becomes an <u>electrical signal</u> again. You <u>tune</u> your radio to that <u>particular carrier wave frequency</u> from the <u>hundreds</u> of <u>radio waves</u> from other stations.

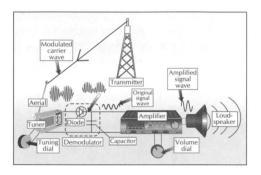

5) A <u>diode</u> in the radio <u>removes half</u> of the modulated carrier <u>wave</u>.

6) A <u>capacitor</u> (see page 246) '<u>filters' out</u> the <u>carrier wave</u>, leaving the <u>original signal</u> from the DJ's voice. This completes the '<u>demodulation</u>'.

7) An <u>amplifier</u> increases the <u>strength</u> of the signal, then it's <u>converted back</u> to a <u>sound wave</u> in the speaker.

## *Different Frequency* Waves travel via *Different Routes*

Radio signals can travel to us via different routes, depending on their frequency.

1) GROUND WAVES  travel in <u>close contact</u> with the <u>ground</u> as they <u>spread out</u> from the transmitter. They travel <u>further over sea</u> than land due to the conductivity of water. Used by <u>LW/MW radio</u> bands (up to <u>3 MHz</u>).

2) SKY WAVES  Frequencies up to about <u>30 MHz</u> (shortwave radio) can <u>reflect</u> off a layer of the atmosphere called the <u>ionosphere</u>. This allows the wave to travel longer distances and deals with the <u>curvature</u> of the Earth. Frequencies <u>above 30 MHz</u> (<u>FM radio and TV</u>) pass straight through the atmosphere and transmissions must be by <u>line of sight</u>.

3) SPACE WAVES  <u>Microwave</u> signals have a very high carrier frequency (<u>over 3000 MHz</u> for satellite TV and telephones). These <u>pass easily</u> through the <u>atmosphere</u> and <u>reflect off satellites</u> orbiting the Earth, enabling the signal to reach <u>distant parts</u> of the planet. Some satellites just <u>retransmit</u> the signal they receive — others are <u>active</u>, <u>sending out</u> <u>signals</u> of their own, e.g. weather monitoring satellites take photos and send the information to Earth.

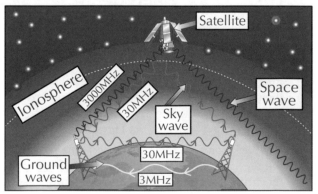

## There's a lot of information to learn here

They might want you to reproduce that diagram to explain the difference between <u>ground waves</u>, <u>sky waves</u> and <u>space waves</u>. Try to remember the shape of all the wiggles too — it will help to remind you which waves are <u>high frequency</u> and which are <u>low frequency</u>.

# Transducers

## There are Three Main Types of Transducer

There are loads of different types or transducers, but they're all the same in one respect:

### TRANSDUCERS TRANSFER ENERGY FROM ONE FORM INTO ANOTHER

## A Loudspeaker changes Electrical Energy into Sound Energy

There's more on electromagnetism on page 38 — go there for the basics.

You need to know about moving-coil loudspeakers — so called because they involve a coil of wire that... well, moves.

1) An <u>alternating electrical signal</u> from an amplifier passes through the <u>wires</u> into the <u>coil</u>.
2) The <u>current in the coil</u> turns it into an <u>electromagnet</u> with a north pole and a south pole. The direction of the current determines which pole is at which end.
3) The <u>coil</u> is <u>attracted</u> to or <u>repelled</u> by the <u>permanent magnet</u> (depending where the poles are).
4) Since the permanent magnet is fixed, the <u>coil moves</u>. (The larger the current, the further it moves.)
5) The coil is connected to the <u>paper cone</u> that moves with it.
6) As the <u>current</u> rapidly <u>changes direction</u>, so does the movement of the <u>coil and cone</u>. The cone vibrates, producing a <u>sound wave</u> of the <u>same frequency</u> as the alternating electrical signal. (How loud the sound is depends on how far the cone moves in and out.)

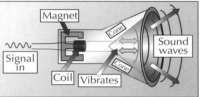

## A Microphone changes Sound Energy into Electrical Energy

A microphone is like a <u>reverse loudspeaker</u>. There are several types of microphone — but you only need to know about the <u>moving-coil microphone</u>:

1) Sound waves make a <u>diaphragm vibrate</u>.
2) The <u>coil</u> is <u>attached</u> to the diaphragm and it moves <u>backwards and forwards</u> past the <u>permanent magnet</u>.
3) This <u>induces</u> a <u>current</u> in the coil. The direction of the current depends on which way the coil is moving. (The higher the frequency of the sound, the more rapidly the current changes direction. The louder the sound, the larger the current induced.)
4) The current <u>signal</u> from the microphone is <u>amplified</u> and then <u>recorded</u> or sent to a <u>loudspeaker</u>.

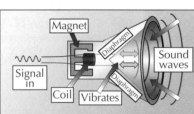

## Magnetic Tape Recorders Use Magnetism

OK, so they're a few decades out of date, but you still need to know how they work. A cassette stores information on a <u>long ribbon</u> of tape containing <u>iron oxide</u> (or similar magnetic substance). In the tape machine there are <u>3 heads</u>:

1) <u>Recording head</u> — The <u>signal</u> from an amplifier passes into the <u>coil</u>, creating a <u>strong magnetic field</u> in the <u>tiny gap</u> in the tape head. As the tape <u>passes</u> the head, the iron oxide becomes <u>magnetised</u> to match the signal. As the size and frequency of the signal varies, the magnetic field varies, producing a <u>pattern of magnetised areas</u> on the tape.

2) <u>Playback head</u> — The <u>magnetised tape</u> passes over the <u>playback head</u> and induces a <u>small current</u> in the coil. The current is passed to a <u>loudspeaker</u> after being amplified.

3) <u>Erase head</u> — The max. frequency recorded on the tape will be less than 20 kHz (the limit of our hearing). A <u>high-frequency</u> (> 50 kHz) alternating <u>current</u> passes through the coil of the erase head. This removes any pattern — it <u>demagnetises</u> the tape.

# Warm-Up and Worked Exam Questions

## Warm-up Questions

1) Which of the following quantities are digital and which are analogue?
   height, a team's score in a football game, shoe size, a runner's time in a race.

2) Which part of a compact disc (CD) player does the equivalent job to the needle on a record player?

3) State one advantage of amplitude modulation over frequency modulation.

4) Is a radio wave of frequency 200 kHz likely to be a ground, sky or space wave?

5) Name three types of information storage system used in electronic devices.

## Worked Exam Questions

1) a) Sketch in the spaces below a typical wave form of an analogue signal and a digital signal.

Analogue                                    Digital

*[2 marks]*

b) Describe what attenuation is, how the problem is usually solved, and what problems the solution itself causes.

*Attenuation is the weakening of a signal the further it has to travel during*

*transmission. The problem can be solved by using repeater stations at regular*

*intervals to amplify the signal. However, this introduces noise and distortion.*

*[3 marks]*

2) a) One music radio station called "Galaxy AM" has the frequency 297 kHz.
   A rival music station called "Planet FM" has the frequency 89 MHz.

   Describe the different ways that the two stations transmit their signals.

   *In amplitude modulation, such as Galaxy AM use, the carrier wave is modulated by*

   *changing its amplitude, so that the amplitude varies at the same rate as the voltage*

   *in the sound signal. In frequency modulation such as Planet FM use, the carrier wave*

   *is modulated by changing its frequency.*

   *[2 marks]*

b) Planet FM claims that it has a better sound quality than Galaxy AM.
   Is that likely to be true? Briefly explain your answer.

   *Yes — FM signals are less susceptible to noise and interference than AM signals.*

   *[1 mark]*

# Exam Questions

1) This question is about tape recorders.

   a) The magnetic tape used to record sound on a tape recorder is made of plastic coated with particles of magnetic material. The diagram below shows the pattern of magnetised particles on the tape after a song has been recorded onto it.

   Which of the diagrams below do you think best shows what the magnetised particles would have looked like before anything was recorded on the tape?

   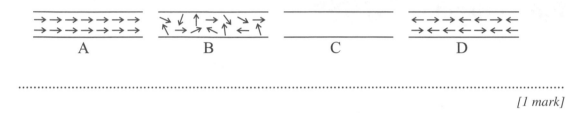

   ......................................................................................................................................

   *[1 mark]*

   b) These three basic principles are used in tape recorders:

   *Principle 1:* A high-frequency alternating current can demagnetise a magnet.

   *Principle 2:* When current flows through a coil of wire, a magnetic field is created in and around the coil.

   *Principle 3:* When a coil of wire experiences a change of magnetic flux, a current is induced in the coil.

   For each of the following three functions of the tape recorder, write down which principle is primarily being used.

   (i) The recording head magnetizes the particles on the tape.

   ...............................................

   (ii) The playback function produces an electric signal that is sent to the amplifier.

   ...............................................

   (iii) The eraser head erases the pattern from the magnetic tape.

   ...............................................

   *[3 marks]*

   c) Which of principles 1, 2 or 3 from part (b) is also used in the operation of a loudspeaker?

   ......................................................................................................................................

# Exam Questions

*[1 mark]*

2) The diagram below shows how a radio receiver picks up a radio transmission and converts it into sound waves.

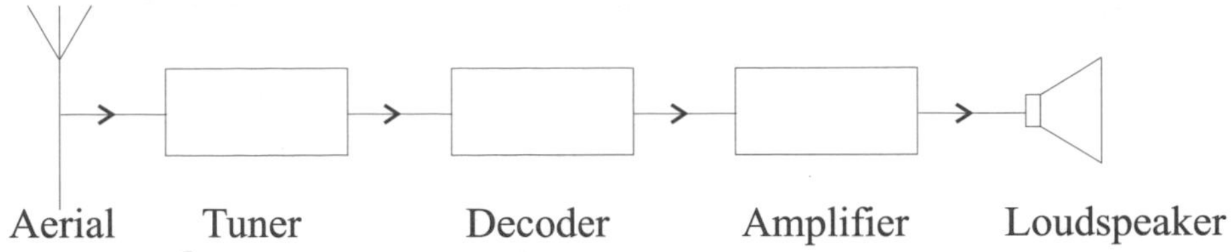

Aerial    Tuner         Decoder      Amplifier      Loudspeaker

a) Which component is a transducer?

.................................................................................................................................................
*[1 mark ]*

b) Which component typically consists of a diode and a capacitor?

.................................................................................................................................................
*[1 mark ]*

c) Which component allows you to choose which frequency you will listen to?

.................................................................................................................................................
*[1 mark ]*

d) Which component is typically controlled by a volume knob on a radio?

.................................................................................................................................................
*[1 mark ]*

e) Communications systems can generally be analysed in terms of eight parts: encoder, modulator, amplifier, transmitter, receiver, decoder, transducer and storage system.

Which two of these parts might involve an aerial?

.................................................................................................................................................
*[1 mark ]*

f) A more private means of sending a signal across a large distance is to send the signal down a cable.

Copper was originally used for these cables, but what is the more modern alternative?

.................................................................................................................................................
*[1 mark ]*

# Lenses and Images

Lenses are usually made of glass or plastic. All lenses change the direction of rays of light by refraction — light slows down when it enters the lens and speeds up when it leaves. (Look back at pages 97-98 to refresh your memory about refraction.)

## Converging and Diverging Lenses do Different Things

There are two main types of lens — converging and diverging. They have different shapes and have opposite effects on light rays:

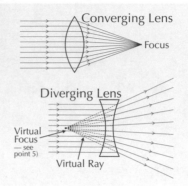

Converging Lens
Focus

Diverging Lens
Virtual Focus — see point 5)
Virtual Ray

 1  A converging lens is convex — it bulges outwards. It causes rays of light to converge (move together) to a focus.

2  A diverging lens is concave — it caves inwards. It causes rays of light to diverge (spread out).

Learn these diagrams — you need to know all of the following:
1) the position of the focus for each lens,
2) that each ray of light changes direction when it enters and when it leaves the lens,
3) that the solid lines are called real rays as they represent where light has really travelled,
4) that the dashed lines on the diverging lens diagram are called virtual rays (as they're not really there),
5) that the virtual rays are drawn to show the point (focus) where the diverging rays appear to come from.

## Taking a Photo Forms an Image on the Film

When you take a photograph of a tree, light from the object (tree) travels to the camera and is focused by the lens, forming an image on the film.

1) The image on the film is a real image because light rays actually meet there.

2) The image is smaller than the object since the lens has focused the rays onto the film by refracting the light.

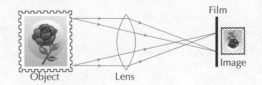

Film
Image
Object
Lens

3) The image is inverted — upside down.

4) The same thing happens in our eye — a real, inverted image forms on the retina.

## A Virtual Image — It's Not Really There

1) A virtual image is different from a real image.

2) A real image is where the light actually hits the screen where the image is formed — like a photo or a movie at the cinema where you see an image of the film on the giant screen.

3) When you look in a mirror you see an image of yourself behind the mirror. This is a virtual image, since light can't possibly have passed through the back of the mirror (check for yourself).

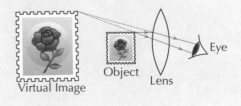

Eye
Object
Lens
Virtual Image

4) A similar thing happens with a converging lens used as a magnifying glass to look at a small stamp.

# Natural Frequency and Resonance

## Everything has a Natural Frequency

EXAMPLES: Give a kid on a swing a push and it'll swing backwards and forwards at a certain rate (frequency).
Gently flick the lip of a decent wine glass and it'll vibrate, 'singing' a particular note.
If a strong gust of wind hits a tall tower it'll vibrate back and forth at a certain frequency.
Everything, given a bit of energy and left to get on with it, will vibrate at its own natural frequency.

## Resonance is when the Natural Frequency is Matched

EXAMPLE: When you're pushing a kid on a swing, you need to push in time with the swing to transfer the maximum energy to the kid. Otherwise it just disrupts the rhythm and the thing just wobbles about all over the place (which wastes energy).

IN GENERAL: When you force an object to vibrate, it's most effective when the driving frequency (the frequency of your 'pushes') matches the natural frequency of the object — then you get resonance.

### RESONANCE happens when DRIVING FREQUENCY = NATURAL FREQUENCY

## Examples of Resonance:

1) Pendulum
   The kid on a swing is a simple pendulum. The period and frequency of the pendulum depends on its length. If you push at the natural frequency of the swing, the amplitude of the oscillation increases rapidly.

2) **Mass on a Spring** 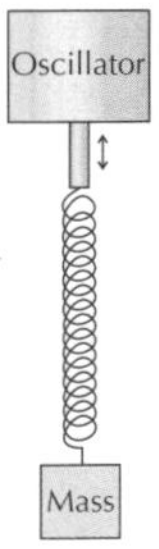
   The diagram shows a system with a natural frequency of 2.5 Hz. A driving oscillator (a machine that makes things vibrate at a chosen frequency) makes it vibrate at a varying frequency rising from 0 to 5 Hz. At low frequencies the mass oscillates with a small amplitude. As the natural frequency is reached the amplitude rises rapidly — the driver is transferring a lot of energy to the mass. Above 2.5 Hz the amplitude falls again.
   *(Changing the mass can change the natural frequency —*
   *increasing the mass decreases the natural frequency.)*

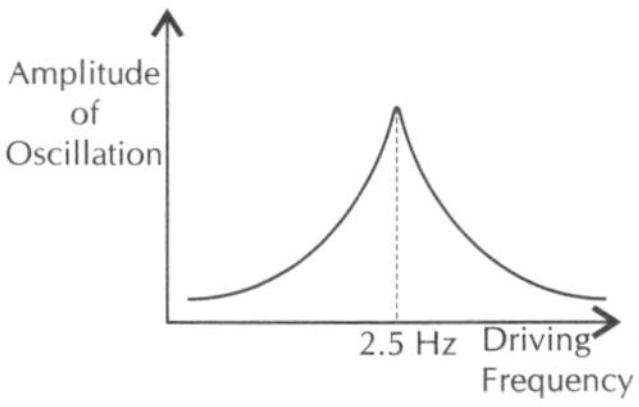

3) Vibrating String 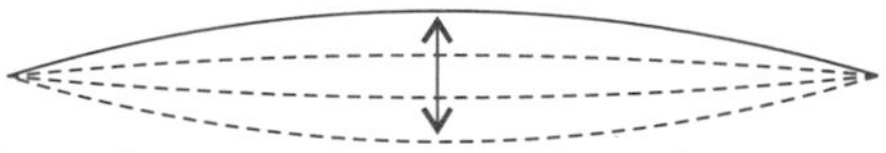
   If the driving oscillator above is attached to a stretched string, a more complex effect occurs because the string has more than one resonant frequency (see next page). When the driver reaches each of these frequencies, the string vibrates in a clear pattern with large amplitude.

4) Column of Air in an Oboe
   The air inside an oboe can be made to vibrate, also with more than one resonant frequency. When you blow on the reed, sound waves of many frequencies travel down the air in the oboe. The natural frequency is 'picked out', resonance occurs and the resonant frequency gets louder — and a musical note is heard. The natural frequency depends on the length of the column of air. *(The longer the column the lower the natural frequency and the lower the pitch of the note.)*

5) Microwaving Food
   Water molecules have a natural frequency matched by the frequency of microwaves. Microwaves make the water molecules in the food resonate, transferring lots of energy to heat the food.

## The Tacoma Narrows Bridge disaster is an example of resonance

In 1940, the Tacoma Narrows Bridge collapsed because the wind made it resonate. Sixty years later they made a similar mistake with the Millennium Bridge — that resonated with the vibrations from people's footsteps.

# Resonance — Harmonics and Musical Notes

A string has several resonant frequencies. If you play a stringed instrument, you may have discovered 'harmonics'. It's not only true of strings, but that's all you need to learn for GCSE Physics.

## Any **Multiple** of the **Natural Frequency** will be **Resonant**

This is the standard experiment. Set up a <u>driving oscillator</u> at one end of a <u>stretched string</u>, and <u>fix</u> the other end. <u>Waves travel</u> down the string and <u>reflect</u> off the fixed end. <u>Vary</u> the <u>frequency</u> of the waves, and as it approaches one of the <u>resonant frequencies</u> you get a '<u>standing wave</u>' like this:

## **Nodes** are **Stationary Points** on the String

At resonant frequencies there are points where the string does not move — called '<u>nodes</u>'. The rest of the string oscillates, making pretty 'loop' <u>patterns</u> too. You can tell <u>which resonant frequency</u> the string is vibrating at by <u>counting</u> how many oscillating <u>loops</u> there are <u>between nodes</u>.

The string on the right is vibrating at the <u>fundamental frequency</u> — the <u>lowest resonant frequency</u>. <u>Half a</u> <u>wavelength</u> fits on the string. The only nodes here are at the <u>ends</u> of the string. There's just <u>one 'loop'</u>.

This is the <u>second harmonic</u> (or <u>first</u> <u>overtone</u>). It occurs at <u>twice</u> the <u>fundamental frequency</u>. <u>One whole</u> <u>wavelength</u> fits on the string. There is now a <u>node</u> in the <u>middle</u> and at the <u>ends</u>, and <u>two 'loops'</u> between.

This is the <u>third harmonic</u> (or <u>second overtone</u>). It occurs at <u>3</u> <u>times</u> the <u>fundamental frequency</u>. 1½ <u>wavelengths</u> fit on the string. There are now <u>four nodes</u> in total, with <u>three 'loops'</u> between them.

## For **Low** Notes you need **Long**, **Heavy**, **Loose** Strings...

Guitars need to be tuned to stop them playing the wrong note (frequency). A guitar player can play a range of notes by choosing <u>different</u> strings or by <u>shortening</u> a string with their fingers. Three things can be changed to <u>alter the natural frequency</u> of a string:

 The longer the string, the lower the note. With longer strings, the half-wavelength you get at the natural frequency is longer, so the frequency is lower. To make a higher pitched note you can shorten the string, which shortens the wavelength and increases the frequency. On a guitar you do this by pressing down on a higher fret.

 The heavier the string, the lower the note. Waves travel slower with a heavier string, so for any given length, a heavier string will have a lower natural frequency.
You can play a higher pitched note on a guitar by plucking a lighter (thinner) string.

③ The looser the string, the lower the note. Waves travel faster with a tighter string.
For any given length a tighter string will have a higher natural frequency.
You can play higher notes on a guitar by tightening the string.

# Interference of Waves

Waves can interfere with each other, e.g. if there are two loudspeakers playing the same thing in a large hall, you can get areas of loud and quiet bits, where the waves have either added to each other or cancelled out.

## When **Waves Meet** they Cause a **Disturbance**

1) All waves cause some kind of <u>disturbance</u> in a medium — water waves disturb water particles, sound waves disturb air particles, electromagnetic waves disturb electric and magnetic fields.
2) When <u>two waves meet</u> at a point they both try to cause their own disturbance.
3) Waves either disturb in the <u>same direction</u> (<u>constructive</u> interference), or in <u>opposite directions</u> (<u>destructive</u> interference).
4) Think of a '<u>pulse</u>' travelling down a slinky spring meeting a pulse travelling in the opposite direction. These diagrams show the <u>possible outcomes</u>:
5) The <u>total amplitude</u> of the waves at that point is the <u>sum</u> of the <u>displacements</u> (you have to take direction into account) of the waves at that point.

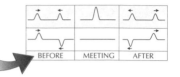

## You get **Patterns** of '**Loud**' and '**Quiet**' Bits

Two speakers both play the same note, starting at <u>exactly</u> the <u>same time</u>, and are arranged as shown: Depending on where you stand in front of them, you might hear a <u>loud sound</u> or <u>almost nothing</u>.

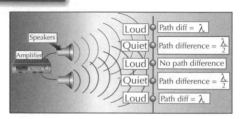

1) At certain points, the sound waves will be <u>in phase</u> — here you get <u>constructive interference</u>. The <u>amplitude</u> of the waves will be <u>doubled</u>, so you'll hear a <u>loud sound</u>.
2) These points occur where the <u>distance travelled</u> by the waves from both speakers is either the <u>same</u> or different by a <u>whole number of wavelengths</u>.
3) At certain other points the sound waves will be exactly <u>out of phase</u> — here you get <u>destructive interference</u> and the waves will <u>cancel out</u>. This means you'll hear almost <u>no sound</u>.
4) These out-of-phase points occur where the difference in the <u>distance travelled</u> by the waves (the "<u>path difference</u>") is ½ wavelength, 1½ wavelengths, 2½ wavelengths, etc.

## Interference helped us **Understand Light** Better

1) <u>Observing interference patterns</u> with water waves, sound waves and microwaves is relatively <u>easy</u> because their wavelengths are <u>quite large</u>.
2) <u>Observing interference</u> effects with <u>light</u> waves is more <u>difficult</u> because their wavelengths are so <u>small</u>. <u>Path differences</u> with light waves have therefore got to be <u>really tiny</u>.
3) A scientist called <u>Young</u> managed it by shining light through a pair of <u>narrow slits</u> that were just a <u>fraction of a millimetre</u> apart. This light then hit a screen in a dark room. There was an interference pattern of <u>light bands</u> (constructive) and <u>dark bands</u> (destructive) on the screen.
4) He was able to calculate the <u>wavelength of the light</u> and help physicists unravel some of its mysteries.

## As usual you need to learn all the diagrams

This is pretty straightforward — you either <u>add waves</u> or <u>take one away</u> from the other. The sound pattern in the middle is a bit more difficult. Start by learning the <u>diagram</u> and it'll make more sense.

# Warm-Up and Worked Exam Question

## Warm-up Questions

1) The blank rectangles A to D represent different optical devices. Which one is a convex lens?

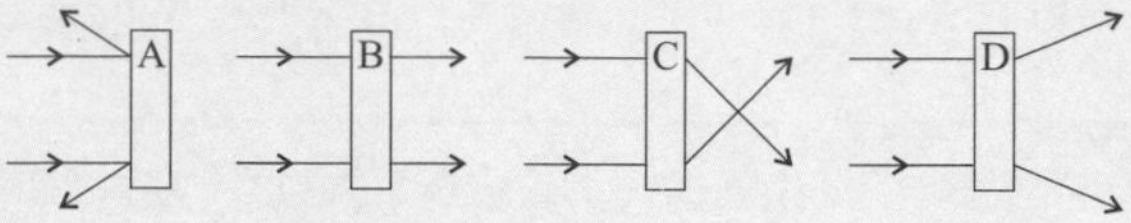

2) Give three examples of real-life systems where resonance can occur.

3) Two pulses are moving inwards from both ends of a cord, as shown below. Which of A, B, C or D shows what happens at the instant when the pulses meet?

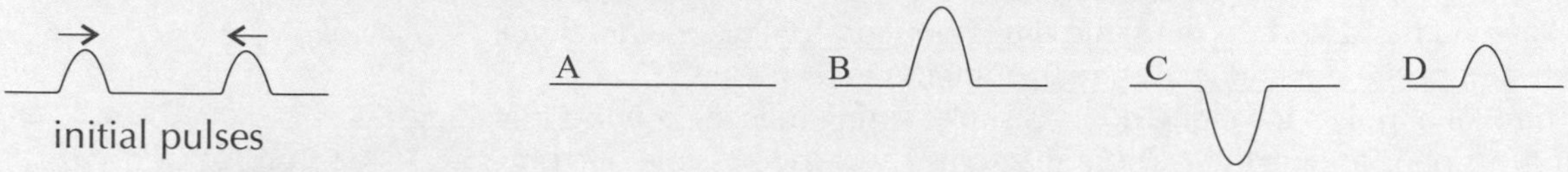

initial pulses

A          B          C          D

4) A guitar player plays a certain note on one of the strings of her guitar. Describe what difference it will make to the note if she (a) presses down on a fret to shorten the string, (b) tightens the string to re-tune it, (c) plucks another string that is heavier.

## Worked Exam Question

1    The diagram shows rays of light travelling towards a lens.

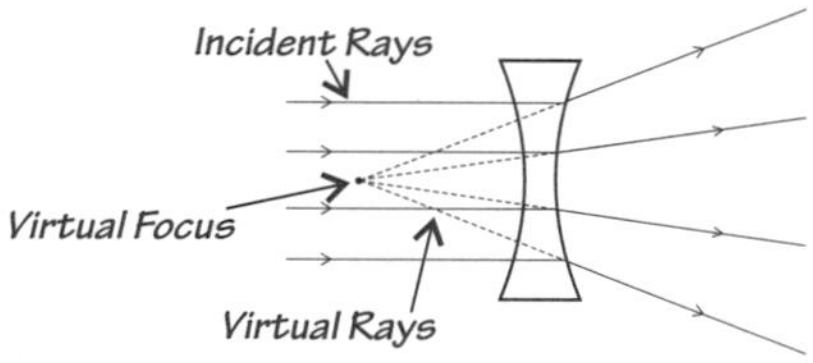

*Incident Rays*

*Virtual Focus*

*Virtual Rays*

a)   Does the diagram above show a converging or diverging lens?
*diverging*
*[1 mark]*

b)   Is it concave or convex?
*concave*
*[1 mark]*

c)   Label the virtual focus, the virtual rays and the incident rays on the diagram.
*[2 marks]*

d)   Which incident ray does not change direction?
*the ray through the centre of the lens*
*[1 mark]*

e)   Would the image formed by this lens be inverted or upright?
*upright*
*[1 mark]*

f)   Explain the meaning of the term "virtual focus".
*the point that the diverging rays appear to come from*
*[1 mark]*

# Exam Questions

1   Seth is pushing Amanda on a swing.  First he gives her one push and then leaves her to swing for a while.  Then he starts pushing in time with the swing, and she starts to swing really high.

a)   Describe what is happening, in terms of the **natural frequency**, the **driving frequency** and **resonance**.

......................................................................................................................................

......................................................................................................................................

*[3 marks]*

b)   At the resonant frequency the maximum amount of energy is transferred from the driving oscillator to the driven oscillator.

Describe the energy transfer that applies to:

(i)      Seth pushing Amanda on a swing,

......................................................................................................................................

(ii)     a microwave oven heating up a bowl of soup.

......................................................................................................................................

*[2 marks]*

c)   A driving oscillator makes a string vibrate at various frequencies.

Sketch on the diagram below how the amplitude of oscillation varies as the driving frequency is changed.  Mark on your graph the natural frequency of the string, 4Hz.

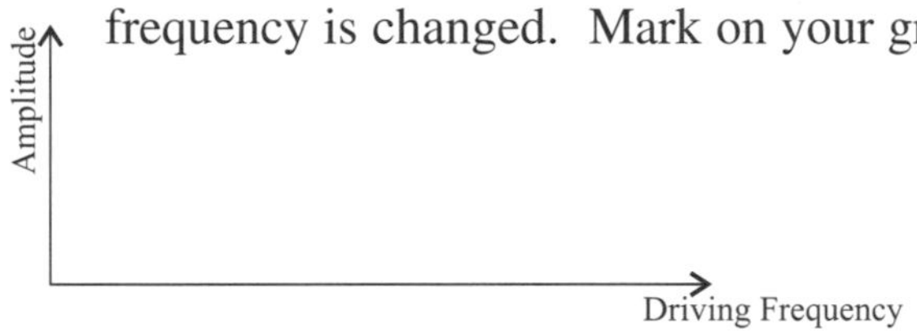

*[2 marks]*

2   a)   The diagram below shows the way a camera forms an image of an object.

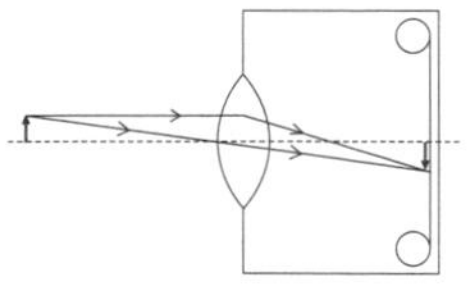

(i)      Label on the diagram the following parts: film, convex lens, object, image.
*[2 marks]*

(ii)     Describe fully the type of image that is formed on the film.

......................................................................................................................................
*[3 marks]*

b)   The diagram shows a naturalist's magnifying glass.The real light rays from the top and the bottom of the feather follow the paths shown.

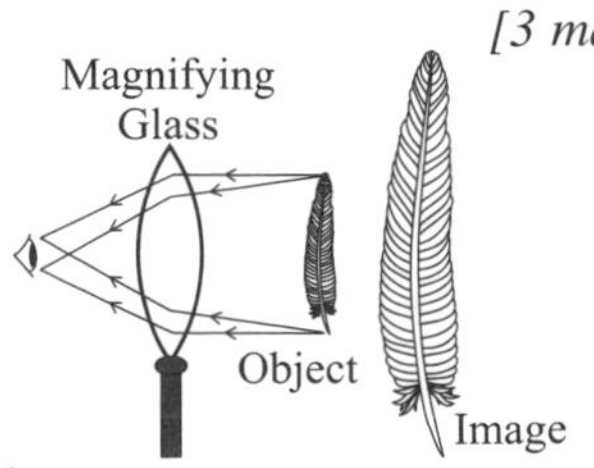

Draw in dotted lines to show how a virtual image is formed.

*[2 marks]*

# Exam Questions

3    A driving oscillator is set up at one end of a stretched string. The string is fixed at the other end, as shown in the diagram. When the oscillator vibrates at the fundamental frequency of the string a standing wave is set up as shown.

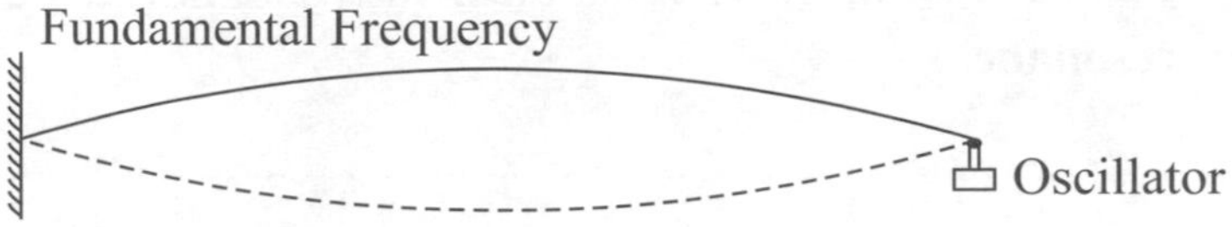

Fundamental Frequency

Oscillator

a)    What is a 'node'?

...................................................................................................................................................

*[1 mark]*

b)    Where do the nodes occur in the diagram?

...................................................................................................................................................

*[1 mark]*

c)    Nodes happen only at specific frequencies. At what frequencies do they occur?

...................................................................................................................................................

*[1 mark]*

d)    How many wavelengths fit on the string when it is vibrating at the fundamental frequency?

...................................................................................................................................................

*[1 mark]*

The diagram below shows the standing wave set up at the third harmonic.

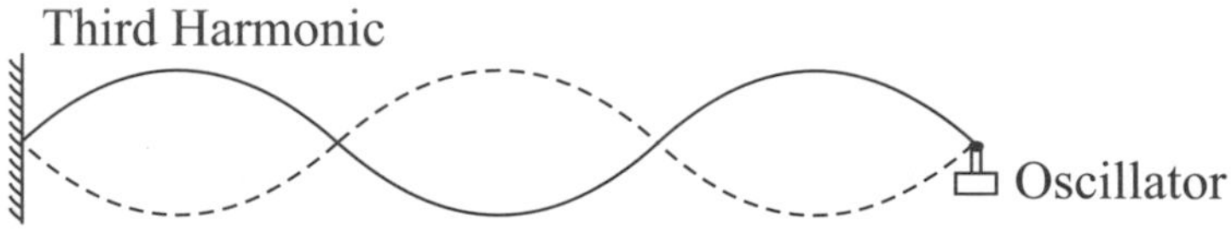

Third Harmonic

Oscillator

e)    (i)     State the frequency as a multiple of the fundamental frequency (F).

...................................................................................................................................................

      (ii)    State the number of nodes.

...................................................................................................................................................

      (iii)   State the number of wavelengths that fit onto the string.

...................................................................................................................................................

*[3 marks]*

# Particles

## Absolute Zero is as Cold as Anything can get — 0 kelvins

1) If you <u>increase</u> the <u>temperature</u> of something, you give its particles more <u>energy</u> — they move about more <u>quickly</u> or <u>vibrate</u> a bit more. In the same way, if you <u>cool</u> a substance down, you're reducing the <u>kinetic energy</u> of the particles.

2) The <u>coldest</u> that anything can ever get is -273 °C — this temperature is known as <u>absolute zero</u>. At absolute zero, atoms have as little <u>kinetic energy</u> as it's <u>possible</u> to get.

3) Absolute zero is the start of the <u>Kelvin</u> scale of temperature.

4) A temperature change of <u>1 °C</u> is also a change of <u>1 kelvin</u>. The two scales are pretty similar — the only difference is where the <u>zero</u> occurs.

5) To convert from <u>degrees Celsius to kelvins</u>, just <u>add 273</u>. And to convert from <u>kelvins to degrees Celsius</u>, just <u>subtract 273</u>.

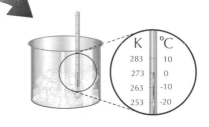

|  | Absolute zero | Freezing point of water | Boiling point of water |
|---|---|---|---|
| Celsius scale | -273 °C | 0 °C | 100 °C |
| Kelvin scale | 0 K | 273 K | 373 K |

*Absolute zero is actually around -<u>273.15 °C</u>, but the 0.15 is usually ignored.*

## Kinetic Energy is Proportional to Temperature

The heading makes this sound more complicated than it actually is...

 **1** If you <u>increase</u> the temperature of a gas, you give its particles <u>more energy</u>.

**2** In fact, if you <u>double</u> the temperature (measured in <u>kelvins</u>), you <u>double</u> the average <u>kinetic energy</u> of the particles.

## Colliding Gas Particles Create Pressure

Remember — gases consist of loads of particles moving about <u>all over the place</u>.

1) As gas particles move about, they <u>bang into</u> each other and anything else that's in their way.

2) Even though gas particles are very light, these collisions cause a <u>force</u> on the object they hit. And if the gas is in a <u>sealed container</u>, the particles hit the container's walls — creating an <u>outward pressure</u>.

3) If the gas is <u>heated</u>, the particles move <u>faster</u>. This increase in kinetic energy means that the particles hit the walls of the container <u>harder</u>, creating more pressure. In fact, temperature and pressure are <u>proportional</u> — if you <u>double</u> the temperature, you'll <u>double</u> the pressure as well.

4) And if you put the <u>same</u> amount of gas in a <u>bigger</u> container, the pressure will decrease, as there will be fewer collisions between the gas particles and the container's walls.

5) In fact, for a <u>fixed mass</u> of gas, this is true:

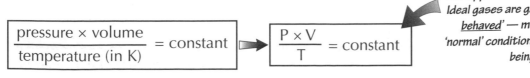

$$\frac{\text{pressure} \times \text{volume}}{\text{temperature (in K)}} = \text{constant} \quad \rightarrow \quad \frac{P \times V}{T} = \text{constant}$$

*This applies to so-called <u>ideal gases</u>. Ideal gases are gases that are '<u>well-behaved</u>' — most gases under 'normal' conditions are quite close to being ideal.*

# Particles

The basic structure of atoms, with negative <u>electrons</u> around a positive <u>nucleus</u>, was discovered by Geiger, Marsden and Rutherford in their 'gold foil and alpha particles' experiment (see page 196). However, it turns out that things are <u>more complicated</u> than that.

## *Electrons* and *Positrons* are *Fundamental Particles*

1) <u>Electrons</u> are very <u>small</u> in size, <u>weigh</u> hardly anything, and are <u>negatively charged</u>.
2) Electrons are also <u>fundamental particles</u> — meaning you can't divide electrons into even <u>smaller</u> particles.
3) Electrons have a positive equivalent called <u>positrons</u>. Positrons are like electrons (they're the <u>same</u> size and mass, for example), but they're <u>positively charged</u>.
4) Positrons are also <u>fundamental</u> particles.

## *Protons* and *Neutrons* are made up of *Smaller Particles*

1) <u>Protons</u> and <u>neutrons</u> are <u>not</u> fundamental particles. They're made up of even smaller particles called <u>quarks</u>. It takes <u>three quarks</u> to make a proton or neutron.
2) There are various kinds of quark, but protons and neutrons consist of just two types — <u>up-quarks</u> and <u>down-quarks</u>.
3) A <u>proton</u> is made of <u>two up-quarks</u> and <u>one down-quark</u>.
4) A <u>neutron</u> is made of <u>two down-quarks</u> and <u>one up-quark</u>.
5) The <u>charges</u> on up and down quarks are shown in the table. When quarks <u>combine</u> to make protons and neutrons, these charges <u>add together</u> to make the overall charges on the proton and neutron.

| Quark | Relative Charge |
|-------|-----------------|
| up    | $\frac{2}{3}$   |
| down  | $-\frac{1}{3}$  |

**Proton**

'up-quark' + 'up-quark' + 'down-quark', so relative charge on proton $= \frac{2}{3} + \frac{2}{3} + \left(-\frac{1}{3}\right) = +1$

*Remember: protons are 'positive' — they're more 'up' than 'down'.*

**Neutron**

'up-quark' + 'down-quark' + 'down-quark', so relative charge on neutron $= \frac{2}{3} + \left(-\frac{1}{3}\right) + \left(-\frac{1}{3}\right) = 0$

## *Quarks Change* — producing *Electrons/Positrons* in the process

Sometimes the number of <u>protons</u> and <u>neutrons</u> in the nucleus can result in the atom being <u>unstable</u>. So to become more <u>stable</u>, the particles in the nucleus may change.

1) Sometimes down-quarks can <u>change</u> into up-quarks, and vice versa.
2) When this happens in a nucleus, a neutron is <u>converted</u> into a proton, or a proton into a neutron.
3) However, the <u>overall charge</u> before and after has to be <u>equal</u>. So when a <u>neutron</u> changes into a <u>proton</u>, a <u>negatively charged</u> particle also has to be produced (so that the overall charge remains <u>zero</u>).
4) The negatively charged particle produced when a neutron changes into a proton is just an <u>electron</u>. This electron is then <u>ejected</u> from the nucleus in a process called β⁻ (<u>beta minus</u>) decay. (This is just normal <u>beta radiation</u>.)
5) When a <u>proton</u> changes into a <u>neutron</u> (i.e., when an <u>up-quark</u> changes into a <u>down-quark</u>), an extra <u>positive</u> charge is needed to keep the overall charge at +1. So the nucleus produces and throws out a <u>positron</u>. This is known as β⁺ (<u>beta plus</u>) decay.

## *Remember — Electrons, Positrons, Quarks are Fundamental Particles*

So... particles can change and become something else — but only if <u>another particle</u> is produced at the same time.

# Particles

It's handy to know whether a particular nucleus is <u>stable</u> or <u>unstable</u> (since unstable ones are <u>radioactive</u>). It all depends on the <u>balance</u> between the number of <u>protons</u> (Z) and the number of <u>neutrons</u> (N).

## *Big Atoms* need *More Neutrons* than *Protons* to be *Stable*

1) The <u>N-Z plot</u> shows whether or not a nucleus with Z protons and N neutrons will be <u>stable</u>.

2) You plot the number of <u>neutrons</u> (N) in an isotope against the number of <u>protons</u> (Z) — if the point lies near the <u>line of stability</u>, it's <u>stable</u>.

3) The <u>line of stability</u> starts off <u>straight</u>, meaning that <u>small</u> atoms need <u>as many</u> neutrons as protons to be stable.

4) But for <u>larger</u> numbers of <u>protons</u>, the line curves upwards. This means that bigger atoms need <u>more neutrons</u> than protons to be stable.

5) If an isotope does <u>not</u> lie near the line, the isotope is <u>unstable</u> and therefore <u>radioactive</u>.

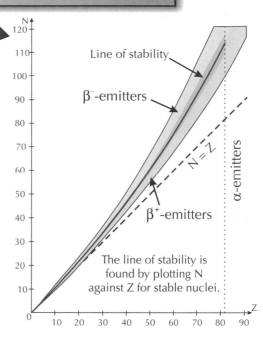

The line of stability is found by plotting N against Z for stable nuclei.

## You can *Predict* the type of *Radioactive Decay*

Using the N-Z plot, you can <u>predict</u> what kind of <u>radiation</u> (if any) an isotope will give off.

1) Isotopes that have <u>more than 82 protons</u> in the nucleus usually try to become more stable by emitting an <u>alpha particle</u>.

2) Isotopes <u>above</u> the line of stability have <u>too many neutrons</u> to be stable. This means that they will undergo β⁻ <u>decay</u> (where a <u>neutron</u> is converted to a <u>proton</u> and an <u>electron</u>).

3) Isotopes <u>beneath</u> the line have <u>too few neutrons</u>. They undergo β⁺ <u>decay</u> (where a <u>proton</u> is converted to a <u>neutron</u> plus a <u>positron</u>).

4) With both β⁻ and β⁺ decay, the particles in the nucleus usually <u>rearrange</u> themselves, and need to get rid of an extra bit of <u>energy</u> as a result — this is done by emitting <u>gamma radiation</u>.

## *Nuclear Equations* need to *Balance*

In radioactive decay, what you start with is the <u>parent element</u>, and what you end up with is the <u>daughter element</u>.

| A typical β⁻ emission |
|---|

$$^{14}_{6}C \rightarrow {}^{14}_{7}N + {}^{0}_{-1}\beta$$

1) A <u>neutron</u> changes into an <u>electron</u> (which gets thrown out) plus a <u>proton</u>.
2) So the total number of particles in the nucleus doesn't change, i.e. the <u>mass number</u> (on top) stays the <u>same</u>.
3) But the <u>extra proton</u> means the atomic number goes <u>up by one</u>.

| A typical β⁺ emission |
|---|

$$^{30}_{15}P \rightarrow {}^{30}_{14}Si + {}^{0}_{+1}\beta$$

1) A <u>proton</u> changes into a <u>positron</u> (which gets kicked out) plus a <u>neutron</u>.
2) So again the total number of particles in the nucleus doesn't change, i.e. the <u>mass number</u> stays the <u>same</u>.
3) But the 'lost' proton means the atomic number goes <u>down by one</u>.

Always check that the <u>mass numbers</u> and <u>atomic numbers</u> in a nuclear equation <u>balance</u>.

## *There's more on Radioactivity in Section Six*
Basically the <u>overall charge</u> and the <u>overall mass</u> is the same <u>before</u> and <u>after</u> a beta decay.

# Particles

Nuclear fission is the process used in nuclear power stations.
It involves splitting a large atom into two smaller ones, which in turn releases energy.

## The Splitting of Uranium-235 needs Neutrons

Uranium-235 (i.e. a uranium atom with a total of 235 protons and neutrons) is used in
some nuclear reactors (and bombs).

1) Uranium-235 (U-235) is actually quite stable, so it needs to be made unstable before it'll split.
2) This is done by firing slow-moving neutrons at the U-235 atom.
3) The neutron joins the nucleus to create an unstable U-236 atom.
4) The U-236 then splits into two smaller atoms,
   plus 2 or 3 fast-moving neutrons.
5) There are different pairs of atoms that U-236 can split
   into — e.g. krypton-90 and barium-144.

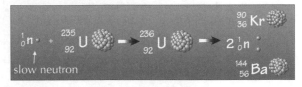

$$_0^1 n + {}_{92}^{235}U \rightarrow {}_{92}^{236}U \rightarrow 2\,{}_0^1 n$$

slow neutron $\quad$ ${}_{36}^{90}Kr$ $\quad$ ${}_{56}^{144}Ba$

## You can split More than One Atom — Chain Reactions

1) To get a useful amount of energy, loads of U-235 atoms have to be split. So neutrons released from previous fissions are used to hit other U-235 atoms.

2) These cause more atoms to (eventually) split, releasing even more neutrons, which hit even more U-235 atoms... and so on and so on. This process is known as a chain reaction.

3) However, neutrons released from the splitting of an atom move too fast to be ideal for converting U-235 into U-236. So a graphite "moderator" is used to slow down the neutrons in a reactor.

4) The fission of an atom of uranium releases loads of energy, in the form of the kinetic energy of the two new atoms (which is basically heat).

## Inside a Gas-Cooled Nuclear Reactor

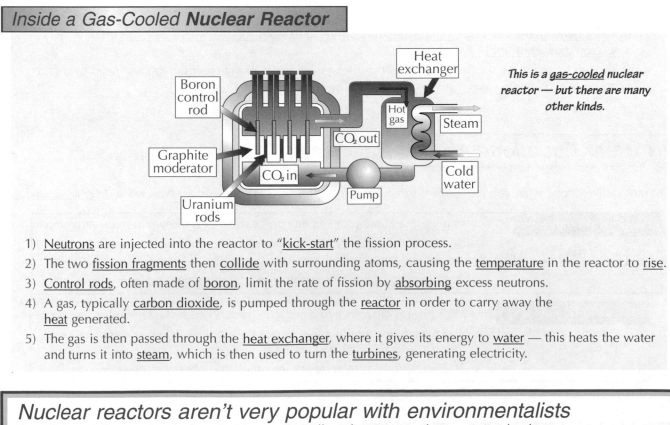

This is a gas-cooled nuclear reactor — but there are many other kinds.

1) Neutrons are injected into the reactor to "kick-start" the fission process.
2) The two fission fragments then collide with surrounding atoms, causing the temperature in the reactor to rise.
3) Control rods, often made of boron, limit the rate of fission by absorbing excess neutrons.
4) A gas, typically carbon dioxide, is pumped through the reactor in order to carry away the heat generated.
5) The gas is then passed through the heat exchanger, where it gives its energy to water — this heats the water and turns it into steam, which is then used to turn the turbines, generating electricity.

## Nuclear reactors aren't very popular with environmentalists

The products left over after nuclear fission are generally radioactive, so they can't just be thrown away.
Sometimes they're put in thick metal containers, which are then placed in a deep hole, which is then filled
with concrete. But some people worry that the materials could leak out after a number of years.

# Particles

Electron guns are used in <u>televisions</u>.  They fire electrons at the screen.

## *Electron Guns use Thermionic Emission*

1) The <u>heater</u> passes <u>energy</u> to the electrons in the <u>cathode</u>.  Once they have enough energy, they "<u>boil off</u>", i.e. they <u>escape</u>.
This process is called <u>thermionic emission</u>.

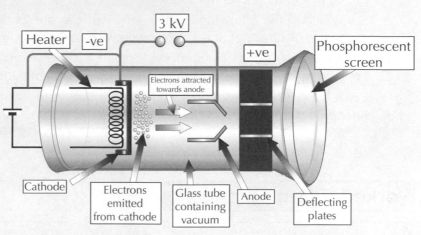

2) The electrons then <u>accelerate</u> as they are pulled towards the (positive) <u>anode</u>.  They then pass through a hole in the anode.

3) An <u>electric field</u> created between two <u>charged</u> metal plates can be used to <u>deflect</u> the electrons.
Using <u>two pairs</u> of plates (the <u>X-</u> and <u>Y-plates</u>), the electron beam can be deflected both <u>up and down</u> (by the Y-plates) and <u>left and right</u> (by the X-plates).

4) In things like television tubes, computer monitors and oscilloscopes, the electrons hit a screen covered in chemicals like <u>phosphorus</u>.  These chemicals emit <u>light</u> when hit by electrons.

5) But electron beams can also be used to produce <u>X-rays</u>.  When the electrons hit a <u>tungsten</u> target, their <u>kinetic energy</u> is converted into <u>X-rays</u>.

## One *Equation* that's *Bound* to be in the *Exam*

① The <u>kinetic energy</u> gained by the electrons is given by:

> **kinetic energy = charge of the electron (q) × accelerating voltage (V)**

$$\frac{KE}{q \times V}$$

② The <u>beam</u> of electrons produced is equivalent to an electrical <u>current</u>.  This means the beam must follow the rules and <u>formulas</u> that are linked to currents, e.g. charge (Q) = current (I) × time (t).

## *Oscilloscopes* are used to *Test* Electrical Signals

An <u>oscilloscope</u> is a device for checking the <u>voltage</u> and <u>frequency</u> of an <u>electrical signal</u>.

1) The screen is divided by equally spaced <u>lines</u> that are used in the same way as <u>gridlines</u> on a graph.

2) The <u>horizontal</u> (x) axis represents <u>time</u>, and the <u>vertical</u> (y) axis represents <u>voltage</u>.

3) The electron beam is made to move in the <u>x-direction</u> by the X-plates.  How quickly the beam moves across the screen is set using the <u>timebase</u> control, which controls the amount of <u>time per division</u>.

4) The <u>vertical</u> (y) axis is set using the <u>volts per division</u> control (which controls the Y-plates). You adjust that according to the <u>electrical signal</u> you're testing.

Time for 1 wave = 40 ms = <u>0.04 s</u>
Frequency is number of waves per second
= 1 ÷ 0.04 = <u>25 Hz</u>

Maximum voltage = <u>4 V</u>

## *Learn the Electron Gun diagram*

There's a lot on this page about electron guns.  But to be absolutely sure you'll be able to answer <u>any</u> question on the subject in the Exam, you need to make sure you know all the normal <u>electricity equations</u> from earlier in the book as well.  They're all there — so take the time to learn them...

# Warm-Up and Worked Exam Question

## Warm-up Questions

1) If the Kelvin temperature of a gas doubles, how will the average kinetic energy of the molecules in the gas change?

2) Say which of the following list of atomic particles are fundamental and which are not: up-quark, electron, neutron, proton, down-quark, positron.

3) State whether (a) a thorium nucleus containing 90 protons, and (b) a cobalt nucleus containing 27 protons, is likely to decay by α or β emission.

4) What happens to phosphorus when it is hit by an electron?   What practical devices use this phenomenon?

## Worked Exam Question

1    The fission process below takes place in nuclear reactors.

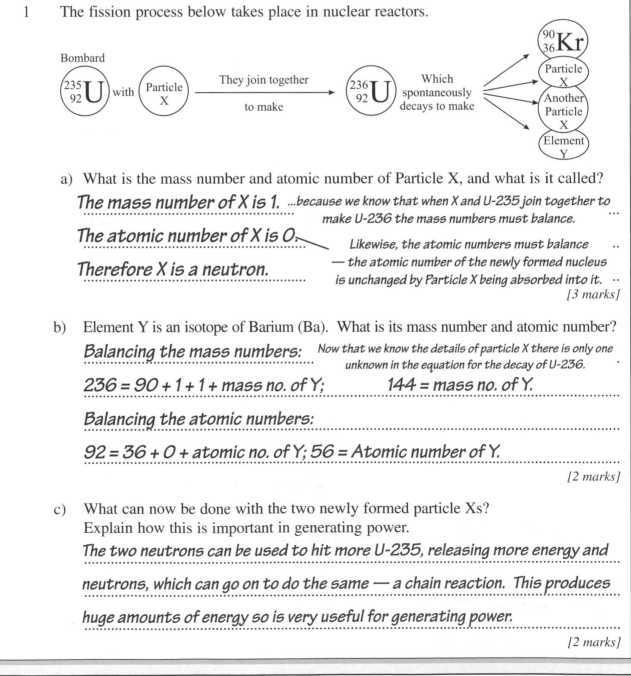

a)  What is the mass number and atomic number of Particle X, and what is it called?

*The mass number of X is 1.*   ...because we know that when X and U-235 join together to
make U-236 the mass numbers must balance. ...

*The atomic number of X is 0.*
Likewise, the atomic numbers must balance
— the atomic number of the newly formed nucleus
*Therefore X is a neutron.*
is unchanged by Particle X being absorbed into it. ..

[3 marks]

b)  Element Y is an isotope of Barium (Ba).  What is its mass number and atomic number?

*Balancing the mass numbers:*   Now that we know the details of particle X there is only one
unknown in the equation for the decay of U-236.

*236 = 90 + 1 + 1 + mass no. of Y;*          *144 = mass no. of Y.*

*Balancing the atomic numbers:*

*92 = 36 + 0 + atomic no. of Y; 56 = Atomic number of Y.*

[2 marks]

c)  What can now be done with the two newly formed particle Xs?
Explain how this is important in generating power.

*The two neutrons can be used to hit more U-235, releasing more energy and*

*neutrons, which can go on to do the same — a chain reaction.  This produces*

*huge amounts of energy so is very useful for generating power.*

[2 marks]

# Exam Questions

1)  a)  What does gas pressure mean in terms of the behaviour of ideal gas particles? Explain how the movement of ideal gas particles changes as the temperature drops towards absolute zero, and what effect this has on the pressure.

    ........................................................................................................................................

    *[3 marks]*

    b)  An ideal gas has a volume of 100 ml and a temperature of 300°C at 5 atmospheres pressure. What will its volume become at a temperature of 21°C and a pressure of one atmosphere?

    ........................................................................................................................................

    *[4 marks]*

2)  The equation below shows a typical beta decay. (Sr stands for strontium, Y for yttrium).

$$^{90}_{38}\text{Sr} \longrightarrow {}^{90}_{39}\text{Y} + {}^{0}_{-1}\beta$$

    a)  Fill in the blanks in this sentence: "One of the 52 _____s within the strontium nucleus was turned into a _____ plus an emitted _____ and gamma rays."
    *[2 marks]*

    b)  An up-quark has a relative charge of +2/3. A down-quark has a relative charge of -1/3. State what type and number of quarks go to make up both a proton and neutron.

    ........................................................................................................................................

    *[2 marks]*

    c)  Describe **beta minus** decay in terms of the changes to a quark within one of the nuclear particles in the parent nucleus. Include what happens to keep the overall charge the same before and after the decay.

    ........................................................................................................................................

    *[2 marks]*

    d)  Describe **beta plus** decay in the same terms as above.

    ........................................................................................................................................

    *[2 marks]*

    e)  Describe the main features of the "line of stability" on a graph of number of protons against number of neutrons for various nuclei.

    ........................................................................................................................................

    *[2 marks]*

# Exam Questions

3) The block diagram shows processes inside a gas-cooled nuclear reactor.

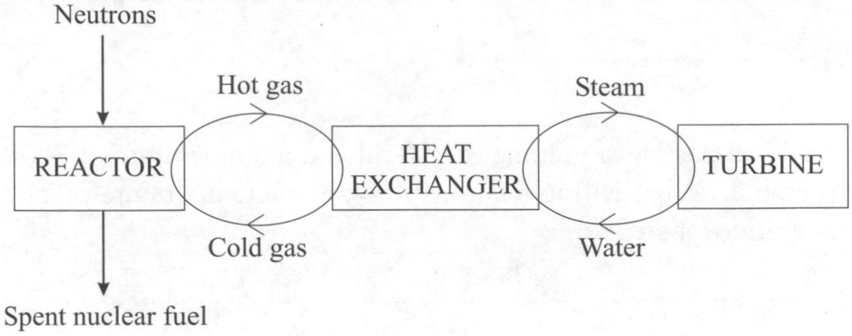

Neutrons

Hot gas          Steam

REACTOR          HEAT
EXCHANGER          TURBINE

Cold gas          Water

Spent nuclear fuel

a) In which component (reactor, heat exchanger or turbine) does fission occur?

.......................................................................................................................................

*[1 mark]*

b) Which component (reactor, heat exchanger or turbine) would also be found in a similar diagram of a coal-fired or oil-fired power station?

.......................................................................................................................................

*[1 mark]*

c) The water and the gas do similar jobs of drawing away heat. You might think that we could do without the middle stage and just use the water from the turbine to cool the reactor. Why would this be a bad idea?

................................................. *Hint — think what happens to substances that* .................................
*come into contact with the reactor.* *[1 mark]*

d) What "kick-starts" the whole process?

.......................................................................................................................................

*[1 mark]*

e) Why can't spent nuclear fuel just be thrown away?

.......................................................................................................................................

*[2 marks]*

f) Match up each substance used in a nuclear power station with its correct role.

Substances:
A — Boron;  B — Uranium or plutonium;  C — Graphite;  D — Carbon dioxide gas.

Roles:
1 — Slows down neutrons that are moving too fast;
2 — Absorbs excess neutrons;   3 — Draws heat off the reactor;   4 — Fuel.

.......................................................................................................................................

*[4 marks]*

# Revision Summary

Do all of the questions relevant to your syllabus.  If you find any difficult, you've got more learning to do.

1)  What is the difference between analogue and digital signals?

2)  Is a vinyl record an example of analogue or digital storage?  Is a CD an example of analogue or digital storage?

3)  How is data stored on a CD?  How are 0s and 1s read during the playback of a music CD?

4)  Will a CD or a vinyl record be more affected by a small scratch?

5)  How is data sent down an optical fibre?
    Give two advantages of sending information down an optical fibre rather than a copper wire.

6)  Noise can be dealt with more easily with digital signals than with analogue.  Why?

7)  Draw diagrams to show how information is encoded on a carrier wave by
    a) amplitude modulation (AM), and b) frequency modulation (FM).

8)  Describe the role of the capacitor in the simple demodulator in a radio.

9)  Describe how the frequency of a radio wave determines whether it travels mainly as a ground wave, a sky wave or a space wave.

10) What is meant by an "active satellite"?

11) What is the function of a transducer?

12) Why does the cone of a loudspeaker vibrate?  What causes a current to be induced in the coil of a microphone?

13) What are the three heads that a tape recorder might have?

14) Describe how a tape recorder records information on magnetic tape.

15) What are virtual rays?

16) List three characteristics of the image that's formed when you take a photo of a building.

17) Use the example of a magnifying glass to explain the meaning of a virtual image.  (Include a ray diagram.)

18) What is meant by the natural frequency of an object?  What is resonance and when does it occur?

19) What happens to the natural frequency of a mass on a spring if the mass is decreased?

20) Draw the pattern formed for the third harmonic on a stretched string.  How many nodes are there?
    What happens at a node?

21) What are the three things that can be done by a guitarist to play a lower note?

22) What is meant by constructive interference?  What is meant by destructive interference?

23) Identical sound waves leave two loudspeakers in phase.  At point A there is constructive interference and at point B there is destructive interference.  Give one possible path difference at each of points A and B.

24) What temperature is absolute zero in °C?  What's the boiling point of water in kelvins?

25) Convert 25 °C into kelvins.

26) What happens to the average KE of gas molecules when the temperature changes from 100 K to 300 K?

27) Describe what happens to the pressure of a gas in a sealed container when the temperature is reduced.

28) Can a positron be split into any smaller particles?  What three particles make up a proton?
    What charge does an up-quark have?

29) Describe the changes of quark that happen when beta minus decay occurs.
    Write out the nuclear equation for the beta minus decay of carbon-14.

30) If a lightweight isotope is plotted on the N-Z stability curve and appears underneath the line of stability, what type of decay is it likely to undergo?

31) How is an atom with 98 protons likely to decay?

32) How do atoms get rid of excess energy after beta decay?

33) What type of particle is U-235 bombarded with to make it split?

34) What is used in a reactor to slow down neutrons which are moving too quickly?

35) Describe how nuclear fission can be used to generate electricity.

36) What term is used to describe how electrons are emitted from the cathode in an electron gun?

37) How can a beam of charged particles in an electron gun be deflected?

38) What is the purpose of having a phosphorescent screen at the end of an oscilloscope tube?

39) Calculate the amount of kinetic energy gained by an electron (charge on an electron = -1.6 × 10$^{-19}$ C)
    when it is accelerated through a voltage of:   a) 10 V        b) 1000 V      c) 5 kV

40) What is the current of a beam of electrons when 0.002 C of charge hit the screen every second?

41) What is represented by the y-axis of an oscilloscope screen?

42) Draw diagrams to show what you would see on an oscilloscope if these signals were put in:
    a) an alternating signal with time period 20 ms and a peak voltage of 2 V
    b) an alternating signal with frequency 50 Hz and a peak voltage of 230 V

# Electronic Systems

There are <u>three</u> main parts of an <u>electronic system</u>. You need to know examples of all three parts, along with the extra circuit components below.

## Electronic Systems = *Input, Process, Output*

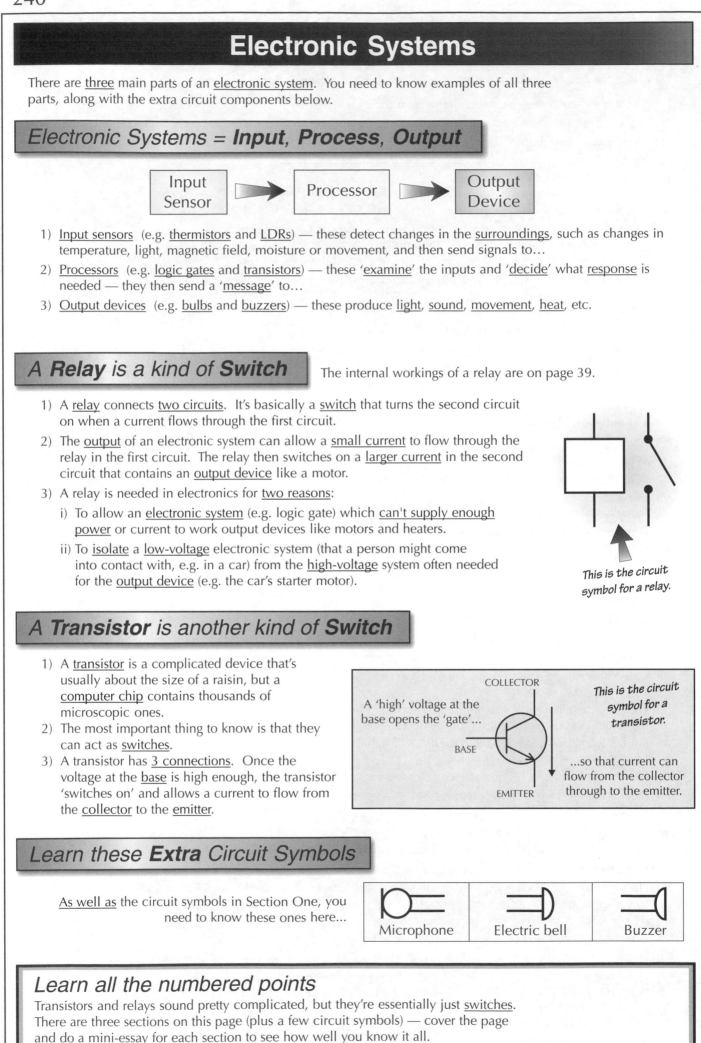

| Input Sensor | → | Processor | → | Output Device |

1) <u>Input sensors</u> (e.g. <u>thermistors</u> and <u>LDRs</u>) — these detect changes in the <u>surroundings</u>, such as changes in temperature, light, magnetic field, moisture or movement, and then send signals to...

2) <u>Processors</u> (e.g. <u>logic gates</u> and <u>transistors</u>) — these 'examine' the inputs and 'decide' what <u>response</u> is needed — they then send a '<u>message</u>' to...

3) <u>Output devices</u> (e.g. <u>bulbs</u> and <u>buzzers</u>) — these produce <u>light</u>, <u>sound</u>, <u>movement</u>, <u>heat</u>, etc.

## A *Relay* is a kind of *Switch*

The internal workings of a relay are on page 39.

1) A <u>relay</u> connects <u>two circuits</u>. It's basically a <u>switch</u> that turns the second circuit on when a current flows through the first circuit.

2) The <u>output</u> of an electronic system can allow a <u>small current</u> to flow through the relay in the first circuit. The relay then switches on a <u>larger current</u> in the second circuit that contains an <u>output device</u> like a motor.

3) A relay is needed in electronics for <u>two reasons</u>:

   i) To allow an <u>electronic system</u> (e.g. logic gate) which <u>can't supply enough power</u> or current to work output devices like motors and heaters.

   ii) To <u>isolate</u> a <u>low-voltage</u> electronic system (that a person might come into contact with, e.g. in a car) from the <u>high-voltage</u> system often needed for the <u>output device</u> (e.g. the car's starter motor).

*This is the circuit symbol for a relay.*

## A *Transistor* is another kind of *Switch*

1) A <u>transistor</u> is a complicated device that's usually about the size of a raisin, but a <u>computer chip</u> contains thousands of microscopic ones.

2) The most important thing to know is that they can act as <u>switches</u>.

3) A transistor has <u>3 connections</u>. Once the voltage at the <u>base</u> is high enough, the transistor 'switches on' and allows a current to flow from the <u>collector</u> to the <u>emitter</u>.

COLLECTOR

A 'high' voltage at the base opens the 'gate'...

BASE

EMITTER

*This is the circuit symbol for a transistor.*

...so that current can flow from the collector through to the emitter.

## Learn these *Extra* Circuit Symbols

<u>As well as</u> the circuit symbols in Section One, you need to know these ones here...

| Microphone | Electric bell | Buzzer |

## Learn all the numbered points

Transistors and relays sound pretty complicated, but they're essentially just <u>switches</u>. There are three sections on this page (plus a few circuit symbols) — cover the page and do a mini-essay for each section to see how well you know it all.

# Logic Gates

## Digital Systems are either On or Off

1) Every connection in a digital system is in one of only <u>two states</u>. They can be either ON or OFF, either HIGH or LOW, either YES or NO, either 1 or 0... you get the picture.

2) In reality a 1 is a <u>high voltage</u> (about 5 V) and a 0 is a <u>low voltage</u> (about 0 V). Any part of the system is in one of these two states — nothing is in between.

3) The state of an output can be shown <u>visually</u> by connecting an LED (with a protective resistor) as shown.

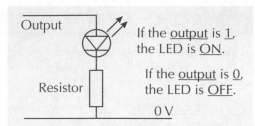

Output

If the <u>output</u> is <u>1</u>, the LED is <u>ON</u>.

If the <u>output</u> is <u>0</u>, the LED is <u>OFF</u>.

Resistor

0 V

## Logic Gates are a type of Digital Processor

Logic gates are small, but they're made up of <u>lots</u> of really small components like <u>transistors</u> and <u>resistors</u>. Each type of logic gate has its own set of <u>rules</u> for converting inputs to outputs, and these rules are best shown in <u>truth tables</u>. The important thing is to list <u>all</u> the possible <u>combinations</u> of input values.

### NOT gate — sometimes called an Inverter

A <u>NOT</u> gate just has <u>one</u> input — and this input can be either <u>1</u> or <u>0</u>, so the truth table has just two rows.

Input        Output

NOT

**NOT**

| Input | Output |
|-------|--------|
| 0 | 1 |
| 1 | 0 |

### AND and OR gates usually have Two Inputs

*Some AND and OR gates have more than two inputs, but you don't have to learn them for GCSE.*

<u>Each input</u> can be 0 or 1, so to allow for <u>all</u> combinations from two inputs, your truth table needs <u>4 rows</u>. There's a certain logic to the names — e.g. an <u>AND</u> gate only gives an output of 1 if both the first input <u>AND</u> the second input are 1. An <u>OR</u> gate just needs either the first <u>OR</u> the second input to be 1.

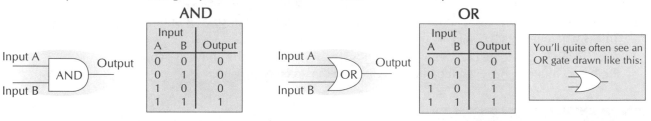

**AND**

Input A

Input B

AND

Output

| Input A | B | Output |
|---------|---|--------|
| 0 | 0 | 0 |
| 0 | 1 | 0 |
| 1 | 0 | 0 |
| 1 | 1 | 1 |

**OR**

Input A

Input B

OR

Output

| Input A | B | Output |
|---------|---|--------|
| 0 | 0 | 0 |
| 0 | 1 | 1 |
| 1 | 0 | 1 |
| 1 | 1 | 1 |

You'll quite often see an OR gate drawn like this:

### NAND and NOR gates have the Opposite Output of AND and OR gates

A <u>NAND</u> gate gives the <u>opposite</u> output to a normal AND gate — e.g. if an AND gate would give an output of 0, a <u>NAND</u> gate would give 1, and vice versa.

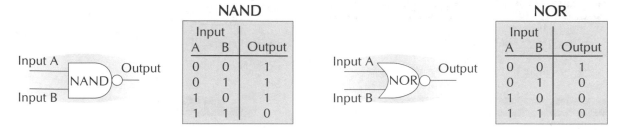

**NAND**

Input A

Input B

NAND

Output

| Input A | B | Output |
|---------|---|--------|
| 0 | 0 | 1 |
| 0 | 1 | 1 |
| 1 | 0 | 1 |
| 1 | 1 | 0 |

**NOR**

Input A

Input B

NOR

Output

| Input A | B | Output |
|---------|---|--------|
| 0 | 0 | 1 |
| 0 | 1 | 0 |
| 1 | 0 | 0 |
| 1 | 1 | 0 |

## Truth tables always come up in GCSE Physics exams

You need to learn the symbol and the truth table for each kind of logic gate. Take it a step at a time.
NOT gates are pretty obvious, and AND and OR aren't so difficult with a bit of thought.
Then for NAND and NOR gates you just need to remember that they're "NOT AND" and "NOT OR".

# Combinations of Logic Gates

You need to be able to construct a <u>truth table</u> for a <u>combination</u> of logic gates.
Approach this kind of thing in an <u>organised</u> way and stick to the rules, and you won't go far wrong.

## *Example — a Greenhouse*

Make sure you understand the following example — a warning system for a <u>greenhouse</u>.
The gardener wants to be warned if it gets <u>too cold</u> or if <u>someone has opened the door</u>.

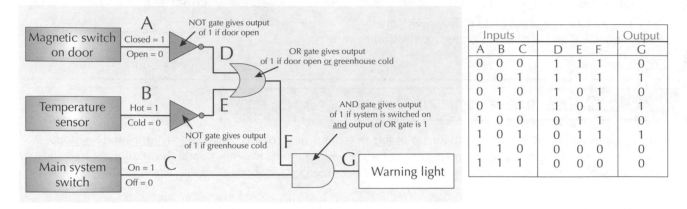

| Inputs | | | | | | Output |
|---|---|---|---|---|---|---|
| A | B | C | D | E | F | G |
| 0 | 0 | 0 | 1 | 1 | 1 | 0 |
| 0 | 0 | 1 | 1 | 1 | 1 | 1 |
| 0 | 1 | 0 | 1 | 0 | 1 | 0 |
| 0 | 1 | 1 | 1 | 0 | 1 | 1 |
| 1 | 0 | 0 | 0 | 1 | 1 | 0 |
| 1 | 0 | 1 | 0 | 1 | 1 | 1 |
| 1 | 1 | 0 | 0 | 0 | 0 | 0 |
| 1 | 1 | 1 | 0 | 0 | 0 | 0 |

The <u>warning light</u> will come on if:   (i)     it is <u>cold</u> in the greenhouse <u>OR</u> if the <u>door</u> is opened,
                                       (ii)       <u>AND</u> the system is switched <u>on</u>.

1) Each connection has a <u>label</u>, and <u>all</u> possible combinations of the inputs are included in the table.
2) What really matters are the <u>inputs</u> and the <u>output</u> — the rest of the truth table is just there to help.

## A *Latch* works like a kind of *Memory* (but is really hard to understand)

1) It's likely that the greenhouse will be too <u>cold</u> in the <u>middle</u> of the night but <u>warm up</u> again by <u>morning</u>.
   This means the warning light will have <u>gone out</u> by the time the gardener gets out of bed.
2) What the gardener needs is some way of getting the warning light to <u>stay on</u> until it is <u>seen</u> and <u>reset</u>.
   This is where the <u>latch</u> comes in.
3) A latch can be made by combining two <u>NOR gates</u> as shown.  In the above system,
   the latch would be <u>between</u> the blue <u>OR</u> gate and the green <u>AND</u> gate.

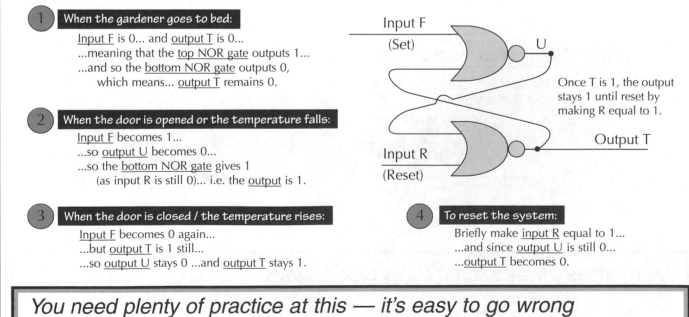

**(1)** **When the gardener goes to bed:**

    <u>Input F</u> is 0... and <u>output T</u> is 0...
    ...meaning that the <u>top NOR gate</u> outputs 1...
    ...and so the <u>bottom NOR gate</u> outputs 0,
       which means... <u>output T</u> remains 0.

**(2)** **When the door is opened or the temperature falls:**

    <u>Input F</u> becomes 1...
    ...so <u>output U</u> becomes 0...
    ...so the <u>bottom NOR gate</u> gives 1
       (as input R is still 0)... i.e. the <u>output</u> is 1.

**(3)** **When the door is closed / the temperature rises:**

    <u>Input F</u> becomes 0 again...
    ...but <u>output T</u> is 1 still...
    ...so <u>output U</u> stays 0 ...and <u>output T</u> stays 1.

**(4)** **To reset the system:**

    Briefly make <u>input R</u> equal to 1...
    ...and since <u>output U</u> is still 0...
    ...<u>output T</u> becomes 0.

## *You need plenty of practice at this — it's easy to go wrong*

<u>Don't rush</u> — draw a truth table, listing all the inputs and outputs, and fill it in ultra-carefully, <u>one bit at a time</u>.

# Warm-Up and Worked Exam Question

## Warm-up Questions

1)

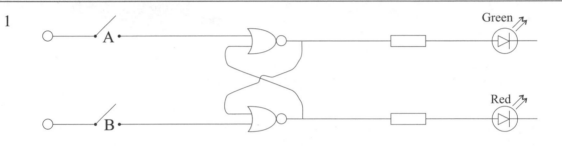

From the list of devices whose symbols are shown above,
write down the names of two of them which are input sensors.

2) Write out the truth tables for an OR gate and a NOR gate.

3) Which logic gate is this the truth table for? ⎯⎯

4) Draw the symbol for a NAND gate.

| A | B | Output |
|---|---|--------|
| 0 | 0 | 0 |
| 0 | 1 | 0 |
| 1 | 0 | 0 |
| 1 | 1 | 1 |

## Worked Exam Question

1

Green

Red

This diagram shows some wiring for a model railway. When a model train comes
onto a certain stretch of track it passes over Switch A, closing it for an instant, and
then a few minutes later passes over Switch B, closing that for an instant.

a) What type of logic gates are used to make this circuit?

*NOR gates*

*[1 mark]*

b) What type of devices are represented by the symbols marked "Green" and "Red"?

*Light Emitting Diodes (LEDs)*

*[1 mark]*

c) Given that initially the green LED is on, what effect does it have when the train
passes over Switch A?

*The green LED goes off and the red LED comes on.*

*[1 mark]*

In the example on page 242, the output U is the green LED and the output T is the red LED.
This is really tricky stuff — probably best to leave questions like this till the end of your exam
so that you don't waste all your time figuring out just one question.

d) What happens when the train then passes over Switch B?

*The red LED goes off and the green LED comes on.*

*[1 mark]*

# Exam Questions

1) a) Mark in the base **B**, collector **C** and emitter **E** of the transistor in the diagram below.

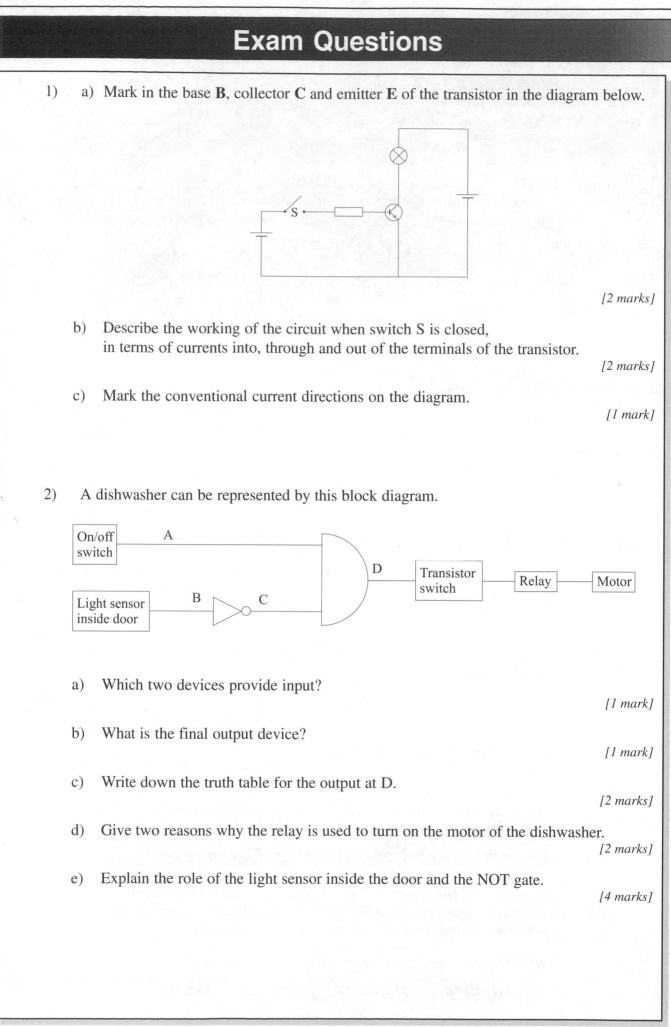

*[2 marks]*

b) Describe the working of the circuit when switch S is closed,
in terms of currents into, through and out of the terminals of the transistor.

*[2 marks]*

c) Mark the conventional current directions on the diagram.

*[1 mark]*

2) A dishwasher can be represented by this block diagram.

a) Which two devices provide input?

*[1 mark]*

b) What is the final output device?

*[1 mark]*

c) Write down the truth table for the output at D.

*[2 marks]*

d) Give two reasons why the relay is used to turn on the motor of the dishwasher.

*[2 marks]*

e) Explain the role of the light sensor inside the door and the NOT gate.

*[4 marks]*

# Potential Dividers

<u>Potential dividers</u> consist of a <u>pair</u> of resistors. One of them usually has a <u>fixed</u> resistance, and the other's resistance <u>varies</u> with temperature, light intensity, etc.

## The *Higher the Resistance*, *the Greater the Voltage Drop*

A voltage across a pair of resistors is 'shared out' according to their <u>relative resistances</u>. The rule is:

### The <u>larger</u> the share of the <u>total resistance</u>, the larger the share of the <u>total voltage</u>.

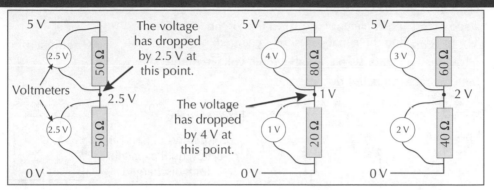

The <u>resistances</u> are <u>equal</u>, so each resistor takes <u>half</u> the <u>voltage</u>.

The top resistor has <u>80%</u> of the total <u>resistance</u>, and so takes <u>80%</u> of the total <u>voltage</u>.

The top resistor has <u>60%</u> of the total <u>resistance</u>, and so takes <u>60%</u> of the total <u>voltage</u>.

## *Potential Dividers* are quite *Useful*

<u>Potential dividers</u> are not only spectacularly interesting — they're <u>useful</u> as well. But first there's an <u>equation</u> you need to learn:

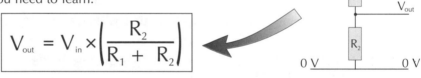

$$V_{out} = V_{in} \times \left( \frac{R_2}{R_1 + R_2} \right)$$

1) Using a <u>thermistor</u> and a <u>variable resistor</u> in a potential divider, you can make a <u>temperature sensor</u> that triggers an output device at a temperature <u>you choose</u>.

2) The sensor on page 242 gave an <u>output of 1</u> when it was hot and an <u>output of 0</u> when it was cold (the <u>output</u> is the point <u>between</u> the two resistors). This is how it works...

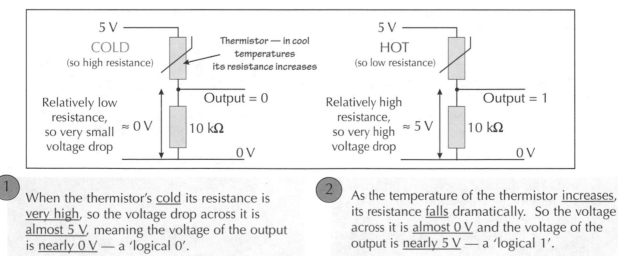

1 When the thermistor's <u>cold</u> its resistance is <u>very high</u>, so the voltage drop across it is <u>almost 5 V</u>, meaning the voltage of the output is <u>nearly 0 V</u> — a 'logical 0'.

2 As the temperature of the thermistor <u>increases</u>, its resistance <u>falls</u> dramatically. So the voltage across it is <u>almost 0 V</u> and the voltage of the output is <u>nearly 5 V</u> — a 'logical 1'.

## *Once again, practice is essential here*

You can change the circuit to make other kinds of sensors too, e.g. you could swap the resistor and the thermistor around in the last diagram above, to get a '<u>cold</u>' sensor that gives a 1 when it's cold and a 0 when it's hot.

# Capacitors

A <u>capacitor</u> is a device that stores charge. The <u>bigger the capacitance</u> of a capacitor, the <u>more charge it can store</u>.

## The **More Charge Stored**, the **Greater the Voltage Drop**

1) You <u>charge</u> a capacitor by connecting it to a source of voltage, e.g. a battery.
2) A <u>current</u> flows around the circuit, and <u>charge</u> gets <u>stored</u> on the capacitor.
3) The <u>more charge</u> that's stored on a capacitor, the <u>larger the</u> <u>potential difference</u> (or voltage) across it.
4) When the voltage across the capacitor is <u>equal</u> to that of the <u>battery</u>, the <u>current stops</u> and the capacitor is <u>fully charged</u>.
5) The voltage across the capacitor <u>won't rise above</u> the voltage of the battery.
6) Without the batteries, the capacitor <u>discharges</u>.

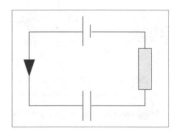

Circuit symbol

<u>Charging</u>

Current flows until capacitor is fully charged

Voltage across capacitor rises as capacitor charges

<u>Discharging</u>

Current flows in opposite direction until capacitor is fully discharged

Voltage falls as capacitor discharges

## The **Charging Time** depends on the **Capacitor** and the **Circuit**

It's possible to <u>control</u> how long it takes for a capacitor to become <u>fully charged</u> (or <u>discharged</u>). There are <u>two</u> important factors — the <u>capacitance</u> of the capacitor and the <u>resistance</u> in the circuit:

1) The <u>greater</u> the <u>capacitance</u>, the more charge it takes to charge it up and so the <u>longer</u> it will take.
2) The <u>larger</u> the <u>resistance</u> of the charging circuit, the <u>lower</u> the <u>current</u> that flows and so the <u>longer</u> it will take to charge up the capacitor (remember — charge = current × time).
3) The rules for capacitor <u>discharge</u> are the same. The <u>larger</u> the <u>capacitance</u> and the <u>resistance</u>, the <u>longer</u> it will take.

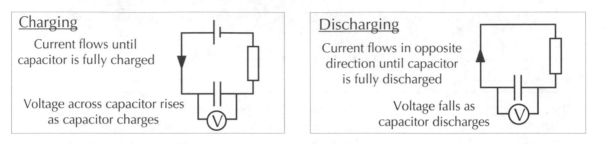

## **Capacitors** are used to cause a **Time Delay**

Capacitors are used in <u>timing circuits</u>, and in input sensors that need a <u>delay</u>. Like on a <u>camera</u> when you want to press the button, and then run round and get in the shot before the picture's taken:

1) The switch is <u>closed</u>. Initially, the capacitor has <u>no charge</u> stored, and so the voltage drop across it is small. This means the <u>voltage drop</u> across the <u>resistor</u> must be <u>big</u>. (And this all means the <u>output voltage</u> will be low.)
2) As the capacitor <u>charges</u>, the <u>voltage drop</u> across it <u>increases</u> (and so the voltage drop across the resistor <u>falls</u>). This all means the voltage at the output <u>increases</u>.
3) The <u>shutter</u> on the camera will <u>open</u> (i.e. the picture will be taken) when the input is <u>close to 5 V</u>.

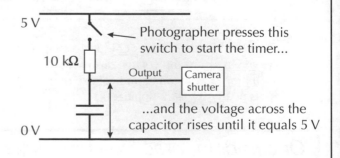

5 V

Photographer presses this switch to start the timer...

10 kΩ

Output  Camera shutter

...and the voltage across the capacitor rises until it equals 5 V

0 V

# More Advanced Circuits

Put the stuff on the last few pages together and you get this circuit. It's an <u>examiner's favourite</u>. So <u>learn</u> it well.

## Learn this circuit — a **Floodlight** that comes on when it's **Dark**

This circuit acts as a <u>light sensor</u> that turns on a <u>floodlight</u> when it gets <u>dark</u>.

This is what happens when it <u>gets dark</u>...

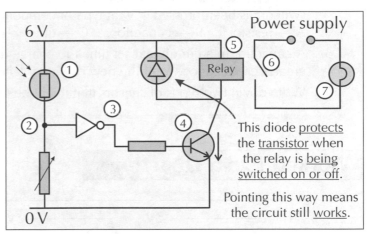

**1** ...the <u>resistance</u> of the LDR becomes <u>high</u>, and so the <u>voltage drop</u> across it <u>increases</u>...

**2** ...so the <u>voltage</u> here becomes close to <u>0 V</u> (a <u>logical 0</u>)...

**3** ...which is converted to a <u>logical 1</u>.

**4** The <u>high input</u> into the <u>transistor</u>...

**5** ...allows <u>current</u> to flow through the <u>relay</u>...

**6** ...which switches on the <u>current</u> in the <u>high-current</u> circuit...

**7** ...which turns on the <u>output</u> device (here, a <u>floodlight</u>).

This diode <u>protects</u> the <u>transistor</u> when the relay is <u>being switched on or off</u>.

Pointing this way means the circuit still <u>works</u>.

## Make sure you know the **Effects** of making **Changes**

Exam questions often ask how the circuit above can be <u>changed</u>, or what the <u>effect</u> of a change might be.

1) The <u>variable resistor</u> can be used to adjust the light level that will trigger the circuit. If its resistance is <u>increased</u>, the light level will have to <u>drop more</u> before the floodlight is switched on.

2) The <u>output device</u> could be anything — a light, a buzzer, a motor, a heater, a lock, etc.

3) If the <u>variable resistor</u> and the <u>LDR</u> are <u>swapped around</u>, then the system will switch on the output device when the light intensity is <u>high</u>.

4) If the LDR is replaced with a <u>thermistor</u>, then the output device will be switched on when it gets <u>cold</u>.

## Learn these other **Applications** of **Electronics**

1) In modern <u>cars</u> there's an array of <u>sensors</u> to improve <u>safety</u> and <u>efficiency</u> — e.g. sensors to <u>prevent skidding</u> when braking, sensors to detect a crash and set off <u>air bags</u>, and sensors to <u>monitor</u> the engine temperature and adjust fuel injection.

2) The development of <u>mobile phone</u> technology has proved very useful for businesses and the general public. They do have potential disadvantages though — there may be <u>health hazards</u> from the radiation given off by the <u>handsets</u> and the <u>masts</u>, and (in theory) you could be <u>tracked</u> by using the signal from your mobile — which is a potential invasion of privacy.

3) The <u>Internet</u> has been at the centre of an information revolution — you can find out about almost anything by pressing a few buttons. It's provided new ways of <u>learning</u> — however it does have the drawback of allowing <u>unsuitable material</u> to be viewed by children.

## Learn the seven steps of the floodlight circuit

The circuit at the top of the page is fairly difficult, but if you can understand that, you should be able to cope with whatever else comes up in the Exam.

# Warm-Up and Worked Exam Questions

## Warm-up Questions

1) In the circuit on the right, if $R_1$ = 140 ohms and $R_2$ = 60 ohms, what is the voltage across $R_2$?

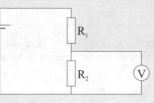

2) When you turn off some radios you will hear the sound continue for a short time after the switch has been pressed. What type of component does this suggest is present in the radio?
A — resistor     B — capacitor     C — timer     D — microphone     E — transistor

3) A capacitor in a circuit takes a time t to charge up. It is then replaced with a capacitor of a greater capacitance. Will the new capacitor take a longer or shorter time to charge up?

4) Write down two types of sensors that might be used in a car.

## Worked Exam Questions

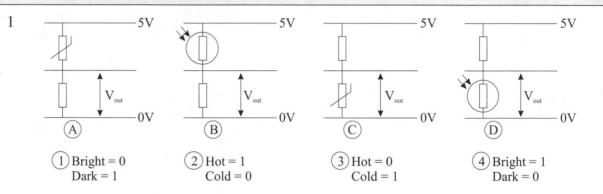

1

① Bright = 0
   Dark = 1

② Hot = 1
   Cold = 0

③ Hot = 0
   Cold = 1

④ Bright = 1
   Dark = 0

a) Correctly match each of the diagrams A to D with the four numbered statements of logic output.

*A matches 2;  B matches 4;  C matches 3;  D matches 1.*

[3 marks]

b) Suggest a practical use for the circuit that could be made by connecting an indicator lamp across the output of diagram A above.

*Any reasonable use where a light indicates an increase in temperature, such as a warning light to indicate that a freezer is beginning to defrost.*

[1 mark]

2  a) Explain in terms of flow of charge and changes to voltages what happens as a battery is connected across a capacitor.

*Current flows round the circuit, depositing charge on the capacitor and increasing the voltage across it.*

[2 marks]

b) When does the capacitor stop charging?

*When the voltage across the capacitor equals the voltage of the battery.*

[2 marks]

# Exam Questions

1) This circuit contains a light dependent resistor (LDR) and a variable resistor set to 9000 Ω.

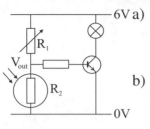

a) In the circuit on the left, what is the output voltage ($V_{out}$) when the LDR has its maximum resistance of 8 kΩ?

*[2 marks]*

b) What is $V_{out}$ when the LDR has its minimum resistance of 200 Ω?

*[2 marks]*

c) A new component is now inserted in the circuit as shown below:

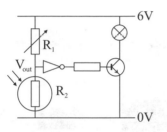

Name this new component, and compare how the circuit responds to changes picked up by the input sensor before and after the new component was added. *[3 marks]*

d) The indicator light is now replaced with a bell, as shown below:

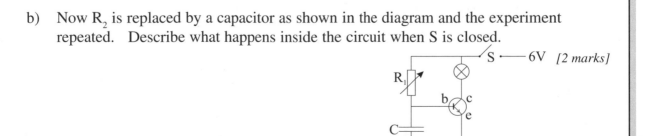

Why is there a need for a relay, and what does the diode do?

*[2 marks]*

2) a) Explain what happens to the circuit on the right when Switch S is closed.

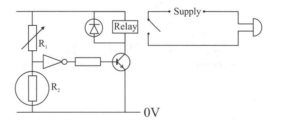

*[2 marks]*

b) Now $R_2$ is replaced by a capacitor as shown in the diagram and the experiment repeated. Describe what happens inside the circuit when S is closed.

*[2 marks]*

c) Which component of the circuit can be adjusted to alter the time the capacitor takes to charge and discharge? *[1 mark]*

# Springs

## Stretching a Steel Spring — Elastic then Plastic

The greater the stretching force on a spring, the greater the <u>extension</u> (i.e. the <u>longer</u> it gets).

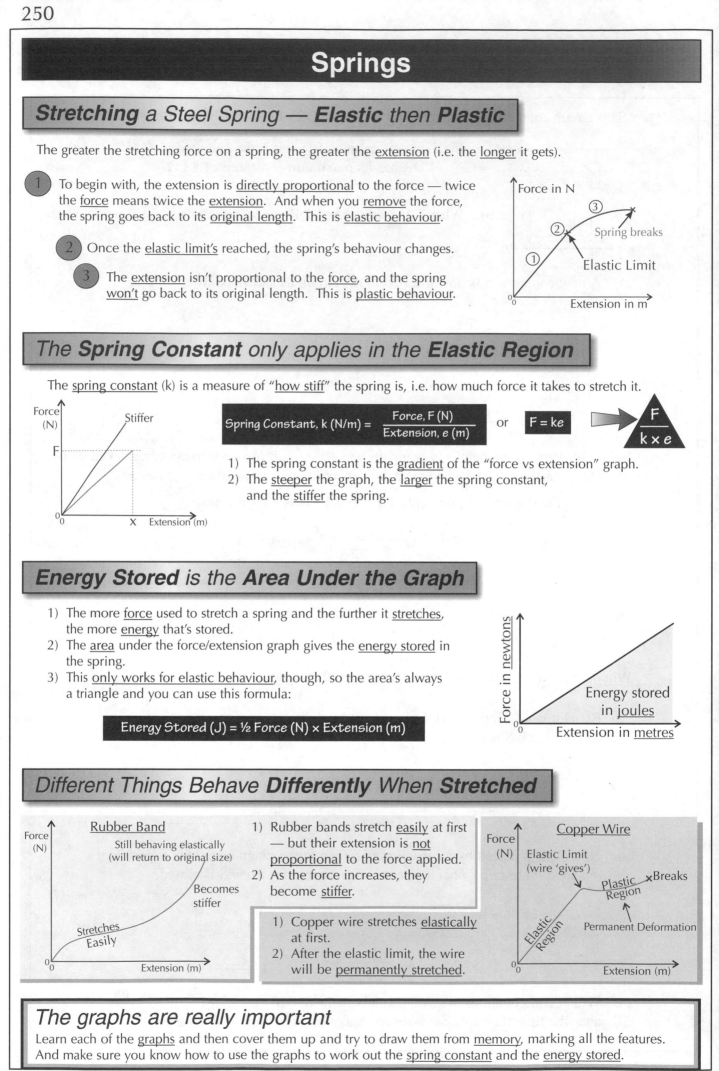

**1** To begin with, the extension is <u>directly proportional</u> to the force — twice the <u>force</u> means twice the <u>extension</u>. And when you <u>remove</u> the force, the spring goes back to its <u>original length</u>. This is <u>elastic behaviour</u>.

**2** Once the <u>elastic limit's</u> reached, the spring's behaviour changes.

**3** The <u>extension</u> isn't proportional to the <u>force</u>, and the spring <u>won't</u> go back to its original length. This is <u>plastic behaviour</u>.

## The Spring Constant only applies in the Elastic Region

The <u>spring constant</u> (k) is a measure of "<u>how stiff</u>" the spring is, i.e. how much force it takes to stretch it.

$$\text{Spring Constant, } k \text{ (N/m)} = \frac{\text{Force, } F \text{ (N)}}{\text{Extension, } e \text{ (m)}} \quad \text{or} \quad F = ke$$

$$\frac{F}{k \times e}$$

1) The spring constant is the <u>gradient</u> of the "force vs extension" graph.
2) The <u>steeper</u> the graph, the <u>larger</u> the spring constant, and the <u>stiffer</u> the spring.

## Energy Stored is the Area Under the Graph

1) The more <u>force</u> used to stretch a spring and the further it <u>stretches</u>, the more <u>energy</u> that's stored.
2) The <u>area</u> under the force/extension graph gives the <u>energy stored</u> in the spring.
3) This <u>only works for elastic behaviour</u>, though, so the area's always a triangle and you can use this formula:

$$\text{Energy Stored (J)} = \tfrac{1}{2} \text{ Force (N)} \times \text{Extension (m)}$$

## Different Things Behave Differently When Stretched

### Rubber Band

1) Rubber bands stretch <u>easily</u> at first — but their extension is <u>not</u> proportional to the force applied.
2) As the force increases, they become <u>stiffer</u>.

### Copper Wire

1) Copper wire stretches <u>elastically</u> at first.
2) After the elastic limit, the wire will be <u>permanently stretched</u>.

## The graphs are really important

Learn each of the <u>graphs</u> and then cover them up and try to draw them from <u>memory</u>, marking all the features. And make sure you know how to use the graphs to work out the <u>spring constant</u> and the <u>energy stored</u>.

# Centre of Mass

## *Density* means How Much *Mass* there is in a Certain *Volume*

$$\text{Density } (\rho) \text{ (kg/m}^3) = \frac{\text{Mass (kg)}}{\text{Volume (m}^3)}$$

*Greek letter 'rho'.*

$$\frac{m}{\rho \times V}$$

Lead is <u>denser</u> than foam, so <u>1 m³</u> of lead <u>weighs more</u> than 1 m³ of foam. And <u>1 tonne</u> of lead takes up <u>less space</u> than 1 tonne of foam.

You don't need to learn the rest of this page if you're doing the <u>Edexcel</u> syllabus.

## The *Centre of Mass* hangs *Directly Below* the *Point of Suspension*

**1** You can think of the <u>centre of mass</u> of an object as the point at which the <u>whole</u> mass is concentrated.

**2** A freely suspended object will <u>swing</u> until its centre of mass is <u>vertically below</u> the <u>point of suspension</u>.

**3** This means you can find the <u>centre of mass</u> of any flat shape like this:

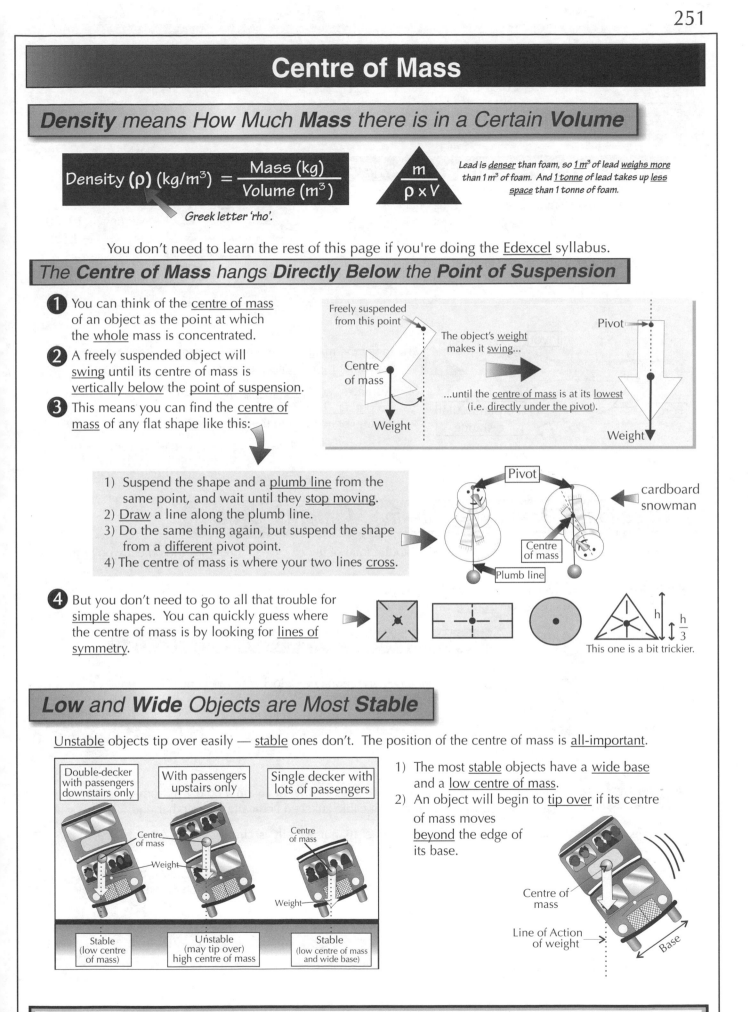

Freely suspended from this point

Centre of mass

Weight

The object's <u>weight</u> makes it <u>swing</u>...

...until the <u>centre of mass</u> is at its <u>lowest</u> (i.e. <u>directly under the pivot</u>).

Pivot

Weight

1) Suspend the shape and a <u>plumb line</u> from the same point, and wait until they <u>stop moving</u>.
2) <u>Draw</u> a line along the plumb line.
3) Do the same thing again, but suspend the shape from a <u>different</u> pivot point.
4) The centre of mass is where your two lines <u>cross</u>.

Pivot

cardboard snowman

Centre of mass

Plumb line

**4** But you don't need to go to all that trouble for <u>simple</u> shapes. You can quickly guess where the centre of mass is by looking for <u>lines of symmetry</u>.

h

$\frac{h}{3}$

This one is a bit trickier.

## *Low* and *Wide* Objects are Most *Stable*

<u>Unstable</u> objects tip over easily — <u>stable</u> ones don't. The position of the centre of mass is <u>all-important</u>.

Double-decker with passengers downstairs only

With passengers upstairs only

Single decker with lots of passengers

Centre of mass

Weight

Centre of mass

Weight

Stable (low centre of mass)

Unstable (may tip over) high centre of mass

Stable (low centre of mass and wide base)

1) The most <u>stable</u> objects have a <u>wide base</u> and a <u>low centre of mass</u>.
2) An object will begin to <u>tip over</u> if its centre of mass moves <u>beyond</u> the edge of its base.

Centre of mass

Line of Action of weight

Base

## *The lower the centre of gravity, the more stable things are*

Make sure you know the equation for <u>density</u>. You also need to be able to describe how to find the <u>centre of mass</u> of any shape. Shut the book and write out the method step by step.

252

# Moments

## A *Moment* is the *Turning Effect* of a Force

MOMENT (Nm) = FORCE (N) × perpendicular DISTANCE (m) between line of action and pivot

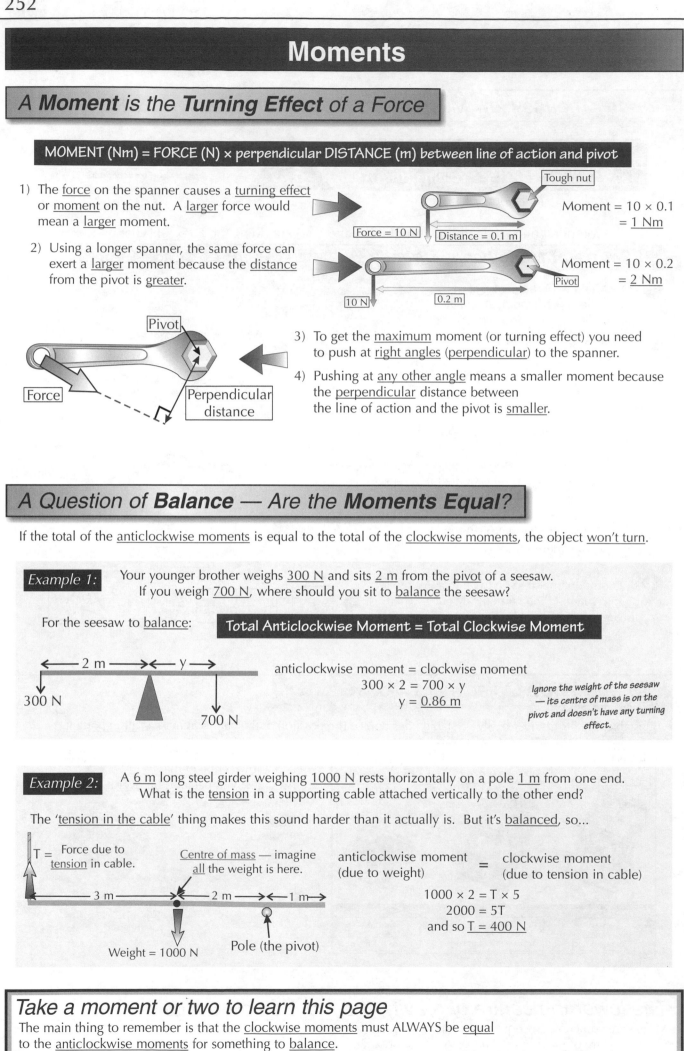

1) The force on the spanner causes a turning effect or moment on the nut. A larger force would mean a larger moment.

2) Using a longer spanner, the same force can exert a larger moment because the distance from the pivot is greater.

Tough nut

Force = 10 N   Distance = 0.1 m

Moment = 10 × 0.1 = 1 Nm

Pivot   10 N   0.2 m

Moment = 10 × 0.2 = 2 Nm

Pivot

Force   Perpendicular distance

3) To get the maximum moment (or turning effect) you need to push at right angles (perpendicular) to the spanner.

4) Pushing at any other angle means a smaller moment because the perpendicular distance between the line of action and the pivot is smaller.

## A Question of *Balance* — Are the *Moments Equal*?

If the total of the anticlockwise moments is equal to the total of the clockwise moments, the object won't turn.

Example 1: Your younger brother weighs 300 N and sits 2 m from the pivot of a seesaw. If you weigh 700 N, where should you sit to balance the seesaw?

For the seesaw to balance:   **Total Anticlockwise Moment = Total Clockwise Moment**

2 m   y

300 N   700 N

anticlockwise moment = clockwise moment
$300 \times 2 = 700 \times y$
$y = 0.86$ m

*Ignore the weight of the seesaw — its centre of mass is on the pivot and doesn't have any turning effect.*

Example 2: A 6 m long steel girder weighing 1000 N rests horizontally on a pole 1 m from one end. What is the tension in a supporting cable attached vertically to the other end?

The 'tension in the cable' thing makes this sound harder than it actually is. But it's balanced, so...

T = Force due to tension in cable.

Centre of mass — imagine all the weight is here.

3 m   2 m   1 m

Weight = 1000 N   Pole (the pivot)

anticlockwise moment (due to weight) = clockwise moment (due to tension in cable)
$1000 \times 2 = T \times 5$
$2000 = 5T$
and so T = 400 N

## Take a moment or two to learn this page

The main thing to remember is that the clockwise moments must ALWAYS be equal to the anticlockwise moments for something to balance.

SEPARATE SCIENCES — ADDITIONAL MATERIAL

# Warm-Up and Worked Exam Question

You could read through this page in a few minutes but there's no point unless you check over any bits you don't know and make sure you understand everything. It's not quick but it's the only way.

## Warm-up Questions

1) A copper wire, a rubber band and a steel spring are all stretched, and the force vs extension graph plotted for each. Which graph(s) will show a straight line up to a certain limit?

2) A plant pot has a mass of 30g. When immersed in water it displaced 12cm³ of water. What is its density in g/cm³?

3) The beam on the right is balanced. What is weight X in newtons?

4) A wellington boot is hung up to dry from a single peg. Where must the centre of mass of the boot be?

   A — the point where the boot is in contact with the peg.

   B — the point on the boot nearest to the ground.

   C — a point directly below the peg.

   D — the tip of the toe of the boot.

0.3m, 0.6m, Pivot, 8N, X

## Worked Exam Question

This worked example is exactly like you'll get in the real exam, except here it's got the answers written in for you already to show you what you should be writing — pretty handy.

1  a)  The "ABC" brand of spring balance is used to weigh objects up to 10N. It has a spring constant, k, of 90 N/m. What force is needed to extend it by 0.06m?

$F = ke = 90 \times 0.06 = 5.4\,N.$

*[1 mark]*

b)  A rival brand, "XYZ", extends 0.03m under a force of 2.5N. Find k for the XYZ balance and state which type of spring is stiffer — the ABC or XYZ.

$k = F \div e = 2.5 \div 0.03 = 83.3\,N/m.$

The ABC is stiffer because it has a higher value of k.

*[2 marks]*

c)  One pupil disobeyed her teacher and suspended a weight of much more than 10N from the ABC spring balance. Explain in terms of elastic and plastic behaviour what this is likely to do to the spring balance.

If it is stretched with more than a certain amount of force the spring may stretch beyond its elastic limit, and its behaviour will change permanently. Its extension will no longer be directly proportional to the force on it, and it won't spring back as far if the force is removed.

*[2 marks]*

# Exam Questions

1) A steel spring is stretched with various different forces. The table below shows the results:

| Force (N) | 0 | 1 | 2 | 3 | 4 | 5 | 6 |
|---|---|---|---|---|---|---|---|
| Length (mm) | 62 | 67 | 72 | 77 | 82 | 91 | 105 |
| Extension (mm) | | | | | | | |

   a) What is the length of the spring when it is not being stretched at all?

   *[1 mark]*

   b) Fill in the values for the next line of the table.

   *[2 marks]*

   c) Plot a graph of force in newtons against extension in millimetres on the grid on the right.

   *[2 marks]*

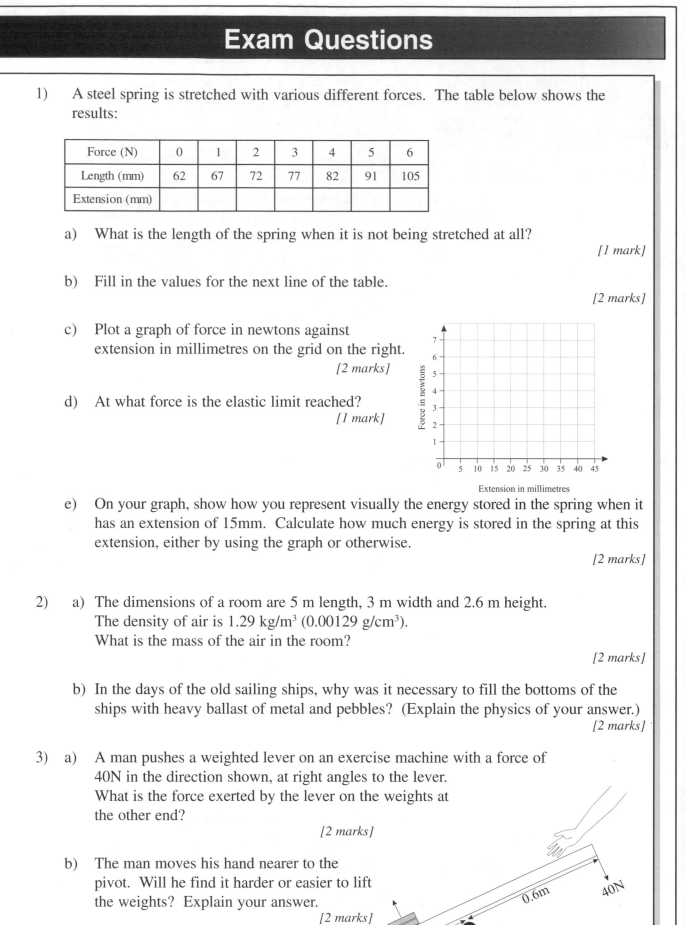

   d) At what force is the elastic limit reached?

   *[1 mark]*

   e) On your graph, show how you represent visually the energy stored in the spring when it has an extension of 15mm. Calculate how much energy is stored in the spring at this extension, either by using the graph or otherwise.

   *[2 marks]*

2) a) The dimensions of a room are 5 m length, 3 m width and 2.6 m height.
   The density of air is 1.29 kg/m$^3$ (0.00129 g/cm$^3$).
   What is the mass of the air in the room?

   *[2 marks]*

   b) In the days of the old sailing ships, why was it necessary to fill the bottoms of the ships with heavy ballast of metal and pebbles? (Explain the physics of your answer.)

   *[2 marks]*

3) a) A man pushes a weighted lever on an exercise machine with a force of 40N in the direction shown, at right angles to the lever.
   What is the force exerted by the lever on the weights at the other end?

   *[2 marks]*

   b) The man moves his hand nearer to the pivot. Will he find it harder or easier to lift the weights? Explain your answer.

   *[2 marks]*

# Momentum

## Momentum = Mass × Velocity

1. The greater the mass of an object and the greater its velocity, the more momentum the object has.

2. Momentum is a vector quantity — it has size and direction (like velocity, but not speed).

$$\frac{\text{momentum}}{\text{mass} \times \text{velocity}}$$

**Momentum (kg m/s) = Mass (kg) × Velocity (m/s)**

## Momentum *Before* = Momentum *After*

Momentum is conserved when no external forces act, i.e. the total momentum after is the same as it was before.

*Example 1:*

Two skaters approach each other, collide and move off together as shown. At what velocity do they move after the collision?

2 m/s — Ed — 80 kg — *Before*
1.5 m/s — Sue — 60 kg
Velocity (v)=? — B — (80+60) kg — *After*

1) Choose which direction is positive. Choose "positive" as "to the right".
2) Total momentum before collision
   = momentum of Ed + momentum of Sue
   = {80 × 2} + {60 × (−1.5)} = 70 kg m/s
3) Total momentum after collision
   = momentum of Ed and Sue together
   = 140 × v
4) So 140v = 70, i.e. v = 0.5 m/s to the right

*Example 2:*

A gun fires a bullet as shown. At what speed does the gun move backwards?

Velocity (v) = ? — 5 kg — 150 m/s — 0.1 kg — *After*

1) Choose which direction is positive. Again, choose "positive" as "to the right".
2) Total momentum before firing
   = 0 kg m/s
3) Total momentum after firing
   = momentum of bullet + momentum of gun
   = (0.1 × 150) + (5 × v)
   = 15 + 5v
4) So 15 + 5v = 0, i.e. v = -3 m/s
   So the gun moves backwards at 3 m/s.

*This is the gun's recoil.*

Rockets work in much the same way — they chuck a load of exhaust gases out backwards, and since momentum is conserved, the rocket moves forwards.

## *Forces* Cause *Changes* in *Momentum*

1) When a force acts on an object, it causes a change in momentum.

2) A larger force means a faster change of momentum (and so a greater acceleration).

$$\text{Force acting (N)} = \frac{\text{Change in Momentum (kg m/s)}}{\text{Time taken for change to happen (s)}}$$

3) Likewise, if someone's momentum changes very quickly (like in a car crash), the forces on the body will be very large (and more likely to cause injury).

4) This is why cars are designed to slow people down over a longer time when they have a crash — the longer it takes for a change in momentum, the smaller the force.

Crumple Zones crumple on impact, increasing the time taken for the car to stop.

Seat Belts stretch slightly, increasing the time taken for the wearer to stop. This reduces the forces acting on the chest.

Air Bags also slow you down more slowly.

## *Momentum is always conserved — but Kinetic Energy usually isn't...*

Momentum is always conserved in collisions and explosions (when no other forces act), but kinetic energy might not be — in fact, the kinetic energy is usually a bit less after a collision. This is because most collisions are inelastic. (If the energy is conserved, it's called an elastic collision.)

# Equations of Motion and Projectiles

## You need to know these Four Equations of Motion

Which of these equations you need to use depends on what you already know, and what you need to find out. But that means you have to know all 4 equations — preferably like the back of your hand.

Altogether, there are 5 things involved in these equations:
u = initial velocity,   v = final velocity,   s = distance,   t = time,   a = acceleration.

**1** $$s = \frac{(u+v)t}{2}$$     **2** $$v = u + at$$     **3** $$s = ut + \frac{1}{2}at^2$$     **4** $$v^2 = u^2 + 2as$$

If you know three things, you can find out either of the other two — if you use the right equation, that is.
And if you use this method twice, you can find out both things you don't know.

How to choose the right equation:
1) Write down which three things you already know.
2) Write down which of the other things you want to find out.
3) Choose the equation that involves all of the things you've written down.
4) Substitute in the numbers, and do the maths.

*Direction is important for velocity, acceleration and distance — so always choose which direction is positive, and stick with it.*

**Example:** *A car going at 10 m/s accelerates at 2 m/s² for 8 s. How far does the car go while accelerating?*

Firstly, choose the "positive" direction — the best option looks like "to the right".
1) You already know u (= 10 m/s), a (= 2 m/s²) and t (= 8 s).
2) You want to find out s.
3) So you need the equation with all these in: u, a, t and s — that's number 3: $s = ut + \frac{1}{2}at^2$.
4) Now substitute in the numbers: $s = (10 \times 8) + \frac{1}{2}(2 \times 8^2) = 80 + 64 = \underline{144 \text{ m}}$

## Projectiles — Deal with Horizontal and Vertical Motion Separately

This is only for OCR students.
It's all about things moving through the air, where the only force acting on them is gravity...

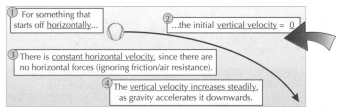
① For something that starts off horizontally...
② ...the initial vertical velocity = 0
③ There is constant horizontal velocity, since there are no horizontal forces (ignoring friction/air resistance).
④ The vertical velocity increases steadily, as gravity accelerates it downwards.

1) Motion can be split into two separate bits — the horizontal bit and the vertical bit.
2) These bits are totally separate — one doesn't affect the other.
3) So gravity (which only acts downwards) doesn't affect horizontal motion at all.

**Example:** *A football's kicked horizontally from a 20 m high wall. How long is it before it lands? Take g = 10 m/s².*

It lands when it's travelled 20 m vertically.
Using $s = ut + \frac{1}{2}at^2$, where u = 0, a = 10 m/s², s = 20 m:

$$20 = (0 \times t) + \frac{1}{2}at^2 = \frac{10t^2}{2}, \text{ i.e. } \underline{t = 2 \text{ s}} \text{ when it lands}$$

*If its horizontal velocity is originally 5 m/s, how far has it travelled when it lands?*    From above.

Using "distance = speed × time", where v = 5 m/s and t = 2 s:
$$s = 5 \times 2 = \underline{10 \text{ m}}.$$

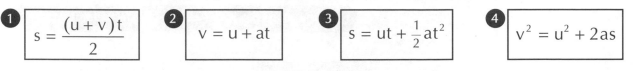
Horizontal velocity | Path of ball | Horizontal velocity constant
Vertical velocity increasing

## Remember to separate horizontal and vertical motion

If you follow these step-by-step instructions to solve motion problems, you really can't possibly go wrong.

# Circular Motion

## Circular Motion — *Velocity* is *Constantly Changing*

1) If an object is travelling at a constant speed in a circle it is constantly changing direction, which means it's accelerating.

2) This means there must be a force acting on it.

3) A force that keeps something moving in a circle is called a centripetal force — it's directed towards the centre of the circle.

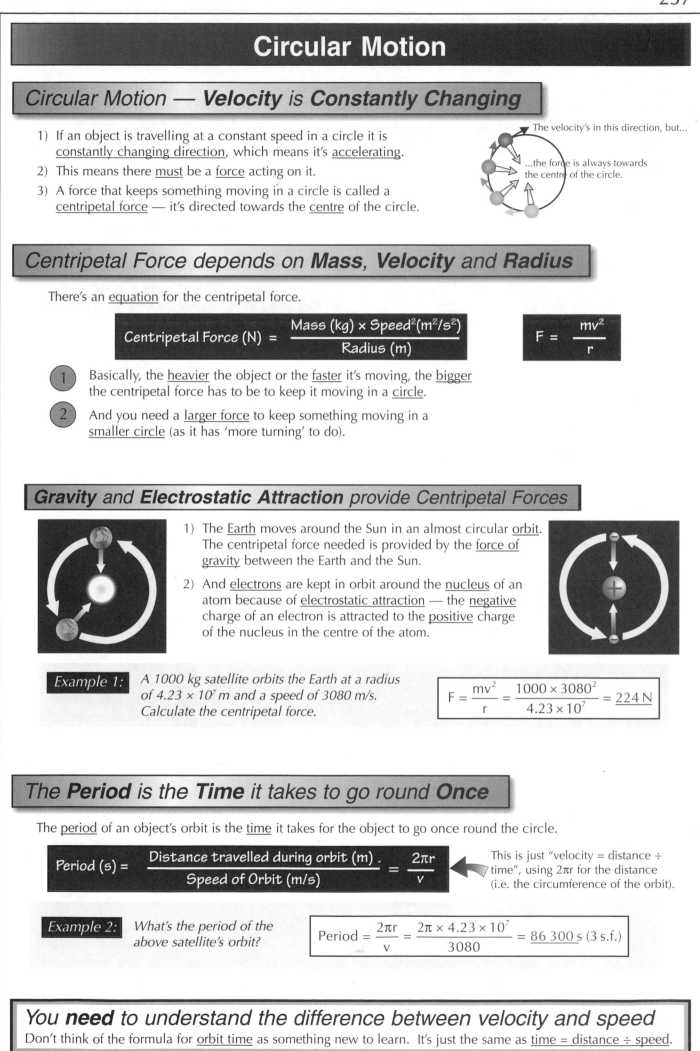

The velocity's in this direction, but...

...the force is always towards the centre of the circle.

## Centripetal Force depends on *Mass*, *Velocity* and *Radius*

There's an equation for the centripetal force.

$$\text{Centripetal Force (N)} = \frac{\text{Mass (kg)} \times \text{Speed}^2 (\text{m}^2/\text{s}^2)}{\text{Radius (m)}}$$

$$F = \frac{mv^2}{r}$$

1 Basically, the heavier the object or the faster it's moving, the bigger the centripetal force has to be to keep it moving in a circle.

2 And you need a larger force to keep something moving in a smaller circle (as it has 'more turning' to do).

## *Gravity* and *Electrostatic Attraction* provide Centripetal Forces

1) The Earth moves around the Sun in an almost circular orbit. The centripetal force needed is provided by the force of gravity between the Earth and the Sun.

2) And electrons are kept in orbit around the nucleus of an atom because of electrostatic attraction — the negative charge of an electron is attracted to the positive charge of the nucleus in the centre of the atom.

**Example 1:** A 1000 kg satellite orbits the Earth at a radius of $4.23 \times 10^7$ m and a speed of 3080 m/s. Calculate the centripetal force.

$$F = \frac{mv^2}{r} = \frac{1000 \times 3080^2}{4.23 \times 10^7} = \underline{224\,\text{N}}$$

## The *Period* is the *Time* it takes to go round *Once*

The period of an object's orbit is the time it takes for the object to go once round the circle.

$$\text{Period (s)} = \frac{\text{Distance travelled during orbit (m)}}{\text{Speed of Orbit (m/s)}} = \frac{2\pi r}{v}$$

This is just "velocity = distance ÷ time", using $2\pi r$ for the distance (i.e. the circumference of the orbit).

**Example 2:** What's the period of the above satellite's orbit?

$$\text{Period} = \frac{2\pi r}{v} = \frac{2\pi \times 4.23 \times 10^7}{3080} = \underline{86\,300\,\text{s}}\ \text{(3 s.f.)}$$

## You **need** to understand the difference between velocity and speed

Don't think of the formula for orbit time as something new to learn. It's just the same as time = distance ÷ speed.

# Specific Heat Capacity and Resistor Colour Codes

There are a couple of other things you might need to know about, depending on which syllabus you're doing. For example, you need to know about specific heat capacity if you're doing the OCR Physics syllabus:

## Use Specific Heat Capacity to calculate Energy Transferred

This equation belongs in the "heat" section — you need to know it.

1) The specific heat capacity (SHC) of a material is the heat energy needed to raise the temperature of 1 kg of it by 1 °C (or 1 K).  (See page 231 for more about the Kelvin scale of temperature.)

2) For example, the specific heat capacity of water is 4200 J/(kg°C).  This means it takes 4200 joules of heat to raise the temperature of 1 kg of water by 1 °C (or 1 kelvin).

3) You can use a material's specific heat capacity to work out how much energy has been transferred.

### Energy transferred = mass × specific heat capacity × temperature change

*Example:* If the SHC of aluminium is 900 J/(kg°C), how much heat is required to raise the temperature of a 1.5 kg aluminium kettle by 5 °C?

*Answer:* Energy transferred = mass × specific heat × temp. change
= 1.5 × 900 × 5 = 6750 J

4) The higher a material's SHC, the more energy it takes to heat it up (and the more heat it can absorb without changing temperature much).

5) This is why in the old days (before fridges), food was stored in dark cupboards on shelves made from concrete or stone.  Concrete has a high specific heat capacity, and so absorbs a lot of heat without changing temperature too much.  This helped keep the cupboard cool and the food fresh.

---

And if you're doing the AQA Physics syllabus, you need to know all about resistor colour codes...

## Know how to Use the Resistor Colour Codes

Resistors have little coloured bands round them showing their resistance. You need to know how to read these resistance 'colour codes'.

| | |
|---|---|
| 0 | black |
| 1 | brown |
| 2 | red |
| 3 | orange |
| 4 | yellow |
| 5 | green |
| 6 | blue |
| 7 | violet |
| 8 | grey |
| 9 | white |

You don't need to remember the values in this table — but you do need to know how to use them.

1) Resistors usually have four coloured bands round them.  The colours of the first three bands each correspond to a value between 0 and 9.

2) The first three bands show the actual resistance of the resistor.
   (i)   the first band is the first digit in the resistance,
   (ii)  the second band is the second digit,
   (iii) the third band shows the number of zeros you'd need to add to these two digits to get the actual resistance.

*Example:* What's the resistance of the resistor on the right?

   (i)  The first band is green (= 5).
   (ii) The second band is blue (= 6).
   (iii)The third band is red (so there are 2 extra zeros).
      So the resistance is 5600 Ω.

3) Finally, the last band is either silver or gold and shows how accurate the resistance actually is — this is called the tolerance.  A gold band means the resistance is accurate to within 5%, while a silver band means it's accurate to within 10%.

So the resistor above is accurate to within 10% — the resistance is 5600 Ω ± 10%. This means the actual resistance could be as little as 5600 – 560 = 5040 Ω, or it could be as big as 5600 + 560 = 6160 Ω.

# Warm-Up and Worked Exam Questions

## Warm-up Questions

1) Write down the equation for calculating momentum.

2) Gravity provides the centripetal force keeping a planet in its orbit around a star.
   What force provides the centripetal force keeping an electron in its orbit around a nucleus?

3) In a Physics experiment, a cannonball is dropped from the top of a high tower. At the same instant an identical cannonball is fired horizontally from a cannon at the top of the same tower. What can you say about the time it will take for each of them to reach the ground (assuming the ground's level)?

4) Which of the following is NOT a vector quantity: momentum, velocity, acceleration, energy, force?

## Worked Exam Questions

1) a) A girl of mass 45 kg steps off a raft onto the riverbank at 3 m/s.
   The raft and the girl were previously at rest, and the raft has a mass of 350 kg.
   How fast does the raft move away?

   *Take direction that raft moves in as positive, and use the principle:*

   *total momentum before = total momentum after = 0*

   *0 = 45 x (-3) + 350 x v = -135 + 350v*

   *135 = 350v;    v = 135/350 = 0.386 m/s*

   *[2 marks]*

   b) When you step off a boat from rest it tends to move slowly away from under your feet. However, if you try to step off a skateboard from rest, it moves away very fast. What is the difference?

   *In both cases your forward momentum will equal the object's reverse momentum.*

   *The boat's mass is much larger than that of the skateboard, so its velocity must*

   *be much smaller than the skateboard's in order to balance your own momentum.*

   *[2 marks]*

2) This question is about resistor colour codes. You will need the following table:

| 0 | 1 | 2 | 3 | 4 | 5 | 6 | 7 | 8 | 9 |
|---|---|---|---|---|---|---|---|---|---|
| black | brown | red | orange | yellow | green | blue | violet | grey | white |

   A resistor has the following colour-coded bands around it: brown, red, brown, silver. What is its maximum and minimum resistance?

   1st two bands give 1st two digits and 3rd     *brown = 1, red = 2, then there's 1 zero, so resistance = 120 Ω*
   band gives number of zeros:

   4th band gives tolerance:     *Silver means ±10%. In this example that's ±12 Ω*

   *So max. resistance = 120 + 12 = 132 Ω; min. resistance = 120 – 12 = 108 Ω.*

   *[2 marks]*

# Exam Questions

1) A car has a mass of 1400 kg. It comes to rest from a speed of 110 km/h in 10 s.

   a) Find the change in its momentum, in kg m/s.

   $M = m \times v$    $m = 1400 \times 110$    *[2 marks]*

   b) Find the size of the average force slowing it down.

   *[1 mark]*

   c) Write down a situation where it is desirable to artificially increase the time taken for a body to undergo a change in its momentum.

   *[2 marks]*

2) A projectile is fired horizontally over the sea from the top of a cliff 11.25 metres high. The projectile has a horizontal velocity of 400 m/s.

   a) How fast is the projectile moving in a vertical direction when it reaches sea level?

   *[2 marks]*

   b) How long will it take for the projectile to fall to sea level?

   *[2 marks]*

   c) How far will it have travelled horizontally by that time?

   *[1 mark]*

3) A cyclist rides round a circular track of radius 159.3 metres at a speed of 21 m/s.

   a) How long does he take to go round once?

   *[2 marks]*

   b) How large is the centripetal force acting on the cyclist, assuming his mass is 55 kg?

   *[2 marks]*

   c) Which direction is the centripetal force on the cyclist acting?
   Choose from the options below:

   A – towards the centre of the track
   B – in the same direction as his velocity at each instant
   C – always towards the front of the bike
   D – downwards
   E – towards the back of the bike

   *[1 mark]*

4) a) The specific heat capacity of copper is 380 J/(kg °C).

   How much energy is needed to raise the temperature of a copper saucepan of mass 1.2kg from room temperature at 20°C to 100°C?

   *[1 mark]*

   b) There is 1.2 kg of watery soup inside the saucepan from part a). It takes much longer for the soup to get hot than the pan.

   What does this tell you about the specific heat capacity of water?

   *[1 mark]*

   c) Why is it useful to make a saucepan out of a substance with a low specific heat capacity?

   *[1 mark]*

# Revision Summary

Practice is the best thing for learning this stuff — so try the questions relevant to your syllabus:

1) Name two types of input sensor. What change in the surroundings does each type respond to?

2) What is the job of a processor in an electronic system?

3) Describe what a relay does. Give two reasons why they are used.

4) What happens when the voltage at the base of a transistor is high enough?

5) Draw the circuit symbols for a) a microphone,  b) an electric bell,  c) a buzzer.

6) Draw the symbol and the truth table for a) an OR gate,  b) an AND gate.

7) Draw the symbol for a NOT gate. What does a NOT gate do?

8) Draw the symbol and the truth table for a NAND gate.

9) When will a NOR gate give an output of 1?

10) Copy and complete the truth table for the logic gate combination shown.

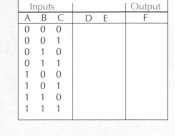

| Inputs | | | | | Output |
|---|---|---|---|---|---|
| A | B | C | D | E | F |
| 0 | 0 | 0 | | | |
| 0 | 0 | 1 | | | |
| 0 | 1 | 0 | | | |
| 0 | 1 | 1 | | | |
| 1 | 0 | 0 | | | |
| 1 | 0 | 1 | | | |
| 1 | 1 | 0 | | | |
| 1 | 1 | 1 | | | |

11) What is a latch used for in an electronic system?

12) Draw a latch made from two NOR gates.

13) For the potential divider on the right calculate:

   a)   the voltage across the 50 Ω resistor

   b)   the voltage across the 200 Ω resistor

   c)   the voltage at the point P.

14) Look at the potential divider on the right. What will happen to the voltage at point X when a bright light shines on the LDR? Explain your answer.

15) What happens to the potential difference (voltage) across a capacitor when more charge is stored on it?

16) In order to increase the time taken to charge a capacitor fully what should you do to:

   a) the capacitance of the capacitor    b) the resistance of the charging circuit?

17) Give one advantage and one disadvantage of:

   a) the development of mobile phone technology   b) the Internet.

18) What is the difference between elastic and plastic behaviour? What is meant by the 'elastic limit' of a spring?

19) Describe how to find a spring constant.

20) How do you find the energy stored from the force/extension graph of an elastic material?

21) Describe how an elastic band acts when stretched by an increasing force.

22) What are the units of density? What is the equation for density?

23) Describe how you would find the centre of mass of an irregularly shaped piece of flat cardboard.

24) Which is more stable — a low, wide object or a tall, thin one?

25) What happens if the line of action of an object's weight lies outside its base?

26) What three things can the engineer do in order to exert a larger moment on the telegraph pole?

27) Calculate the moment exerted about the nut by the force on this spanner.

28) What must the clockwise moments be equal to, for an object to balance?

29) "Momentum is usually lost in collisions and explosions." True or False?

30) An acrobat (mass 50 kg) jumps off a wall and hits the floor at a speed of 4 m/s. She bends her knees and stops moving in 0.5 s. What is the average force acting on her?

31) Explain why air bags, seat belts and crumple zones reduce the risk of serious injury in a car crash.

32) Write down the four equations of motion. Which equation should you use to find the acceleration if you know the initial velocity, the final velocity and the time taken?

33) If a bullet is fired horizontally, what happens to its horizontal and vertical velocities?

34) Explain why an object's velocity is constantly changing if it is moving in a circle.

35) What three things does a centripetal force depend on?

36) Calculate the centripetal force needed on an athletics hammer (mass 3 kg) moving in a circle of radius 1.5 m at a speed of 15 m/s. Calculate the time taken by the hammer to do a complete circle.

37) What provides the centripetal force needed to keep artificial satellites from drifting off into space?

38) Explain what specific heat capacity is. How much heat energy is required to increase the temperature of 3 kg of water by 7 °C? (*Take the SHC of water to be 4200 J/(kg°C).*)

39) The bands on a resistor are blue, grey, brown and gold (in that order). What are the maximum and minimum possible resistances of the resistor?

# Practice Exam

Once you've been through all the questions in this book, you should feel pretty confident about the exam. As final preparation, here is a **practice exam** to really get you set for the real thing. It's split into **two sections** — one for if you're doing Double Science and an <u>extra</u> section for if you're taking GCSE Physics (i.e. Separate Sciences). The paper is designed to give you the best possible preparation for the differing question styles of the actual exams, whichever syllabus you're following. If you're doing Foundation then you won't have learnt every bit — but it's still good practice.

## General Certificate of Secondary Education

## GCSE Science: Double Award
## GCSE Physics

| Centre name | | | | |
|---|---|---|---|---|
| Centre number | | | | |
| Candidate number | | | | |

## Paper 1

| Surname |
|---|
| Other names |
| Candidate signature |

**Time allowed**:

<u>GCSE Science: Double Award</u>

**Section A**               1 hour 15 minutes

<u>GCSE Physics</u>

**Sections A and B**        1 hour 45 minutes

### Instructions to candidates
- If you are taking **GCSE Science: Double Award** answer **all** of the questions in **Section A**. Do **not** answer Section B.
- If you are taking **GCSE Physics** answer **all** of the questions in **Sections A and B**.
- Write your name and other details in the spaces provided above.
- Answer **all** questions in the spaces provided.
- Do all rough work on the paper.

### Information for candidates
- The marks available are given in brackets at the end of each question or part-question.
- Marks will not be deducted for incorrect answers.
- In calculations show clearly how you work out your answers.
- There are **13** questions in **Section A** of this paper. There are **4** questions in **Section B** of this paper.
- There are no blank pages.

### Advice to candidates
- Work steadily through the paper.
- Don't spend too long on one question.
- If you have time at the end, go back and check your answers.

## SECTION A
**This section is for candidates taking GCSE Double Science and candidates taking GCSE Physics.**
**There is 1 hour and 15 minutes available for this section.**

1    The central heating system in a house pumps hot water around the building.
The temperature inside is monitored and controlled using a sensing circuit.
This controls a switch for the pump.
The sensing circuit contains a thermistor.
The graph shows how the resistance of the thermistor varies with temperature.

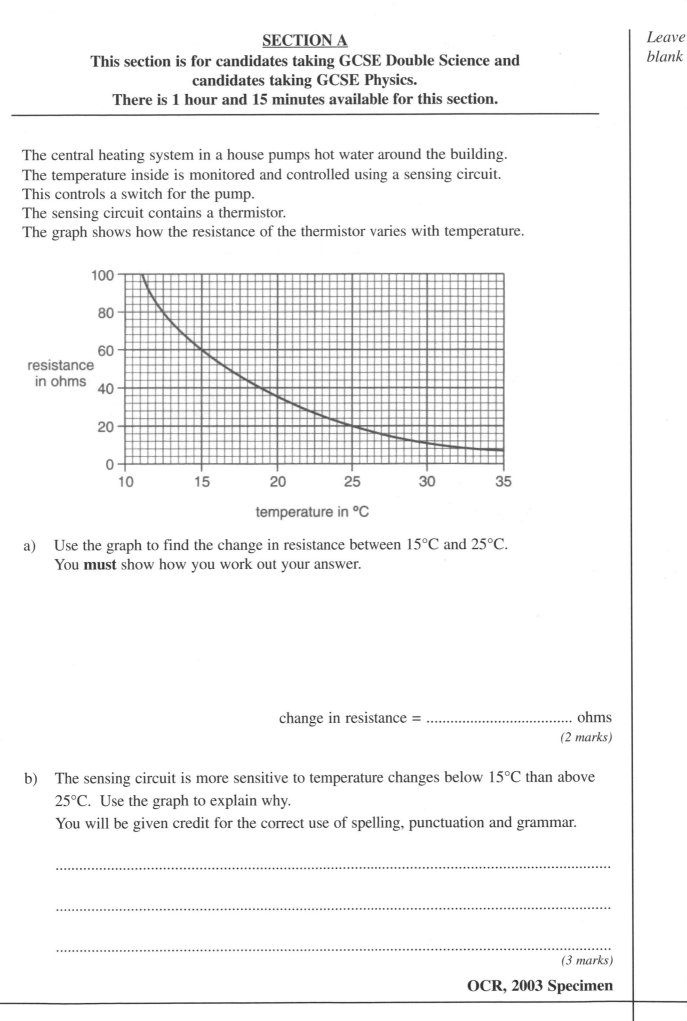

a)   Use the graph to find the change in resistance between 15°C and 25°C.
You **must** show how you work out your answer.

change in resistance = ..................................... ohms

*(2 marks)*

b)   The sensing circuit is more sensitive to temperature changes below 15°C than above 25°C. Use the graph to explain why.
You will be given credit for the correct use of spelling, punctuation and grammar.

.................................................................................................................................

.................................................................................................................................

.................................................................................................................................

*(3 marks)*

**OCR, 2003 Specimen**

2  a)  Describe, in as much detail as you can, the orbit of a planet in the Solar System.

.......................................................................................................................................

.......................................................................................................................................

.......................................................................................................................................

.......................................................................................................................................

.......................................................................................................................................

.......................................................................................................................................

*(3 marks)*

b)  Give **two** different uses of artificial satellites which orbit the Earth.

1.  ..............................................................................................................................

.......................................................................................................................................

2.  ..............................................................................................................................

.......................................................................................................................................

*(2 marks)*

c)  Comets are rarely seen.  Explain this in terms of their orbits.

.......................................................................................................................................

.......................................................................................................................................

.......................................................................................................................................

.......................................................................................................................................

.......................................................................................................................................

*(3 marks)*

**AQA, 2000**

3    This question is about the movements of the Earth's plates and earthquake waves. The diagram represents a zone where an ocean plate meets a continental plate.

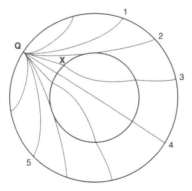

a)  In the diagram the ocean plate is slowly moving downwards under the continental plate.  What happens to the rock in the ocean plate as it moves into the mantle?

...................................................................................................................................

*(1 mark)*

b)  Why does the ocean plate move under the continental plate?

...................................................................................................................................

...................................................................................................................................

*(1 mark)*

c)  The diagram shows a cross-section of the Earth.  There is an earthquake at place **Q**. Sensitive detectors at positions **1**, **2**, **3**, **4** and **5** are designed to pick up two types of earthquake waves.  These types are called **P** and **S** waves.

(i)  Finish these sentences.

**P** waves are .......................................... waves which can be detected at

positions .......................................... .

**S** waves are .......................................... waves which can be detected at

positions .......................................... .

*(4 marks)*

**QUESTION 3 CONTINUED OVERLEAF**

(ii) The path of one of the earthquake waves changes direction at the point marked **X**. What is the name given to this change of direction of waves?

..................................................................................
*(1 mark)*

(iii) Explain why the wave path changes direction at **X**.

..................................................................................

..................................................................................

..................................................................................
*(2 marks)*

**OCR, 2003 Specimen**

4    A student carries out an experiment with a steel ball bearing and a tube of thick oil.
The diagram shows the apparatus used.
The student releases the ball bearing and it falls through the oil.

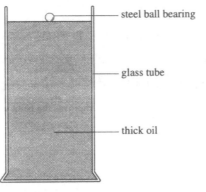

a)    Two forces **X** and **Y** act on the ball bearing as it falls through the oil.
This is shown on the diagram.

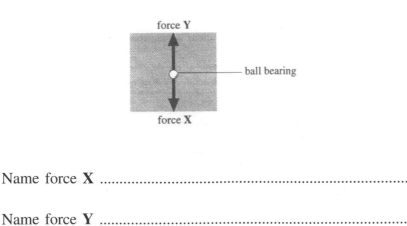

Name force **X** ...................................................................

Name force **Y** ...................................................................
*(2 marks)*

b) The graph shows how the speed of the ball bearing changes as it falls through the oil.

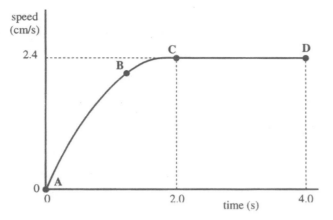

(i) What is happening to the speed of the ball bearing between points **A** and **B**?

......................................................................................................................

......................................................................................................................
*(1 mark)*

Explain, in terms of forces **X** and **Y**, why this happens ...........................................

......................................................................................................................

......................................................................................................................

......................................................................................................................

......................................................................................................................
*(1 mark)*

(ii) What is happening to the speed of the ball bearing between points **C** and **D**?

......................................................................................................................

......................................................................................................................
*(1 mark)*

Explain, in terms of forces **X** and **Y**, why this happens ...........................................

......................................................................................................................

......................................................................................................................

......................................................................................................................
*(3 marks)*

**QUESTION 4 CONTINUED OVERLEAF**

c) Use the graph to help you to calculate the distance travelled by the ball bearing between points **C** and **D**.

........................................................................................................................

........................................................................................................................

........................................................................................................................

Distance ................................................ cm

*(2 marks)*

**AQA, 2000**

5 a) Power stations burn fuel to generate electricity.

Write down the name of one fuel they burn to generate electricity.

Answer ..................................................

*(1 mark)*

b) Look at the diagram of the power station.

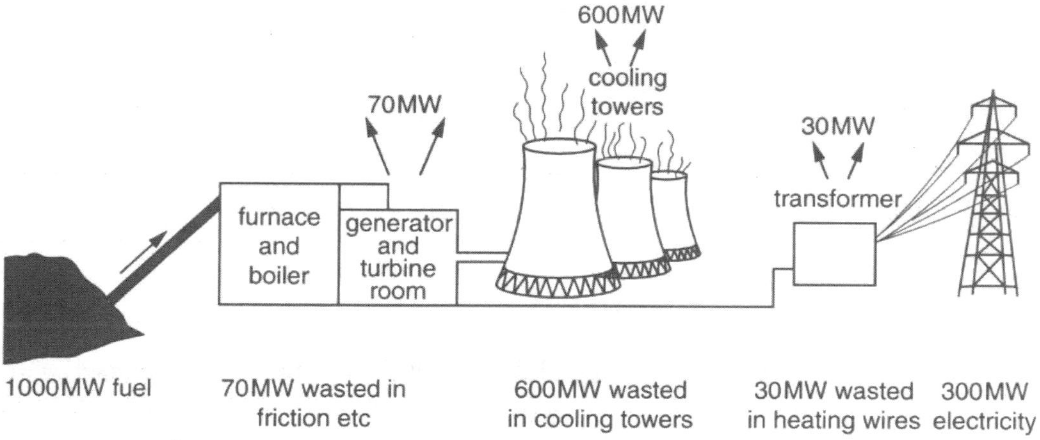

Calculate the efficiency of the power station.

$$\% \text{ efficiency} = \frac{\text{useful output}}{\text{input}} \times 100$$

........................................................................................................................

........................................................................................................................

........................................................................................................................

........................................................................................................................

Answer ..................................................... %

*(2 marks)*

**OCR, 2000**

6   This question is about electromagnetism.
    Graham makes a simple electric bell.

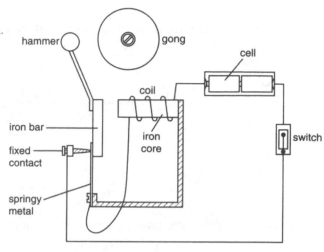

He closes the switch.

a)   The hammer moves to the right and hits the gong.
     Explain why.

     ......................................................................................................................

     ......................................................................................................................

     ......................................................................................................................
                                                                    *(2 marks)*

b)   The hammer now moves back to the left.
     Explain why.

     ......................................................................................................................

     ......................................................................................................................

     ......................................................................................................................
                                                                    *(2 marks)*

                                                        **OCR, 2003 Specimen**

7   a)   The diagram shows the passage of light beam **A** travelling down an optical fibre.

         (i)   State the name of the process that takes place as the light **A** beam travels down the
               optical fibre.

               ......................................................................................................................
                                                                    *(1 mark)*

         (ii)  Complete the diagram to show the passage of light beam **B** down the same optical
               fibre.

                                                                    *(1 mark)*

                    **QUESTION 7 CONTINUED OVERLEAF**

(iii) Suggest why beam **B** will take slightly longer to travel down the fibre than beam **A**.

.......................................................................................................................

.......................................................................................................................

.......................................................................................................................

.......................................................................................................................

*(2 marks)*

b) Optical fibres are used to carry information. The information is carried by the light beam in the form of a digital signal.

(i) Draw a diagram to show what is meant by a digital signal.

*(1 mark)*

(ii) The signal from a microphone is an analogue signal. How does an analogue signal differ from a digital signal?

.......................................................................................................................

.......................................................................................................................

*(1 mark)*

c) When signals are sent through optical fibres they lose energy.

(i) State what happens to the brightness of the light beam as it loses energy.

.......................................................................................................................

*(1 mark)*

(ii) State **one** disadvantage of losing energy as the light beam travels through the optical fibre.

.......................................................................................................................

.......................................................................................................................

*(1 mark)*

**Edexcel, 2003 Specimen**

8    The table gives some information about the three main types of nuclear radiation.
     Use the information in the table to answer the questions that follow.

| Name | Consists of | Penetration | Ionising ability | Charge: +/−/0 |
|------|-------------|-------------|------------------|----------------|
| **Alpha** (α) | two protons and two neutrons | stopped by thin paper or several cm of air | strong | .................. |
| **Beta** (β) | electron | stopped by thick aluminium or thin lead | weak | .................. |
| **Gamma** (γ) | short-wavelength electromagnetic radiation | reduced by thick lead or concrete | very weak | .................. |

a)   Complete the last column in the table.

*(3 marks)*

b)   Smoke alarms use a source of alpha radiation in a small chamber.
     The alpha radiation ionises air particles, causing a current between the electrodes.
     When smoke enters the chamber, the ionisation current stops and the alarm sounds.

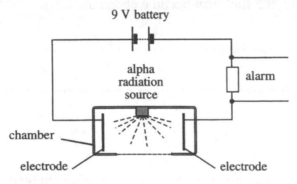

(i)   State **two** reasons why alpha radiation is more suitable than either gamma or beta
      radiation for use in a smoke alarm.

      1 ...........................................................................................................................

      ...............................................................................................................................

      2 ...........................................................................................................................

      ...............................................................................................................................

*(2 marks)*

(ii)  Suggest **one** way in which a "false alarm" could sound and suggest how this
      could be prevented.

      ...............................................................................................................................

      ...............................................................................................................................

      ...............................................................................................................................

*(2 marks)*

**Edexcel, 2001**

9    This question is about static electricity.

a)   Static electricity can be a problem when fuel is put into an aircraft.
     There are two types of electric charge.
     Write down the two types of electric charge.

     ................................................. and ................................................................
                                                                                                        *(2 marks)*

b)   Fuel is pumped along the plastic pipe.
     The pipe may become charged.

     (i)   Explain why this may be dangerous.

           ........................................................................................................

           ........................................................................................................
                                                                                                        *(1 mark)*

     (ii)  Explain how the fuel may become charged.

           ........................................................................................................

           ........................................................................................................

           ........................................................................................................
                                                                                                        *(2 marks)*

     (iii) The engineer can stop the aeroplane becoming charged.
           Suggest how.

           ........................................................................................................

           ........................................................................................................
                                                                                                        *(1 mark)*

                                                                                        **OCR, 2000**

10    The radioactive isotope, carbon-14, decays by beta (β) particle emission.

a)    Plants absorb carbon-14 from the atmosphere.  The graph shows the decay curve for 1 g of carbon-14 taken from a flax plant.

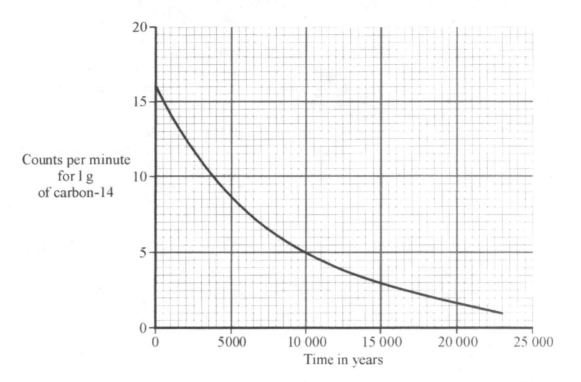

Use the graph to find the half-life of carbon-14.  You should show clearly on your graph how you obtain your answer.

Half-life = ............................................ years.

*(2 marks)*

b)    Linen is a cloth made from the flax plant.  A recent exhibition included part of a linen shirt, believed to have belonged to St Thomas à Becket, who died in 1162.  Extracting carbon-14 from the cloth would allow the age of the shirt to be verified.

If 1 g of carbon-14 extracted from the cloth were to give 870 counts in 1 hour, would it be possible for the shirt to have once belonged to St Thomas à Becket?  You must show clearly the steps used and reason for your decision.

........................................................................................................................

........................................................................................................................

........................................................................................................................

........................................................................................................................

*(3 marks)*

**AQA, 2003 Specimen**

11  a)  Water slides at leisure pools are very popular with children.

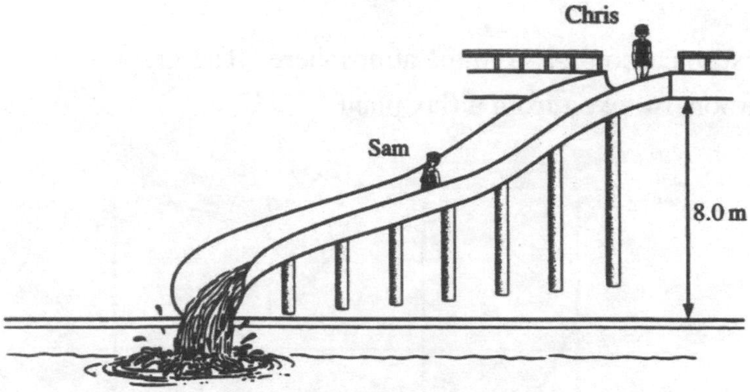

(i)  Chris has climbed to the top of the slide.  What type of energy has she gained?

........................................................................................................................

*(1 mark)*

(ii)  Sam is shown moving through a point halfway down the slide.  State TWO types of energy she has.

1.  ...............................................................................................................

2.  ...............................................................................................................

*(2 marks)*

(iii)  Describe what happens to Sam's energy as she enters the water.

........................................................................................................................

........................................................................................................................

........................................................................................................................

*(2 marks)*

b)  Water is pumped from the pool to the top of the slide, a height of 8.5 m.

(i)  How much work is done when 450 N of water is pumped to the top of the slide?

........................................................................................................................

........................................................................................................................

*(3 marks)*

(ii)  It takes 15 seconds to pump 450 N of water to the top of the slide.  Calculate the useful power output of the pump.

........................................................................................................................

*(2 marks)*

**Edexcel, 2000**

12  a)  Complete the following sentences about waves.

Waves travelling across the surface of water are ................................................... waves.

The disturbance in the water is ........................................ the direction in which the wave is travelling.

Sound waves in air are ...................................................................... waves.

The disturbance in the air is ...................................... the direction in which the wave is travelling.

.............................................................. waves can travel through a vacuum.

When a wave moves through a gap it spreads out.  This is called ...................................
*(6 marks)*

b)  The diagram shows apparatus used to find the position of a flaw in a metal casting. The cathode ray oscilloscope is producing a visual display to show how the ultrasonic waves are partly reflected at different boundaries.

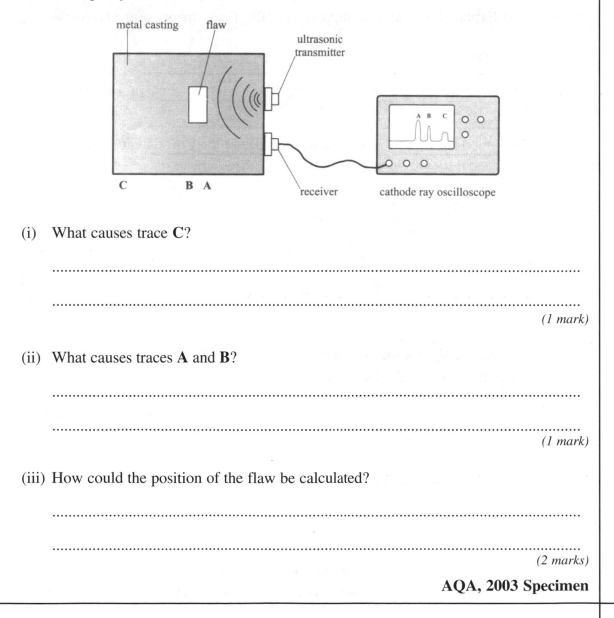

(i)   What causes trace **C**?

..................................................................................................................................

..................................................................................................................................
*(1 mark)*

(ii)  What causes traces **A** and **B**?

..................................................................................................................................

..................................................................................................................................
*(1 mark)*

(iii) How could the position of the flaw be calculated?

..................................................................................................................................

..................................................................................................................................
*(2 marks)*

**AQA, 2003 Specimen**

13  a)  The diagrams show the line spectra from a nearby galaxy and a distant galaxy.

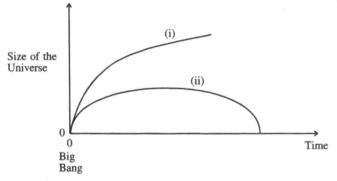

(i)  Indicate **on the diagram** which end of the spectrum is red.

*(1 mark)*

(ii)  What does the shift of the line spectra for the distant galaxy suggest about its motion relative to the nearby galaxy?

.................................................................................................................................

.................................................................................................................................

*(1 mark)*

(iii)  Explain how this shift supports the 'Big Bang' theory of the Universe.

.................................................................................................................................

.................................................................................................................................

*(1 mark)*

b)  The graph shows two ways in which the size of the Universe might change over time.

Explain the significance of curves (i) and (ii) and the different predictions they make about the future of the Universe.

.................................................................................................................................

.................................................................................................................................

.................................................................................................................................

.................................................................................................................................

*(2 marks)*

**Edexcel, 2001**

**END OF SECTION A**

**GCSE PHYSICS CANDIDATES ONLY:  NOW START SECTION B**

**SECTION B**
**This section is for GCSE Physics candidates only.**
**There are 30 minutes available to complete this section.**

1    Radio waves can travel from the transmitter to a receiver in three different ways.
     **Space waves** can only be received if the receiver is within sight of the transmitter.
     **Sky waves** are reflected by the ionosphere.
     **Ground waves** follow the Earth's curvature.
     The diagrams show these waves.

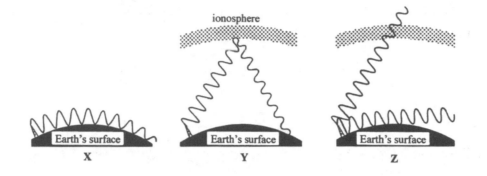

a)   Which type of radio wave is used for communications with satellites?
     Give a reason for your choice.

     ................................................................................................................................

     ................................................................................................................................

     ................................................................................................................................
                                                                                            *(2 marks)*

b)   In 1901, Marconi sent a radio message from Canada to Cornwall.

     (i)   Explain why a space wave cannot travel from Canada to Cornwall.

           ..........................................................................................................................

           ..........................................................................................................................
                                                                                            *(1 mark)*

     (ii)  Suggest why the first radio transmissions were called "wireless".

           ..........................................................................................................................

           ..........................................................................................................................
                                                                                            *(1 mark)*

                                                                            **Edexcel, 2000**

2  a)  The atoms $^{14}_{7}\text{N}$ and $^{15}_{7}\text{N}$ are isotopes of nitrogen.
Write down **one** similarity and **one** difference between the nuclei of these isotopes.

similarity ...............................................................................................................

difference .............................................................................................................

*(2 marks)*

b)  The graph shows the relationship between the number of neutrons and the number of protons in a nucleus.

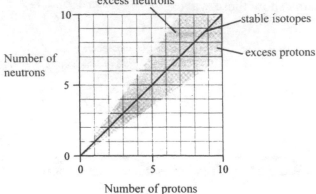

(i)  What is the relationship between the number of protons and the number of neutrons in a stable nucleus?

.......................................................................................................................

*(1 mark)*

(ii)  Use an **X** to mark the position of $^{15}_{7}\text{N}$ on the graph.

*(1 mark)*

(iii)  Give the reason why $^{15}_{7}\text{N}$ is unstable.

.......................................................................................................................

*(1 mark)*

(iv)  The diagram shows the decay of $^{15}_{7}\text{N}$.

What changes take place when the nucleus of $^{15}_{7}\text{N}$ decays?

.......................................................................................................................

.......................................................................................................................

.......................................................................................................................

*(3 marks)*

**Edexcel, 2002**

3  a) An experiment was carried out to measure how the pressure of a gas changes with temperature.

The volume and mass of the gas were kept constant throughout the experiment.

The following results were obtained.

| Pressure in kPa | 90 | 96 | 103 | 110 | 117 | 123 |
|---|---|---|---|---|---|---|
| Temperature in °C | 0 | 20 | 40 | 60 | 80 | 100 |

(i) Use the grid to draw a graph of these results.

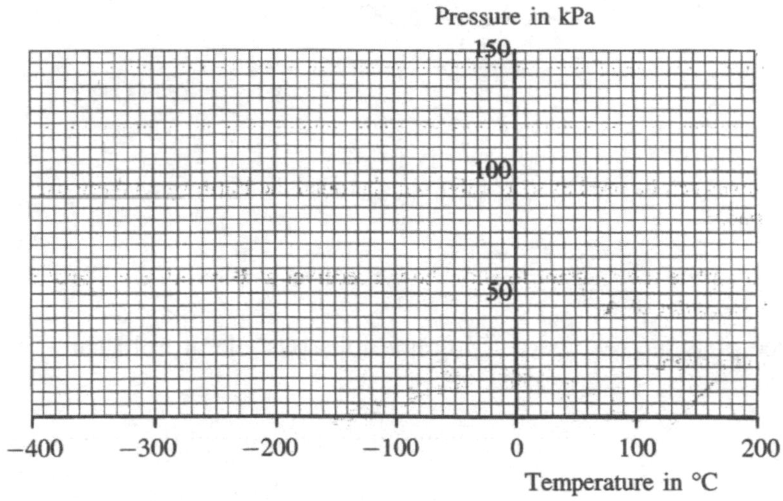

Pressure in kPa

Temperature in °C

*(3 marks)*

(ii) Describe, using the graph, how the pressure of the gas changes with temperature.

..............................................................................................................................

*(1 mark)*

(iii) Use your graph to estimate the temperature at which the gas exerts no pressure.

..............................................................................................................................

*(2 marks)*

(iv) What name is given to the temperature at which a gas exerts no pressure?

..............................................................................................................................

*(1 mark)*

b) At the start of a journey, the temperature and pressure inside a car tyre were 17°C and 300 kPa.

During the journey, the pressure was 350 kPa.

Calculate the temperature during the journey.

..............................................................................................................................

..............................................................................................................................

..............................................................................................................................

..............................................................................................................................

*(4 marks)*

**Edexcel, 2001**

280

4    The diagram shows the paths of alpha particles from a radioactive source directed at thin gold foil.

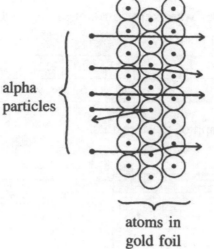

alpha particles

atoms in gold foil

a)   Describe **three** possible results of firing an alpha particle at gold foil.

1    ...................................................................................................................................

     ...................................................................................................................................

2    ...................................................................................................................................

     ...................................................................................................................................

3    ...................................................................................................................................

     ...................................................................................................................................

*(3 marks)*

b)   Explain what the diagram shows about the sign of the charge on the nucleus.

     ...................................................................................................................................

     ...................................................................................................................................

     ...................................................................................................................................

*(2 marks)*

c) Describe how the results of the experiment led to Rutherford's nuclear model of the atom.

..........................................................................................................................................

..........................................................................................................................................

..........................................................................................................................................

..........................................................................................................................................

..........................................................................................................................................

..........................................................................................................................................

*(4 marks)*

**Edexcel, 2001**

**END OF SECTION B**

## Page 4 (Warm-Up Questions)

1) a) V  b) t  c) Q  d) m  e) I  f) F

2) a)  ampere, A      b)  metres per second, m/s   c)  joule, J

   d)  second, s      e)  coulomb, C              f)  pascal, Pa

3) b) P = F/A  c) P = VI

4) a) P = E/t  b) m = DV  c) A = F/P

## Page 5 (Exam Questions)

1) a)  F = ma = 50 × 0.9 = 45N  *(2 marks)*

   b)  m = F/a = 45/0.5 = 90kg  *(2 marks)*

   c)  Pressure = Force / Area = 660 / 0.03 = 22 000 Pa = 22 kPa  *(2 marks)*

2) a)  f = 1/T = 1/0.00357 = 280 Hz  *(1 mark)*

   b)  v = f × λ
       λ = v/f = 330/280 = 1.18 m  *(2 marks)*

3)     K.E. = ½ mv²
       K.E. = ½ × 1500 × (97 × 1000 /3600)²
            = ½ × 1500 × 26.94²
            = ½ × 1500 × 726
            = 544,502 J = 544 kJ  *(3 marks)*

4) a)  the temperature is constant, so use P(1)V(1) = P(2)V(2)
       P(1)V(1) = 1 × 120 = 120
       P(2)V(2) = 3.1 × V(2)
       So  120 / 3.1 = V(2)
       V(2) = 38.7 cm³  *(2 marks)*

   b)  the temperature is constant, so use P(1)V(1) = P(2)V(2)
       50 × 0.0005 = P(2) × 0.0003
       50 × 0.0005 / 0.0003 = P(2)
       P(2) = 83.3 kPa  *(2 marks)*

## Page 12 (Warm-up Questions)

1)     An electric current is a flow of ELECTRONS round a circuit.  These
       particles carry a NEGATIVE electric charge.  They are pushed away from
       the NEGATIVE terminal of the cell and round to the POSITIVE terminal,
       the opposite direction to CONVENTIONAL current.  No current will flow
       through a circuit unless the circuit is COMPLETE.

2)  a)                b)                c)                d)

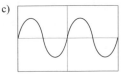

3)     The current decreases.

4)     The resistance increases.

5)     Any sensible answers, e.g.
       LED – used in a stereo system; the light that flashes red when the volume
       goes above a certain level.
       Light Dependent Resistor – used in an alarm set to go off when a drawer is
       opened.
       Thermistor – used in a temperature sensor in a car engine.

## Page 13 (Exam Questions)

1) a)  i)  electrons  *(1 mark)*

       ii) Negatively charged electrons are repelled from the negative pole of the
           cell and pushed along the wire towards the positive pole.  *(2 marks)*

       iii) The current increases.  *(1 mark)*

   b)  When a voltage is applied, the positively charged ions move towards the
       negative electrode, and the negatively charged ions move towards the
       positive electrode, producing an electric current.  *(2 marks)*

   c)

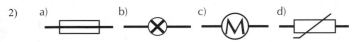

       *(1 mark)*

2) a)  (i)  Component C is the Light-Dependent Resistor  *(1 mark)*

       (ii) Component A is the cell.  *(1 mark)*

       (iii) Component B is the Light Emitting Diode  *(1 mark)*

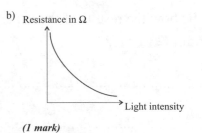

b)  Resistance in Ω → Light intensity

       *(1 mark)*

   c)  It will emit light.  *(1 mark)*

   d)  No current would flow because the diode (LED) lets current flow through
       it in one direction only.  *(2 marks)*

## Page 19 (Warm-up Questions)

1)     2V

2) i)  Parallel

   ii) Series

   iii) Parallel

3)     "Series circuits are not much used in real life because if one component
       does not work, the whole circuit does not work."  Or "Series circuits are
       not much used in real life because the components cannot be switched off
       independently." (or similar)

## Page 20 (Exam Questions)

1) a)  Series  *(1 mark)*

   b)  2 Ω  *(1 mark)*

   c)  5V  *(1 mark)*

   d)  No, because the same current flows through all points of a series circuit
       (unless it was connected in voltmeter branch of circuit).  *(1 mark)*

   e)  Zero  *(1 mark)*

2) a)  12V  *(1 mark)*

   b)  4 Ω  *(1 mark)*

   c)  5A  *(1 mark)*

   d)  2.5V  *(1 mark)*

   e)  14.5V (Hint: add the voltage across R₁ to the voltage across the parallel
       components.)  *(2 marks)*

## Page 24 (Warm-up Questions)

1)     Friction

2) a)  Proton

   b)  Electron

3)     Inkjet printer and photocopier.

4)     Any two of lighting, and/or a spark causing an explosion in any of the
       following situations: near a fuel tank or pipe, near paper rollers, near a
       grain chute, or near dusty or fume-filled places generally.

## Page 25 (Exam Questions)

1) a)  Electrons have been scraped off the duster onto the rod.  *(2 marks)*

   b)  The charge on the duster is now positive due to the negative electrons
       having left it for the polythene rod.  *(2 marks)*

   c)  Unlike the polythene rod, metal is a conductor.  That means that any
       charge transferred to it will be conducted away easily.  *(3 marks)*

2) a)  More electrons than protons.  *(1 mark)*

   b)  Electrons within each grain of toner are repelled from the negative charge
       on the drum, inducing a net positive charge on the side of the grain nearest
       the drum.  This attracts the grains of toner closer to the drum.  *(2 marks)*

   c)  If the paper were charged its attraction would help transfer the ink from
       the drum onto its surface.  *(2 marks)*

   d)  Sanjay could rub the balloon with the glove to charge it.  Then he could
       bring the balloon near to the toner in the saucer and show how the toner is
       attracted to the balloon.  Finally he could roll the blackened part of the
       balloon on the paper, transferring the toner onto the paper.  *(2 marks)*

## Page 29 (Warm-up Questions)

1) a) Lamp or lightbulb or LED
   b) Cell or battery

2) a) $1.3 \times 0.5 = 0.65$ kWh   b) $1300 \times 0.5 \times 60 \times 60 = 2\,340\,000$ J

3) a) 1026.4 units   b) £51.32

4) (d)

## Page 30 (Exam Questions)

1) a) i) Sound *(1 mark)*
      ii) Heat *(1 mark)*
      iii) Heat *(1 mark)*
      iv) Kinetic *(1 mark)*
      v) Light *(1 mark)*
   b) i) $E = QV$ *(1 mark)*
      ii) The charge loses electrical energy. *(1 mark)*

2) A = no. of units used = $16969 - 14303 = 2666$
   B = $2666 \times 4.68 = 12476.88$p = £124.77
   C = 124.77 + 8.64 = £133.41
   D = £133.41 × 0.175 = £23.35
   E = £133.41 + £23.35 = £156.76
   *(5 marks available — 1 for each)*

3) a) $Q = It$ *(1 mark)*
      $Q = 3.3 \times 60 \times 60 = 11\,880$ C *(1 mark)*
   b) $E = QV$ *(1 mark)*
      $E = 11\,880 \times 12 = 142\,560$ J *(1 mark)*

## Page 35 (Warm-up Questions)

1) Any three of: frayed cables, too-long cables, cables in contact with something hot or wet, pets chewing cables, cables within reach of children.

2) a) Blue
   b) Green and yellow
   c) Brown

3) It means that the appliance has a plastic casing and no metal parts showing.

4) a) $P = IV$
   b) $P = I^2R$

5) Alternating current, 400kV.

## Page 36 (Exam Questions)

1) a) The blue wire should be connected to the neutral terminal, not the earth. The green and yellow wire should be connected to the earth terminal, not the neutral. *(2 marks —1 mark for each wire)*

   b) The 5A fuse is too low. It would blow every time the kettle was used. The current through the kettle is given by $I = P/V = 2000/240 = 8.33$A, which is greater than 5A. A 13A fuse would be suitable.
   *(2 marks — 1 mark for correct fuse and 1 mark for correct reasoning)*

   c) Plastic is an insulator. *(1 mark)*

   d) If the live wire touches the case then the case itself becomes live and anyone touching it could get an electric shock. *(1 mark)*

   e) The earth wire must be connected to the case. *(1 mark)*
   This provides a low resistance path to earth, meaning that if the case becomes accidentally connected to the live terminal, a big current will surge from the live wire, through the case, and out down the earth wire. *(1 mark)*
   The surge in the current will blow the fuse and isolate the live supply. The fuse must be placed in the live wire because that is the one we need to isolate from the appliance. *(1 mark)*

2) a) Power supplied is given by $P = VI$. So to transmit a lot of power, either the voltage must be high or the current must be high. *(1 mark)*
   Power lost due to resistance is given by $P = I^2R$. *(1 mark)*
   Because power loss increases so steeply with the current, we want to keep the current low — and that means the voltage must be high. *(1 mark)*
   [3 marks available in total]

   b) Alternating current is used so that a transformer can be used to change the voltage. A transformer does not work with direct current. *(2 marks)*

   c) The step-down transformer allows us to reduce the high voltage in the power lines to a low voltage suitable for domestic use. *(1 mark)*

## Page 40 (Warm-up Questions)

1) A solenoid is a coil of wire.

2) Strength of electric current; what the core is made of; number of turns in the solenoid.

3) It means that once magnetised, steel retains its magnetism.

4) Any three of: scrap yard/car breaker's yard electromagnet, circuit breaker, relay switch ("reed relay" or just "relay" is also acceptable), electric bell, loudspeaker.

## Page 41 (Exam Questions)

1) a) The north pole. *(1 mark)*
   b) By changing the direction of the current. *(1 mark)*
   c) The field would be weaker. *(1 mark)*
   d) After putting it in the solenoid with a DC supply, turn off the supply and remove the rod. *(1 mark)*
   e) You would need an AC supply. Put the rod in the solenoid, turn the supply on, then remove the rod while the current is still flowing. *(2 marks)*

2) The correct order is D, B, A, F, C, E. *(5 marks)*

## Page 48 (Warm-up Questions)

1) thuMb: Motion (or force); First finger: Field (i.e. magnetic field); seCond finger: Current (i.e. electric current).

2) More current, more turns on the coil, stronger magnetic field, a soft iron core in the coil.

3) Electromagnetic induction is the creation of a voltage (and maybe a current) in a wire which is experiencing a change in magnetic field. The induced voltage is proportional to the rate of change of flux through the wire.

4) speed of movement of coil, number of turns on coil, area of coil, strength of magnetic field.

5) 20V

## Page 49-50 (Exam Questions)

1) a)

|  | Input | Output |
|---|---|---|
| Motor | Electrical | Kinetic |
| Generator | Kinetic | Electrical |

*(2 marks)*

   b) Slip rings and brushes *(1 mark)*

   c) A generator rotates a coil in a stationary magnetic field. A dynamo rotates a magnet, causing the magnetic field through a stationary coil to change. *(2 marks)*

   d) (iv) *(1 mark)*

   e) and f)

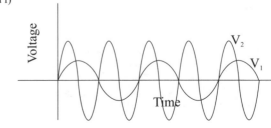

*(2 marks for each curve)*

2) a)

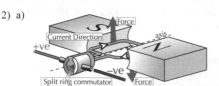

*(4 marks)*

b) (iii) *(1 mark)*

c) Without the split ring commutator the forces acting on the side arms of the coil would only act so as to make the coil rotate for half a cycle. Because the current direction is changed as the coil reaches the vertical position, however, the coil continues to rotate. *(2 marks)*

d) There would be no effect — the two changes would cancel each other out. *(1 mark)*

3) a) A *(1 mark)*

b) Using $V_P I_P = V_S I_S$

$I_P = V_S I_S / V_P = 400,000 \times 0.1/20,000 = 2A$ *(2 marks)*

c) Efficient power transmission requires high voltages and low currents. Transformers are used to step up the voltages produced by the power stations to a very high value for efficient transmission, and also to step down the voltages to safe useable levels at the other end. *(3 marks)*

d) Energy is wasted when eddy currents form in the iron core of the transformer and heat it up. To minimize eddy currents the core is laminated with layers of insulation. *(2 marks)*

e) B *(1 mark)*

## Page 57 (Warm-up Questions)

1) Gravity (or gravitational attraction).

2) The astronaut's mass stays the same.

3) Increase the force or increase the perpendicular distance of the force from the pivot.

4) a)

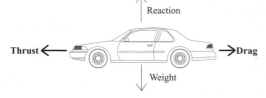

b) i) The reaction and weight forces will be balanced.

ii) The thrust force and drag force are unbalanced (thrust greater than drag, thus allowing acceleration).

5) a) Any examples where useful energy is lost due to friction, e.g. in machinery where surfaces are sliding over each other, such as inside a car's engine, in the wheel hub of a bicycle, on the hinge of a door, etc.

b) Any examples where friction helps something to work, e.g. in the brakes of a vehicle, when we strike matches, when we walk, to hold knots in string, to hold nails and screws in wood, etc.

## Page 58-59 (Exam Questions)

1) a) In icy conditions, the maximum force of friction between the tyres and the road is less than it is in dry conditions *(1 mark)*, therefore the braking force must be reduced so that it does not exceed the maximum force of friction *(1 mark)*.

b) The aerofoil makes the lorry's shape more streamlined *(1 mark)* reducing the drag force on the lorry *(1 mark)* and therefore saving fuel *(1 mark)*.

c) The lorry has to do more work against friction as it goes faster *(1 mark)* so the lorry's engine works less hard and uses less fuel if it is driven at a lower speed *(1 mark)*.

2) a) Weight = mass × g = 85 kg × 10 N/kg = 850N *(1 mark)*

b) i) The mass of the satellite is the same everywhere = 85kg. *(1 mark)*
ii) Weight = mass × g = 85kg × 6N/kg = 510N *(1 mark)*

c) At the height of the satellite's orbit there will be little or no air *(1 mark)*, so the drag force on the satellite will be very small or non-existent. *(1 mark)*.

3) a) Jenny's dad can move closer to the pivot *(1 mark)* or Jenny can move further away from the pivot *(1 mark)*.

b) Jenny's clockwise moment = 200 × 3.0 = 600 Nm *(1 mark)*

Dad's anticlockwise moment = 800 × D = 800D Nm *(1 mark)*

For equilibrium, the total clockwise moment = the total anticlockwise moment *(1 mark)*. Hence, 800D = 600, D = 600/800 = 0.75m *(1 mark)*.

c) Jenny's clockwise moment = 200 × 0.3 = 600 Nm *(1 mark)*

Alex's clockwise moment = 400 × 2.0 = 800 Nm *(1 mark)*

Therefore, the total clockwise moment = 800 + 600 = 1400Nm *(1 mark)*

Dad's anticlockwise moment = 800E Nm *(1 mark)*

Therefore, for equilibrium, 800E = 1400, E = 1400/800 = 1.75m *(1 mark)*

d) Weight = mass × g, therefore mass = weight/g = 200/10

= 20kg *(1 mark)*

e) No, they would sit in exactly the same places *(1 mark)* because the gravitational field strength will affect both sides of the seesaw in the same way *(1 mark)*.

## Page 63 (Warm-up Questions)

1) a) balanced

b) unbalanced (the Moon is travelling in a circle and, therefore, has a constantly changing velocity)

c) balanced

d) unbalanced (the coin will accelerate towards the ground)

2) a) F = ma = 900kg × 3m/s² = 2700N

b) a = F/m = 18N/3kg = 6m/s²

3) From Newton's Third Law the trailer will exert a force of 100N on the tractor.

## Page 64-65 (Exam Questions)

1) a) Weight = mass × g = 2 700 000kg × 10N/kg = 27 000 000N (=2.7 × 10⁷N) *(1 mark)*

b) Each engine produces 7 000 000N of thrust, therefore the total thrust = 5 × 7 000 000N = 35 000 000 N (=3.5 × 10⁷N) *(1 mark)*

The resultant force = thrust – weight = 35 000 000 – 27 000 000N

= 8 000 000N (= 8.0 × 10⁶N) *(1 mark)*

c) Acceleration = resultant force/mass of rocket = 8 000 000N/2 700 000kg *(1 mark)* = 2.96m/s² ≈ 3.0m/s² *(1 mark)*

d) The acceleration of the rocket will increase *(1 mark)* because the mass of the rocket will decrease as fuel is burnt *(1 mark)*. (Another possible, but less good, explanation is that the weight of the rocket decreases since the gravitational field strength decreases as the rocket rises above the Earth.)

e) The rocket exerts a force on its exhaust gases backwards *(1 mark)*. From Newton's Third Law, the exhaust gases exert an equal force on the rocket forwards, causing it to accelerate *(1 mark)*.

f) The rocket will travel at a constant velocity *(1 mark)*. This happens because the resultant force on the rocket is zero *(1 mark)*. This fact arises from Newton's First Law *(1 mark)*.

2) a)

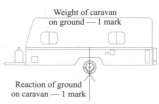

*(1 mark for each correct force)*

The two forces are equal in size *(1 mark)*

b) If the car and caravan are travelling at a constant velocity, the resultant force on the caravan must be zero *(1 mark)* from Newton's First Law. Therefore the force of the car on the caravan is equal and opposite to the drag force on the caravan = 2300N *(1 mark)*.

c) The resultant force on the car and caravan is zero *(1 mark)*. Therefore the engine force = drag force on car + drag force on caravan.

= 1800N + 2300N = 4100N *(1 mark)*

d) The extra force needed to accelerate the car and caravan = total mass × acceleration = (1000kg + 800kg) × 2m/s² = 3600N *(1 mark)*

Therefore, the new thrust force = total drag force + acceleration force = 4100N + 3600N = 7700N *(1 mark)*

3) a) i)

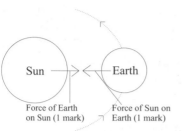

Sun → ← Earth

Force of Earth on Sun (1 mark)    Force of Sun on Earth (1 mark)

ii) From Newton's Third Law, these forces are the same size. *(1 mark)*

b) i) When you kick the ball, you apply a force to the ball. This force is unbalanced *(1 mark)* and so, according to Newton's Second Law, the ball accelerates away *(1 mark)*.

ii) When your foot exerts a force on the ball, from Newton's Third Law *(1 mark)*, the ball exerts an equal and opposite force on your foot, causing you to feel the ball *(1 mark)*.

## Page 70 (Warm-up Questions)

1)    s = d/t = 280/8 = 35m/s

2)    a = Δv/t = 15/3 = 5m/s²

3) a) d = s × t = 16 × 6 = 96m

   b) K.E. = ½ mv² = ½ × 900 × 16² = 115 200J

4) a) Any two from: how fast you're going, how alert (or not) you are; how bad the visibility is.

   b) Any two from: how fast you're going; how heavily loaded the vehicle is; how good your brakes are; how good the grip is.

## Page 71 (Exam Questions)

1) a) i) Either: a = Δv/t *(1 mark)* = (24 − 6)/12 *(1 mark)* = 1.5m/s² *(1 mark)*

   or a = gradient = vertical/horizontal *(1 mark)* = (24 − 6)/12 *(1 mark)* = 1.5m/s² *(1 mark)*

   ii) F = ma = 1200kg × 1.5m/s² *(1 mark)* = 1800N *(1 mark)*

   iii)

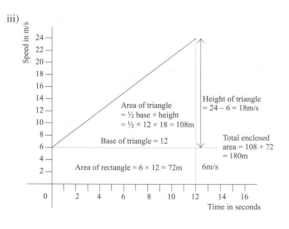

Area of triangle = ½ base × height = ½ × 12 × 18 = 108m

Height of triangle = 24 − 6 = 18m/s

Base of triangle = 12

Total enclosed area = 108 + 72 = 180m

Area of rectangle = 6 × 12 = 72m    6m/s

Distance travelled  = area under graph between 0s and 12s

= area of rectangle + area of triangle *(1 mark)*

= (6 × 12) + (½ × 12 × 18) = 72 + 108 *(1 mark)*

= 180m *(1 mark)*

b) i) Resultant force must be zero *(1 mark)* because the car is not accelerating, and hence the forces must be balanced *(1 mark)*.

   ii) The kinetic energy = ½ mv² *(1 mark)* = ½ × 1200 × 24² *(1 mark)* = 345 600J *(1 mark)*

   iii) The work done by the brakes in stopping the car = the kinetic energy transferred *(1 mark)*. Therefore forces of brakes × braking distance = 345 600J; therefore force = 345 600/50 *(1 mark)* = 6912N *(1 mark)*

iv) The force calculated above = the total opposing force *(1 mark)* which will include the drag force as well as the braking force *(1 mark)*.

v) The car's braking distance would be greater *(1 mark)* because the car's extra mass would mean that it would have more kinetic energy and, therefore, would require more work to be done in order to stop it *(1 mark)*.

## Page 75 (Warm-up Questions)

1) a)   10mm    b)   15mm

2) a)   40cm³    b)   160cm³

3)    A 40N force produces an extension of 5cm. Therefore, every cm of extension needs a force of 8N.

   a)   16N (A length of 12cm means an extension of 2cm. Therefore, the force needed is 2 × 8N = 16N).

   b)   24N

4)    The correct answer is (b).

5)    48cm³ (Use $P_1V_1 = P_2V_2$)

## Page 76-77 (Exam Questions)

1) a)

| Force in N | Ruler reading in mm | Extension in mm |
|---|---|---|
| 0 | 120 | 0 |
| 2 | 126 | 6 |
| 4 | 132 | 12 |
| 6 | 138 | 18 |
| 8 | 144 | 24 |

*(1 mark if all of the extensions are correctly calculated)*

b)

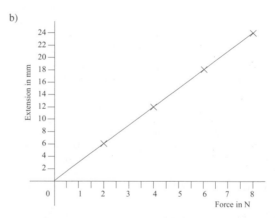

*8 marks total —*
1 mark for a sensible scale on each axis *(2 marks)*;
1 mark for having a label and correct unit on each axis *(2 marks)*;
*3 marks* for all points plotted correctly (deduct 1 mark for each incorrectly plotted point);
1 mark for a straight line passing through all of the points.

c) Force to produce an extension of 15mm = 5N. *(1 mark)*

d) 10N produces a reading of 55mm; 80N produces a reading of 90mm. An extra 70N stretched the spring by 90 − 55 = 35mm *(1 mark)* The force needed to stretch spring by 1mm = 70/35 = 2N *(1 mark)*. The unknown weight produces an extension of 75 − 55mm = 20mm and, therefore, must have an extra force of 20mm × 2N/mm = 40N *(1 mark)*. Therefore, the unknown weight = 10N + 40N = 50N *(1 mark)* *(4 marks for the correct answer, irrespective of method. Another way of doing the question is to notice that an extra 10N is needed for each 5mm that the spring stretches).*

e) i) The elastic limit is the point above which the spring will not return to its original length when the weights are removed *(1 mark)* and below which the spring will return to its original length when the weights are removed *(1 mark)*. *(2 marks maximum for any reasonable explanation)*

   ii) Some reference should be made to the fact that, above the elastic limit, the extension is no longer proportional to the load or Hooke's law is no longer obeyed or spring is permanently stretched *(1 mark)* resulting in the weighing scales being inaccurate from then on (for all weights) *(1 mark)*.

2) a)

| Pressure in kPa | Volume in m³ | 1/Volume in m⁻³ |
|---|---|---|
| 300 | 0.00017 | 6000 |
| 250 | 0.00020 | **5000** |
| 200 | 0.00025 | **4000** |
| 150 | 0.00033 | 3000 |
| 100 | 0.00050 | **2000** |

*1 mark* for each correct value of l/volume calculated *(3 marks maximum)*

b)

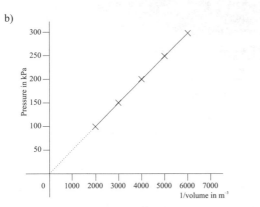

Graph: 1 mark for a suitable scale on each axis *(2 marks)*; 1 mark for a correct label and a correct unit on each axis *(2 marks)*; award *1 mark* if all of the points are plotted correctly; *1 mark* for a straight line that passes through all of the points and which can be extrapolated through the origin.

c) The graph is a straight line that passes through the origin *(1 mark)*, showing that the pressure is directly proportional to l/volume *(1 mark)* which is the same as saying that pressure is inversely proportional to volume, which is Boyle's law *(1 mark)*.

d) Either (i) the student could multiply together each pair of values for pressure and volume *(1 mark)* and show that the same answer is obtained each time *(1 mark)*, or (ii) the student could plot a graph of pressure versus volume *(1 mark)* and take readings to show that when the pressure is doubled the volume is halved and vice versa *(1 mark)*.

e) At 75 kPa, the value for l/volume is 1500m⁻³ *(1 mark)*, so the volume = 1/1500 = 0.0006666m³ = 0.000 67m³ *(1 mark)*

f) The student paused for two reasons:

i) To allow the temperature to return to room temperature (a gas cools down when it expands and heats up when it is compressed *(1 mark)*.

ii) To allow the oil to flow down the sides of the glass tube so that the correct reading of the volume could be taken *(1 mark)*.

g) i) 63kPa (62.5kPa) *(2 marks)*

ii) 42kPa (41.7kPa) *(2 marks)*

(Use $P_1V_1 = P_2V_2$ in each case, or read off graph. 1 mark (maximum) should be deducted if the unit is missing or incorrect in either answer).

## Page 81 (Warm-up Questions)

1) A = Wavelength, B = Amplitude.

2) transverse, e.g. light, radio waves, water waves, microwaves longitudinal, e.g. sound, shock waves

3) They can cause ionisation/damage to cells.

## Page 82 (Exam Questions)

1) a) 20 waves pass a given point each second. *(1 mark)*

b) T = 1/f = 1/20 s = 0.05 s *(2 marks)*

c) Two from: light waves, radio waves, microwaves, X-rays, gamma rays, waves on strings, water waves, etc. *(2 marks)*.

2) a)

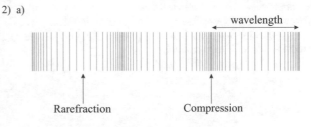

*(3 marks — 1 mark for each correct label)*

b) Any reasonable answer, e.g. shock waves, sound waves. *(1 mark)*

## Page 88 (Warm-up Questions)

1) vibrating, longitudinal, solid, gas, vacuum.

2) a) 35 000 vibrations (waves) each second.

b) Too high for people to hear. Not too high for dogs to hear.

3) Drum — 100Hz; Squeak — 8000Hz; Ultrasound — 40 000Hz.

## Page 89-90 (Exam Questions)

1) a) B (biggest amplitude, therefore loudest).

b) D (longest time period).

c) C

d) A & D (same "height").

*(4 marks available — 1 mark for each correct answer)*

2) a) Sound waves *(1 mark)*, with a frequency too high for humans to hear *(1 mark)*.

b) X-rays might damage foetus — ultrasound is much safer. *(1 mark)*

c) Waves are reflected from boundaries between different media. Reflected waves are used to produce an image. *(2 marks)*

d) Two of: cleaning, breaking kidney stones, quality control, sonar *(2 marks)*

e) 35kHz (only answer bigger than 20 000 Hz). *(1 mark)*

## Page 93 (Warm-up Questions)

1) v = fλ = 2500 × 13.2 / 100 = 330 m/s

2) f = v/λ = 3 × 10⁸ ÷ 1500 = 200 000 Hz = 200 kHz

3) distance = speed × time = 330 × 0.8 = 264 m

4) s = vt = 1400 × 0.7 = 980m, so sea is 980/2 = 490m deep

## Page 94 (Exam Questions)

1) a) wavelength = speed/frequency
   = 1400/28 000   *(1 mark)*
   = 0.05m   *(1 mark)*

b) distance = speed × time
   = 1400 × 0.2   *(1 mark)*
   = 280m *(1 mark)* this is distance there and back
   So, distance to seabed is 280/2 = 140m *(1 mark)*

## Page 101 (Warm-up Questions)

1) refraction, dispersion, total internal reflection.

2) violet

3) The largest angle of incidence at which light will still be refracted.

## Page 102-103 (Exam Questions)

1) a) *a* = Angle of Incidence, *b* = Angle of Refraction. *(2 marks)*

b) Refraction. *(1 mark)*

c) i)

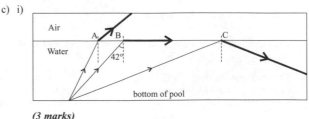

*(3 marks)*

ii) total internal reflection *(1 mark)*

iii) Light travels slower in water than in air, so at A the wave speeds up. Because it meets the air at an angle, it speeds up on one side first, so the direction changes. *(1 mark)*

2) a) Any two of: direction, wavelength and speed *(2 marks)*

b) air into glass: the ray bends towards the normal *(1 mark)*
glass into air: the ray bends away from the normal *(1 mark)*

c) Frequency stays the same. *(1 mark)*

3) a) 1 — Red; 2 — Orange; 3 — Yellow; 4 — Green; 5 — Blue; 6 — Indigo; 7 — Violet. *(2 marks)*

b) Each frequency of light is travelling at a slightly different speed to start with, so each is bent by a slightly different amount. *(1 mark)*

c)

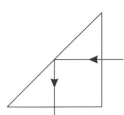

*(1 mark for 45°-45°-90° prism, 1 mark for rays drawn correctly)*

## Page 109 (Warm-up Questions)

1) They give better quality reflection than a mirror, they're easier to hold in place and they're more robust.

2) signal doesn't need boosting as often, can carry more information, more secure, less interference

3) Analogue examples include: clock with hands, thermometer, speedometer, dimmer switch
Digital examples include: ordinary light switch, any other on/off switch, digital clocks, electronic scales

4)
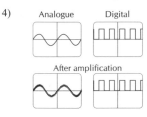

## Page 110-111 (Exam Questions)

1) a) Signal can take only two different values ('on' and 'off') *(1 mark)*

b) Analogue: dimmer switch, thermometer, speedometer *(1 mark)*
Digital: electronic scales, light switch *(1 mark)*

c) It gets less and less like the original and there will be more noise *(2 marks)*

d)  *(2 marks)*

e) E.g. mobile phones, cable TV, digital radio *(1 mark)*

2) a) 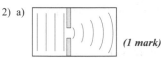 *(1 mark)*

b) Diffraction *(1 mark)*

c) 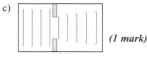 *(1 mark)*

d) The waves would diffract less — they would be less curved. *(1 mark)*

e) Long wavelength radio waves. *(1 mark)*

## Page 115 (Warm-up Questions)

1) Ultraviolet.

2) a) E.g. Cooking, satellite signals.

b) E.g. Night vision, remote controls, toasters.

c) E.g. Kill cancer cells, kill bacteria.

3) Radio waves.

## Page 116-117 (Exam Questions)

1) a) Radio waves have a longer wavelength / lower frequency. *(1 mark)*

b) They have different frequencies (shorter for cooking). *(1 mark)*

c) The microwaves are absorbed by living tissue and the heat will damage/kill the cells. *(1 mark)*

2) a) i) Danger — Skin cancer; Use — Fluorescent tubes/security marking.

ii) Danger — Cancer; Use — Detecting broken bones.

iii) Danger — Burns; Use — Night vision/remote controls.

iv) Danger — Blindness; Use — Fibre-optic communication/endoscopes.
*(8 marks)*

b) They all travel at the same speed in a vacuum. *(1 mark)*

## Page 121 (Warm-up Questions)

1) S-waves are transverse, P-waves are longitudinal.

2) The Earth is made up of <u>four</u> layers. These layers are the <u>crust</u>, the <u>mantle</u>, the <u>outer</u> core and the <u>inner</u> core.

3) Iron and nickel.

## Page 122 (Exam Questions)

1) a) Y is the mantle, Z is the core. *(2 marks)*

b) P-waves are longitudinal. *(1 mark)*

c) The outer core is not solid (hence it must be liquid). *(1 mark)*

d) P-waves travel faster through the middle of the core. *(1 mark)*

e) The paths curve due to the increasing density of the mantle (the waves are refracted). *(1 mark)*

## Page 126 (Warm-up Questions)

1) They look like they can fit together like a jigsaw, suggesting they were once joined together but have moved apart.

2) E.g. Matching fossils, identical rock sequences, same living creatures.

3) He couldn't give a convincing reason why it happened, and he wasn't a qualified geologist.

## Page 127 (Exam Questions)

1) a) The line along which two or more tectonic plates (or parts of tectonic plates) meet. *(1 mark)*

b) When tectonic plates rub together, there is a lot of friction, meaning the plates often get stuck *(1 mark)*. This results in a build up of force and then a sudden movement of one or both of the plates *(1 mark)*.

2) a) The sediments were pushed together so that they folded up, creating mountains. *(1 mark)*

b) The plates are still moving together. This pushes the mountains higher by a few cm each year *(1 mark)*.

c) As the plates move apart, magma keeps rising up to fill the gap. The water cools this magma, leaving a ridge of new crust under the Atlantic Ocean. *(2 marks — 1 mark for each of the main points)*

## Page 133 (Warm-up Questions)

1) Pluto.

2) Bodies (e.g. lumps of rock) that enter the Earth's atmosphere, sometimes colliding with the Earth.

3) Light from the Sun reflects off the Moon, and that's what we can see from Earth.

4) Their orbits look strange compared with the constellations, which makes it look like they just 'wander' randomly amongst the stars seen at night.

5) Venus, Mars, Jupiter or Saturn (any two).

6) Any phrase giving the correct order, e.g. My Very Energetic Maiden Aunt Just Swam Under North Pier.

## Page 134 (Exam Questions)

1) a) Mercury, Venus, Earth and Mars (in any order). *(1 mark)*

   b) Gravity attracts the Earth towards the Sun. The Earth is travelling fast enough not to get pulled towards the Sun, but isn't fast enough to escape the pull of gravity altogether, so instead it stays in a steady orbit around the Sun. *(2 marks)*

   c) Slightly elliptical or oval shaped. *(1 mark)*

   d) Large lumps of rock (roughly 1-1000km diameter) orbiting the Sun, mostly in a belt between the orbits of Mars and Jupiter. *(1 mark)*

   e) It's now too far away to be seen, and it has no tail because it isn't near the Sun any more. *(2 marks)*

## Page 137 (Warm-up Questions)

1) The moon.

2) E.g. monitoring weather, communications, space research, spying.

3) There is no atmosphere in the way to blur the picture.

4) It has increased.

5) Some of the signal is lost into space so could in theory be picked up by aliens, if they had the correct equipment.

## Page 138 (Exam Questions)

1) a) Has an orbit of 24 hours, is situated over the equator and travels eastwards. *(3 marks)*

   b) Nearer to the Earth to see detail, travels over the whole of the Earth. *(2 marks)*

2) a) Go to other places, land and analyse samples or send back pictures. *(1 mark)*

   b) Any two from: Look for intelligent signals coming from space. Look for chemical changes in the atmospheres of other planets. Study the light coming from planets in space to find out what the surface is like. *(2 marks)*

## Page 142 (Warm-up Questions)

1) moon, planet, sun, solar system, galaxy, universe.

2) Nuclear fusion (usually of hydrogen to produce helium).

3) a) An independent group or collection of stars (held together by gravity).

   b) gravity

   c) rotate/spin

4) It has such large gravity, that even light cannot escape (so it looks black).

## Page 143 (Exam Questions)

1) Points to include: <u>hydrogen in core will run out</u>; it will <u>expand</u> and form a <u>red giant</u>; it will <u>cool and contract</u> to become a <u>white dwarf</u> *(1 mark each underlined point)*.

2) a) Either a neutron star or a black hole. *(1 mark)*

   b) Heavier elements are made in the final stages of a massive star, and during the supernova explosion. The supernova pushes these into space. They are brought together again when a new solar/star system is made. *(3 marks — 1 mark for each point)*

## Page 148 (Warm-up Questions)

1) All directions.

2) Microwave.

3) Universe has expanded (and cooled).

4) The Universe is the same everywhere, and always has been.

5) Matter is being created to fill in the gaps as it expands (so it still looks the same).

6) Explosion that started the Universe.

7) Accept answer between 12 and 16 billion years.

8) Gravity.

## Page 149 (Exam Questions)

1) a) Red shift or Doppler effect. *(1 mark)*

   b) It's moving away from us. *(1 mark)*

   c) It's moving away faster than the near one. *(1 mark)*

   d) It's expanding. *(1 mark)*

   e) By how quickly the Universe is expanding. *(1 mark)*

## Page 155 (Warm-up Questions)

1) a) sound

   b) kinetic

   c) thermal / heat

2) a) 105J

   b) The Principle of Conservation of Energy.

3) a) Chemical to Elastic Potential.

   b) Elastic Potential to Kinetic.

   c) Kinetic to Gravitational Potential.

## Page 156 (Exam Questions)

1) a) electrical *(1 mark)*

   b) 3000 J *(1 mark)*

   c) Efficiency = Useful Energy Output ÷ Total Energy Input
      = 2000 ÷ 5000 = 0.4 or 40%. *(2 marks)*

   d) **Chemical**    Electrical
      *(2 marks)*

   e) 12% efficient, so useful output is 12% of 2000 J = 240 J *(2 marks)*

## Page 162 (Warm-up Questions)

1) a) Force

   b) Potential Energy

   c) ½, Velocity$^2$

2) a) Power

   b) watts

3) a) Yes

   b) No

   c) Yes

## Page 163-164 (Exam Questions)

1) a) Work Done = Force × Distance
      = 2000 × 30
      = 60 000J *(2 marks)*

   b) Power = Work Done ÷ Time taken
      = 60 000 ÷ 120
      = 500W *(2 marks)*

   c) Potential Energy = Mass × g × Height
      = 800 × 10 × 4
      = 32 000 J *(2 marks)*

   d) Some of the work done is wasted in overcoming friction and producing heat energy. *(1 mark)*

2) a) Potential Energy = Mass × g × Height
      = 0.5 × 10 × 50 = 250J *(2 marks)*

   b) 250J (because the potential energy lost is changed into kinetic energy). *(1 mark)*

   c) Kinetic Energy = ½ × Mass × Velocity$^2$
      250 = ½ × 0.5 × Velocity$^2$
      250 = 0.25 × Velocity$^2$
      250 ÷ 0.25 = Velocity$^2$
      1000 = Velocity$^2$
      Velocity = √1000 = 31.6m/s *(3 marks)*

   d) If it loses 99% of its energy it will also lose 99% of its height *(1 mark)* so the height of the first bounce will be 0.5m *(1 mark)*.

## Page 172 (Warm-up Questions)

1) White is a poor absorber of heat, so less heat from the Sun will enter the houses.

2) a) radiation,   b) conduction, c) convection.

3) The high initial cost means that it will take a long time for the energy savings to pay back the money spent installing it.

## Page 173-174 (Exam Questions)

1) a) Free electrons in the metal gain energy *(1 mark)*. They move quickly through the metal carrying the energy with them *(1 mark)*.

   b) Heat loss by radiation *(1 mark)* is increased by the dark, dull surface *(1 mark)*.

   c) Convection. *(1 mark)*

   d) Air is a very poor conductor of heat *(1 mark)*. It cannot transfer heat by convection because the air is trapped in pockets and cannot move *(1 mark)*.

   e) The cost of installing cavity wall insulation is high *(1 mark)* so the payback period is likely to be high, even with the saving of £65 per year *(1 mark)*.

2) a) Element is black, which is a good radiator of heat *(1 mark)*. Fins create a large surface area *(1 mark)*.

   b) A metal *(1 mark)*. Because it will quickly conduct the heat away (from the coolant to the air/into the fins) *(1 mark)*.

   c) Heat transferred by electromagnetic waves or infrared waves. *(1 mark)*

   d) The silver lining. *(1 mark)*

   e) The outer particles gain heat energy from the surroundings, giving them extra kinetic energy/making them vibrate more *(1 mark)*. Some of the energy is passed on when they collide with neighbouring particles *(1 mark)*.

   f) Any sensible answer, e.g. trapping air bubbles in the plastic or replacing with foam containing air bubbles. *(1 mark)*

## Page 179 (Warm-up Questions)

1) Renewable — will never run out.
Non-renewable — will run out.
Renewable — any four of: wind, waves, tides, hydroelectric, solar, geothermal, food, biomass (wood).
Non-renewable — coal, oil, natural gas, nuclear fuels (uranium and plutonium).

2) They are often expensive, many are unreliable as they depend on the weather, they can harm the environment visually (e.g. wind turbines).

3) Sun → light energy → plants → photosynthesis → BIOMASS.
Sun → heats atmosphere → creates WIND → and therefore WAVES.
Sun → heats seawater → clouds → rain → HYDROELECTRICITY.

4) It has a boiler containing water which is heated to produce steam; the steam is used to drive a turbine; the turbine is attached to a generator. The generator turns to produce the electricity.

5) Carbon dioxide is released, adding to the Greenhouse Effect, which is causing global warming.
Burning coal and oil releases sulphur dioxide which causes acid rain.
Coal mining makes a mess of the landscape.
Oil spillages cause serious environmental problems.

## Page 180 (Exam Questions)

1) a) Any three of: wind, waves, tides, biomass. *(3 marks)*

   b) Any sensible answers, e.g. improve the insulation of their buildings (double glazing, roof insulation, draught excluders, cavity wall insulation) *(1 mark)*; turn off lights and other electrical items when they are not needed *(1 mark)*.

   c) They produce solid waste which stays dangerously radioactive for thousands of years *(1 mark)*. The power stations contain a lot of radioactive materials so dismantling them is dangerous and expensive *(1 mark)*. There is the risk of a major accident or catastrophe *(1 mark)*.

2) a) They are cheaper *(1 mark)*, they are reliable (don't depend on weather) *(1 mark)*.

   b) Carbon dioxide is released, adding to the Greenhouse Effect, which is causing global warming.
Burning coal and oil releases sulphur dioxide which causes acid rain.
Coal mining makes a mess of the landscape.
Oil spillages cause serious environmental problems. *(4 marks)*

## Page 185 (Warm-up Questions)

1) This involves putting wind turbines in exposed places. Each wind turbine has its own generator inside it so the electricity is generated directly from the wind turning the blades.

2) A dam is built to capture rainwater and flood a valley. Water is allowed out through turbines, which drive a generator. Diagram should include all features on the diagram at the bottom of page 181.

3) Pumped storage is a method of storing energy during times of low energy demand. Water is pumped up from a low reservoir to a higher one while demand is low, then when demand is high the water can be released to flow through the turbines, increasing electricity supply.

4) Wave power uses energy from waves, which are caused by the wind. Wave converters around the coast rock up and down, pushing air through a turbine.
Tidal power uses the energy from the tides, which are caused by the gravitational attraction of the Moon. A dam is built over an estuary, which fills up when the tide comes in, turning the turbines as it goes. The water is then allowed back through turbines at a controlled speed.

## Page 186 (Exam Questions)

1) a) wave power *(1 mark)*

   b) tidal power *(1 mark)*

   c) a turbine *(1 mark)*

   d) tidal power *(1 mark)*

2) a) wave; hydroelectric; tidal; wind *(4 marks)*

   b) the mountain valley in Wales *(1 mark)*

## Page 191 (Warm-up Questions)

1) Water is pumped in pipes down to hot rocks and returns as steam to drive a generator.

2) cost is the main drawback — it's very expensive to drill down far enough

3) It doesn't cause a problem with the greenhouse effect because the $CO_2$ released in the burning of the wood is balanced out by the $CO_2$ used up as the trees were growing in the first place. The trees are grown as quickly as they are burnt so they will never run out.

4) Burning rainforests is bad for the environment because the trees take a very long time to grow and cannot be replaced as quickly as they are burned. This results in an increase in $CO_2$ levels, contributing to the greenhouse effect. It's also bad for the various species of plants and animals, whose habitat is destroyed.

5) Solar cells — creating electric currents directly from the sunlight.
Solar panels — heating water in water pipes, e.g. to feed directly into hot water tanks.
Solar furnaces — using mirrors to focus light to heat water to very high temperatures, producing steam to drive a turbine

## Page 192-193 (Exam Questions)

1) a) (i) mostly nuclear energy (radioactive decay) *(1 mark)*

      (ii) It is extremely expensive to drill down far enough to heat the water sufficiently. *(1 mark)*

   b) Advantages include: quite far north so high number of daylight hours in summer
Disadvantages include: it's often cloudy in Scotland, so at certain times of the year there will be little energy produced. Also fewer daylight hours in winter.
*(2 marks for a reasonable advantage and disadvantage)*

2) a) There are many possible answers — here are two examples:
Burning coal — produces many pollutant gases which can cause acid rain OR increases carbon dioxide levels, adding to the greenhouse effect etc.
Nuclear Fission — increases levels of background radiation in areas around the power station OR poses risk of major disaster if radiation leaks etc. *(4 marks — 1 mark for each example and 1 mark for each effect)*

b) Points include:
Most renewables have a bad visual impact, e.g. wind farms, tidal barrages. The amount of electricity produced is often very low compared with that gained from non-renewable plants.
The equipment required to harness renewable energy is often expensive for relatively small gain.
Energy production from renewables can be unreliable, as it relies on the weather, time of year, etc.
*(2 marks for 2 sensible arguments against renewable energy)*

c) Reasons include:
The $CO_2$ produced when the trees are burned is removed by newly growing trees. Because the trees are grown at the same rate as they are burned, there is no overall increase in $CO_2$ levels. When fossil fuels are burned, they increase the level of $CO_2$ and other pollutants, as there is nothing to absorb the $CO_2$. Also fossil fuels are not renewable, whereas trees are. *(2 marks for making 2 sensible points)*

d) geothermal or solar *(1 mark)*

## Page 201 (Warm-up Questions)

1) C

2) F

3) Because certain rocks, which aren't distributed equally across the UK, are naturally more radioactive.

4) A gamma source.

## Page 202-203 (Exam Questions)

1) a) A — Because most particles went straight through, most of the gold atom must have been empty space. *(1 mark)*

B — Some particles were deflected a lot, implying they collided with something. This suggests most of the mass is concentrated in a small part of the whole volume (the 'nucleus'). *(1 mark)*

C — Some of the (positively charged) α-particles were repelled slightly, implying that the nucleus must also be positively charged. *(1 mark)*

b) No. Alpha radiation does not penetrate very far into metals, so all of it would have been blocked by a sheet of gold as thick as a sheet of paper, making the experiment impossible. *(2 marks)*

c) beta radiation *(1 mark)*

2) a) Because the unstable isotopes will decay to become other, stable elements. *(1 mark)*

b) When unstable nuclei decay they give out radiation which can be detected. *(1 mark)*

c) Alpha radiation is more strongly ionising, because of the size of its particles. Being relatively large, they are more likely to hit atoms and knock electrons off than a gamma ray is. *(2 marks)*

d) While it is true that you are shielded from cosmic rays down a mine, they are not the only source of background radiation. There may well be radiation coming from the rocks that are all around you underground. *(2 marks)*

## Page 208 (Warm-up Questions)

1) a) gamma   b) beta   c) gamma

2) a) becquerel = one radioactive nucleus decaying per second

b) Geiger counter = device used to detect radiation

c) ionisation = atom gaining or losing electrons

3) Any from:
Keep in lead-lined box.
Replace as soon as experiment complete.
Point source away from body.
Keep source at arm's length.
Handle with tongs.
Keep students at a safe distance.

## Page 209 (Exam Questions)

1) a) Some beta particles will pass through the paper. *(1 mark)* But the amount passing through will be affected by the thickness of the paper. *(1 mark)*

b) The amount of radiation detected will decrease. *(1 mark)*

c) The amount of beta particles emitted by the source is decreasing. *(1 mark)*
The device makes the paper thinner to keep the amount of beta particles reaching the detector constant. *(1 mark)*

d) Use a source with a much longer half-life. *(1 mark)*

e) Because the metal sheet will absorb almost all of the beta particles. *(1 mark)*

f) gamma *(1 mark)*

## Page 214 (Warm-up Questions)

1) a) Uranium-238  b) Carbon-14  c) Carbon-14

2) a) Beta  b) Alpha  c) Alpha  d) Gamma  e) Gamma  f) Beta

## Page 215 (Exam Questions)

1) a) background radiation   *(1 mark)*

b)

| Corrected Count Rate in Counts per Minute |
|---|
| 400 |
| 283 |
| 200 |
| 142 |
| 100 |

*(2 marks for all values correct, 1 mark for 3 or 4 values correct)*

c)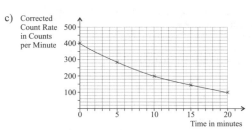

*(2 marks for all points correct, 1 mark for 3 or 4 points correct)*

d) half-life is 10 minutes   *(1 mark)*
This is the time it takes for the activity of the sample to halve. *(1 mark)*

e) 40 minutes   *(1 mark)*
Because the sample must go through another two half-lives, making a total of four since the beginning of the experiment. *(1 mark)*

## Page 221 (Warm-up Questions)

1) Digital: team's score, shoe size
Analogue: height, runner's time in race

2) diode laser (or just 'laser')

3) AM signals have a longer range than FM signals.

4) A ground wave.

5) Any three of: CD, hard drive, magnetic tape, floppy disk, vinyl record or any other reasonable answer.

## Page 222-223 (Exam Questions)

1 a) B *(1 mark)*

b) (i) Principle 2  (ii) Principle 3  (iii) Principle 1   *(3 marks)*

c) Principle 2 *(1 mark)*

2 a) The loudspeaker *(1 mark)*

b) The decoder *(1 mark)*

c) The tuner *(1 mark)*

d) The amplifier *(1 mark)*

e) Transmitter and receiver *(1 mark)*

f) An optical fibre *(1 mark)*

## Page 228 (Warm-up Questions)

1) C

2) Examples include:
- A driving oscillator making a mass on a string vibrate at its natural frequency.
- A column of air inside an oboe made to vibrate when someone blows down the oboe.
- In a microwave oven, the microwaves making the water molecules in the food vibrate at their natural frequency.
- A pendulum being pushed. (The pendulum could be a child on a swing.)
- A singer hitting a high note, causing a wine glass to vibrate or even crack.
- Bridges and towers made to resonate by gusts of wind.

3) B

4) It will make the note (a) higher, (b) higher, (c) lower.

## Page 229-230 (Exam Questions)

1) a) The rate at which Amanda's swing oscillates having been given one push and then left is the natural frequency of the swing. The rate at which Seth pushes the swing is the driving frequency. When the driving frequency matches the natural frequency of the swing then resonance occurs and the amplitude of the oscillation increases. *(3 marks)*

   b) (i) The person pushing the swing transfers energy from his or her muscles to make the swing move. *(1 mark)*

   (ii) The microwaves transfer some of their energy to make the water molecules in the soup vibrate, heating up the soup. *(1 mark)*

   c) 
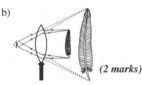
   *(2 marks)*

2) a) (i) 
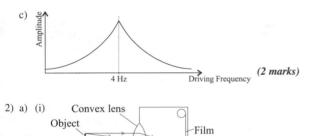
   *(2 marks)*

   (ii) Real, diminished and inverted. *(3 marks)*

   b) 

   *(2 marks)*

3) a) A node is a point where the string does not move in the standing wave pattern. *(1 mark)*

   b) There are two nodes — one where the string is fixed to the wall and one at the oscillator. *(1 mark)*

   c) Nodes occur at frequencies that are multiples of the fundamental frequency. *(1 mark)*

   d) Half a wavelength. *(1 mark)*

   e) (i) frequency = 3F *(1 mark)*

   (ii) 4 nodes *(1 mark)*

   (iii) 1½ wavelengths *(1 mark)*

## Page 236 (Warm-up Questions)

1) It will double as well. The Kelvin temperature of a gas is proportional to the average KE of the molecules.

2) Fundamental: up-quark, electron, down-quark, positron.
Non-fundamental: neutron, proton.

3) The thorium is more likely to decay by alpha emission because it has more than 82 protons in the nucleus.
The cobalt is more likely to decay by beta emission because it has fewer than 82 protons.

4) It emits light; this effect is used in the phosphorescent screens of computers, oscilloscopes and TVs.

## Page 237-238 (Exam Questions)

1) a) Pressure on the walls of a container arises from the force that the particles exert on the walls when they collide with them. As the temperature drops and the particles move more slowly this force decreases, tending towards an idealized limit at absolute zero where they would not move or collide at all and the pressure would be zero. *(3 marks)*

   b) First convert the temperatures from Celsius to kelvin by adding 273.
   $T_1 = 273 + 300 = 573K$. $T_2 = 273 + 21 = 294K$.
   $P_1V_1/T_1 = P_2V_2/T_2$.
   $5 \times 100/573 = 1 \times V_2/294$
   $294 \times 5 \times 100/573 = V_2 = 256.5$ ml
   *(4 marks for correct answer, otherwise 1 for each correct line of working)*

2) a) One of the 52 **neutrons** within the strontium nucleus was turned into a **proton** plus an emitted **electron** and gamma rays. *(2 marks)*

   b) Two up-quarks and one down-quark are present in a proton.
   Two down-quarks and one up-quark are present in a neutron. *(2 marks)*

   c) One down-quark from a neutron inside the parent nucleus is turned into an up-quark. That means a change in relative charge of +1, but this is balanced by the creation of an electron with its relative charge of -1. *(2 marks)*

   d) One up-quark within a proton is turned into a down-quark. That means a change in relative charge of -1, but this is balanced by the creation of a positron with its relative charge of +1. *(2 marks)*

   e) The line starts off straight (with a gradient of 1), as for small atoms to be stable the numbers of protons and neutrons must be about equal. For larger atoms to be stable more neutrons than protons are necessary, so the line curves upwards. *(2 marks)*

3) a) The reactor. *(1 mark)*

   b) The turbine. *(1 mark)*

   c) Anything that passes through the reactor, like the gas does, will become slightly radioactive. The gas must therefore be kept isolated from the general environment, unlike the steam that turns the turbines. *(1 mark)*

   d) The injection of neutrons into the reactor. *(1 mark)*

   e) The spent fuel is radioactive and dangerous to health. It remains so for many years. Therefore it must be sealed in containers and buried. *(2 marks)*

   f) A — 2; B — 4; C — 1; D — 3. *(4 marks)*

## Page 243 (Warm-up Questions)

1) Any two of thermistor, light-dependent resistor (LDR) or microphone.

2)

| A | B | A OR B |   | A | B | A NOR B |
|---|---|--------|---|---|---|---------|
| 0 | 0 | 0 |   | 0 | 0 | 1 |
| 0 | 1 | 1 |   | 0 | 1 | 0 |
| 1 | 0 | 1 |   | 1 | 0 | 0 |
| 1 | 1 | 1 |   | 1 | 1 | 0 |

3) An AND gate.

4) 

## Page 244 (Exam Questions)

1) a) and c)

(diagram of circuit)

*(2 marks for labels and 1 mark for direction)*

   b) When the switch is pressed it completes the circuit and a small current flows into the base of the transistor. This switches the transistor on and enables a large current to flow from the collector to the emitter. The lamp comes on. *(2 marks)*

c) see diagram part a)

2) a) On/off switch, light sensor inside door. *(1 mark)*

b) The motor. *(1 mark)*

c)

| A | B | C | D |
|---|---|---|---|
| 0 | 0 | 1 | 0 |
| 0 | 1 | 0 | 0 |
| 1 | 0 | 1 | 1 |
| 1 | 1 | 0 | 0 |

*(2 marks)*

d) Firstly because a logic gate or transistor does not supply enough power to run the motor. Secondly because the relay isolates the high-current motor circuit from the circuit that people come into contact with. *(2 marks)*

e) To avoid floods the machine should not start washing unless the door is closed. A light sensor inside the door gives output ON if it is receiving light and OFF if it is in darkness. The NOT gate swaps this round so that the output is ON if the sensor is in darkness, as it will be if the door is closed. *(4 marks)*

## Page 248 (Warm-up Questions)

1) $V = V_{in} \times R_2 \div (R_1 + R_2)$
$= 12 \times 60 \div (140 + 60) = 12 \times 60 \div 200 = 12 \times 0.3 = 3.6$ V

2) B

3) a longer time

4) Any two reasonable answers, e.g. sensors to prevent skidding when braking, sensors to detect impact and set off an airbag, sensors to monitor engine temperature, sensors to adjust fuel injection, or any other reasonable gauge or sensor.

## Page 249 (Exam Questions)

1) a) $V_{out} = V_{in} \times R_2 \div (R_1 + R_2)$
$V_{out} = 6 \times 8000 \div (9000 + 8000)$
$V_{out} = 48000 \div 17000 = 2.8$ V
*(2 marks for correct answer, otherwise 1 for working)*

b) $V_{out} = V_{in} \times R_2 \div (R_1 + R_2)$
$V_{out} = 6 \times 200 \div (9000 + 200)$
$V_{out} = 1200 \div 9200 = 0.13$ V
*(2 marks for correct answer, otherwise 1 for working)*

c) The new component is a NOT gate (or inverter). Before, the indicator light came on when the light shining on the LDR fell below a certain intensity. Now, with the NOT gate present, the indicator light comes on when the light shining on the LDR goes above a certain intensity.
*(3 marks — 1 for naming the component and 2 for the comparison)*

d) The bell would need more power and a higher current than the transistor can provide. *(1 mark)* The diode protects the transistor when the relay is being switched on or off. *(1 mark)*

2) a) If $R_1$ is sufficiently low, then when the switch is closed the output voltage will be enough to make a current flow into the base of the transistor, switching on a larger current from the collector to the emitter. Hence the lamp will light. *(2 marks)*

b) When the switch is closed charge flows onto the plate of the capacitor and gradually builds up the voltage across it until it reaches the threshold voltage of the transistor. Then the transistor comes on and the lamp lights. *(2 marks)*

c) The variable resistor. *(1 mark)*

## Page 253 (Warm-up Questions)

1) copper wire and steel spring

2) density = mass ÷ vol = 30 ÷ 12 = 2.5 g/cm³

3) balancing moments gives:
$8 \times 0.3 = X \times 0.6$
so $X = 2.4 \div 0.6 = 4$ N

4) C

## Page 254 (Exam Questions)

1) a) 62mm *(1 mark)*

b)

| Force (N) | 0 | 1 | 2 | 3 | 4 | 5 | 6 |
|---|---|---|---|---|---|---|---|
| Length (mm) | 62 | 67 | 72 | 77 | 82 | 91 | 105 |
| Extension (mm) | 0 | 5 | 10 | 15 | 20 | 29 | 43 |

*(2 marks)*

c) and e) *2 marks for plotting the line in the graph below*.

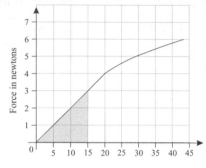

Extension in millimetres

d) about 4 N *(1 mark)*

e) See the completed graph in answer (c). The shaded area under the graph shows the energy used. Energy stored = $1/2 \times 3 \times 0.015 = 0.0225$ J. (Note the units of the extension must be converted to metres so that the answer will be in joules.)
*(2 marks)*

2) a) Volume = $5 \times 3 \times 2.6 = 39$ m³.
m = V × ρ = $39 \times 1.29 = 50.31$ kg. *(2 marks)*

b) To keep the ship upright. An object will tip over if it tilts far enough so that a vertical line through its centre of mass falls outside the edge of the base of the object; it requires a much steeper angle of tilt for this to happen when the centre of mass is near the bottom. Putting heavy ballast at the base of the ship moved the centre of mass as low as possible. *(2 marks)*

3) a) Clockwise moment = anticlockwise moment.
$F \times 0.2 = 40 \times 0.6$
$F = 40 \times 0.6 \div 0.2 = 120$N *(2 marks)*

b) Harder, because with a smaller distance between the line of action of the force and the pivot the turning effect of his push will be less and hence he will need more force. *(2 marks)*

## Page 259 (Warm-up Questions)

1) momentum = mass × velocity

2) electrostatic attraction

3) The two cannonballs will reach the ground simultaneously.
The vertical motion of a projectile is not affected by its horizontal motion.

4) energy

## Page 260 (Exam Questions)

1) a) First convert speed to m/s so final answer will come out in kg m/s.
110 km/h = 110,000 / (60 × 60) = 30.6 m/s
Change in momentum is from (1400 × 30.6) kg m/s to zero.
Change in momentum = 42 800 kg m/s (to 3 s.f.) *(2 marks)*

b) Force = change in momentum / time taken for change
Force = 42,800 / 10 = 4,280 N. *(1 mark)*

c) Any situation where the forces acting on a body need to be reduced by increasing the time for a change in momentum, such as in a car crash, where crumple zones, seat belts and airbags are designed to slow down the deceleration, or in transporting delicate objects such as eggs in boxes designed to crumple on impact. *(2 marks)*

2) a) Considering vertical motion only:
$v^2 = u^2 + 2as = 0 + (2 \times 10 \times 11.25) = 225$
v = 15 m/s *(2 marks)*

b) Considering vertical motion only:
t = (v −u) /a = (15 − 0) / 10
t = 1.5 s *(2 marks)*

c) Considering horizontal motion only:
velocity × time = distance
Distance = 400 × 1.5 = 600 m *(1 mark)*

3) a) Total distance covered going round circuit once = 2 × π × r
= 2 × 3.14 × 159.3 = 1001 m
Time to go round once = distance / speed = 1001 / 21 = 47.7 s *(2 marks)*

b) Centripetal force = $mv^2$ /r = (55 × 21²) / 159.3 = 152.3 N *(2 marks)*

c) A *(1 mark)*

4) a) Energy needed = mass × specific heat capacity × change in temperature
Energy needed = 1.2 × 380 × 80 = 36 480 J *(1 mark)*

b) That water has a very high specific heat capacity, so it takes more energy to heat it up. *(1 mark)*

c) So as to need as little energy as possible to heat the pan, since that is only a means to the end of heating whatever you are cooking inside it. *(1 mark)*

# EXAM PAPER ANSWERS

*We have used the following conventions:*

*Where units are shown in brackets, e.g. 247.6 (m/s), it means that you can get full marks even if you miss off the units.*

*Forward slashes separate alternative answers, e.g. circular / oval / egg shaped.*

## Section A

1 a) $R_{15}$ = 60 ohms, $R_{25}$ = 20 ohms *(1 mark)*
Change = $R_{15} - R_{25}$ = 40 ohms *(1 mark)*
*(Full marks for correct answer)*

b) You need to make reference to steeper gradient *(1 mark)*, and the idea of a bigger change in resistance per equal rise in temperature *(1 mark)* at temperatures lower than 15°C.
*(3 marks available in total if your answer addresses the question and your description makes the meaning clear)*

2 a) orbits circular / oval / egg shaped *(1 mark)*
*OR* slightly squashed circles / ellipses *(2 marks)*
sun at centre of orbit / orbit the sun / attracted by sun's gravity *(1 mark)*
correct reference to orbital time (longer if further from sun) *(1 mark)*
*(3 marks maximum)*

b) Any two of:
communication (between different places on Earth)
monitoring (something relevant, e.g. weather)
observing the universe / planets / stars / galaxies
*(2 marks for 2 reasonable answers)*

c) Award 1 mark for each of these points:
long thin ellipse / far from circular *(1 mark)*
rarely near the sun / much closer to sun at some times than others / sun not at centre of orbit *(1 mark)*
only seen when near the sun *(1 mark)*
(ice melts) when near to sun to form tail / tail reflects sunlight *(1 mark)*
comet moves faster when near the sun / moves slower when away from sun *(1 mark)*
*(3 marks maximum)*
*Remember — you need to make three distinct points for the three marks.*

3 a) Accept any of:
becomes semi-molten / gets hotter / dissolves / melts / melts into the mantle *(1 mark)*

b) Accept any of:
higher density / pushed by other plates / convection currents *(1 mark)*

c) (i) Accept any of:
primary / push-pull / pressure / longitudinal *(1 mark)*,
1, 2, 3, 4, 5 *(1 mark)*
Accept any of:
secondary / shake / shear / sideways / transverse *(1 mark)*,
1, 2, 5 *(1 mark)*

(ii) refraction *(1 mark)*

(iii) wave changes speed *(1 mark)*,
change in density / going from solid to liquid *(1 mark)*

4 a) force X = weight / gravity *(1 mark)*
force Y = friction / (viscous) drag / resistance *(1 mark)*
(Don't accept buoyancy, upthrust or air resistance)

b) (i) speed is increasing / acceleration / it (the ball bearing) is accelerating *(1 mark)*
force X is greater than force Y (likewise if X and Y are named) *(1 mark)*

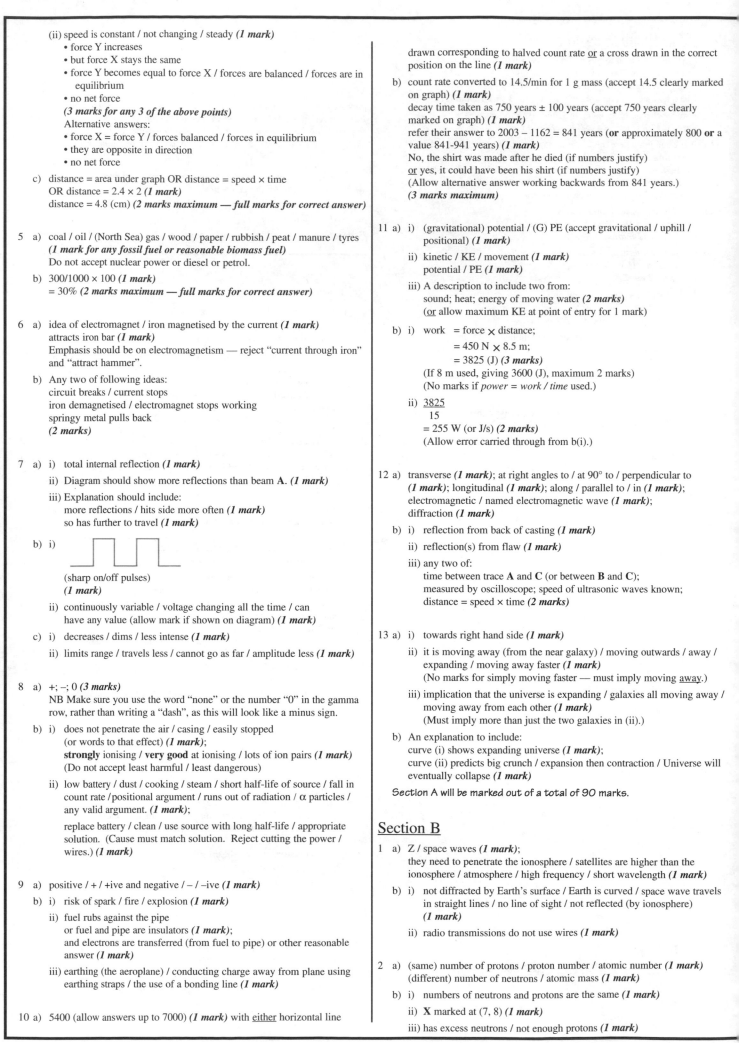

(ii) speed is constant / not changing / steady *(1 mark)*
- force Y increases
- but force X stays the same
- force Y becomes equal to force X / forces are balanced / forces are in equilibrium
- no net force

*(3 marks for any 3 of the above points)*
Alternative answers:
- force X = force Y / forces balanced / forces in equilibrium
- they are opposite in direction
- no net force

c) distance = area under graph OR distance = speed × time
OR distance = 2.4 × 2 *(1 mark)*
distance = 4.8 (cm) *(2 marks maximum — full marks for correct answer)*

5 a) coal / oil / (North Sea) gas / wood / paper / rubbish / peat / manure / tyres
*(1 mark for any fossil fuel or reasonable biomass fuel)*
Do not accept nuclear power or diesel or petrol.

b) 300/1000 × 100 *(1 mark)*
= 30% *(2 marks maximum — full marks for correct answer)*

6 a) idea of electromagnet / iron magnetised by the current *(1 mark)*
attracts iron bar *(1 mark)*
Emphasis should be on electromagnetism — reject "current through iron" and "attract hammer".

b) Any two of following ideas:
circuit breaks / current stops
iron demagnetised / electromagnet stops working
springy metal pulls back
*(2 marks)*

7 a) i) total internal reflection *(1 mark)*
ii) Diagram should show more reflections than beam **A**. *(1 mark)*
iii) Explanation should include:
more reflections / hits side more often *(1 mark)*
so has further to travel *(1 mark)*

b) i)
(sharp on/off pulses)
*(1 mark)*
ii) continuously variable / voltage changing all the time / can have any value (allow mark if shown on diagram) *(1 mark)*

c) i) decreases / dims / less intense *(1 mark)*
ii) limits range / travels less / cannot go as far / amplitude less *(1 mark)*

8 a) +; –; 0 *(3 marks)*
NB Make sure you use the word "none" or the number "0" in the gamma row, rather than writing a "dash", as this will look like a minus sign.

b) i) does not penetrate the air / casing / easily stopped
(or words to that effect) *(1 mark)*;
**strongly** ionising / **very good** at ionising / lots of ion pairs *(1 mark)*
(Do not accept least harmful / least dangerous)

ii) low battery / dust / cooking / steam / short half-life of source / fall in count rate /positional argument / runs out of radiation / α particles / any valid argument. *(1 mark)*;

replace battery / clean / use source with long half-life / appropriate solution. (Cause must match solution. Reject cutting the power / wires.) *(1 mark)*

9 a) positive / + / +ive and negative / – / –ive *(1 mark)*
b) i) risk of spark / fire / explosion *(1 mark)*
ii) fuel rubs against the pipe
or fuel and pipe are insulators *(1 mark)*;
and electrons are transferred (from fuel to pipe) or other reasonable answer *(1 mark)*
iii) earthing (the aeroplane) / conducting charge away from plane using earthing straps / the use of a bonding line *(1 mark)*

10 a) 5400 (allow answers up to 7000) *(1 mark)* with either horizontal line

drawn corresponding to halved count rate or a cross drawn in the correct position on the line *(1 mark)*

b) count rate converted to 14.5/min for 1 g mass (accept 14.5 clearly marked on graph) *(1 mark)*
decay time taken as 750 years ± 100 years (accept 750 years clearly marked on graph) *(1 mark)*
refer their answer to 2003 – 1162 = 841 years (**or** approximately 800 **or** a value 841-941 years) *(1 mark)*
No, the shirt was made after he died (if numbers justify)
or yes, it could have been his shirt (if numbers justify)
(Allow alternative answer working backwards from 841 years.)
*(3 marks maximum)*

11 a) i) (gravitational) potential / (G) PE (accept gravitational / uphill / positional) *(1 mark)*
ii) kinetic / KE / movement *(1 mark)*
potential / PE *(1 mark)*
iii) A description to include two from:
sound; heat; energy of moving water *(2 marks)*
(or allow maximum KE at point of entry for 1 mark)

b) i) work  = force × distance;
= 450 N × 8.5 m;
= 3825 (J) *(3 marks)*
(If 8 m used, giving 3600 (J), maximum 2 marks)
(No marks if *power = work / time* used.)

ii) $\dfrac{3825}{15}$
= 255 W (or J/s) *(2 marks)*
(Allow error carried through from b(i).)

12 a) transverse *(1 mark)*; at right angles to / at 90° to / perpendicular to *(1 mark)*; longitudinal *(1 mark)*; along / parallel to / in *(1 mark)*; electromagnetic / named electromagnetic wave *(1 mark)*; diffraction *(1 mark)*

b) i) reflection from back of casting *(1 mark)*
ii) reflection(s) from flaw *(1 mark)*
iii) any two of:
time between trace **A** and **C** (or between **B** and **C**);
measured by oscilloscope; speed of ultrasonic waves known;
distance = speed × time *(2 marks)*

13 a) i) towards right hand side *(1 mark)*
ii) it is moving away (from the near galaxy) / moving outwards / away / expanding / moving away faster *(1 mark)*
(No marks for simply moving faster — must imply moving away.)
iii) implication that the universe is expanding / galaxies all moving away / moving away from each other *(1 mark)*
(Must imply more than just the two galaxies in (ii).)

b) An explanation to include:
curve (i) shows expanding universe *(1 mark)*;
curve (ii) predicts big crunch / expansion then contraction / Universe will eventually collapse *(1 mark)*

Section A will be marked out of a total of 90 marks.

## Section B

1 a) Z / space waves *(1 mark)*;
they need to penetrate the ionosphere / satellites are higher than the ionosphere / atmosphere / high frequency / short wavelength *(1 mark)*

b) i) not diffracted by Earth's surface / Earth is curved / space wave travels in straight lines / no line of sight / not reflected (by ionosphere)
*(1 mark)*
ii) radio transmissions do not use wires *(1 mark)*

2 a) (same) number of protons / proton number / atomic number *(1 mark)*
(different) number of neutrons / atomic mass *(1 mark)*

b) i) numbers of neutrons and protons are the same *(1 mark)*
ii) **X** marked at (7, 8) *(1 mark)*
iii) has excess neutrons / not enough protons *(1 mark)*

iv) Any three from:
- one extra proton
- one less neutron
- electron / ß⁻ particle emitted / formed / produced / ß decay (or words to that effect)
- down quark changes to up quark

*(3 marks)*

(NB allow 2 marks for "a neutron changes to a proton".
Accept a correct nuclear equation:

$$^{15}_{7}\text{N} \rightarrow \,^{15}_{8}\text{O} + \,^{0}_{-1}\beta \quad \text{for full marks.)}$$

3 a) i)

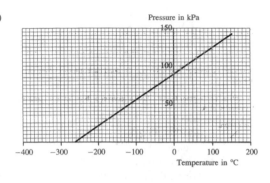

Points plotted correctly; *(allow 1 mark for 4 correct)*
Straight line. *(3 marks for correct graph)*

ii) Pressure increases with temperature *(1 mark)*

iii) line back tracked;
-260 ± 10°C *(2 marks)*

iv) absolute zero / 0 K *(1 mark)*

b) A calculation to include either:
- temperature changed to kelvin;

- $\dfrac{300}{290} = \dfrac{350}{T_2}$;

- $T_2 = \dfrac{290 \times 350}{300}$;

- 338 (K);

**OR**

- $\dfrac{300}{17} = \dfrac{350}{T_2}$;

- $T_2 = \dfrac{17 \times 350}{300}$;

- 19.8 / 20 (°C) *(4 marks)*

4 a) A description to include:
- pass through undeviated;
- be deflected;
- be "back-scattered" / deflected or turned through large angle / reflected

*(3 marks)*

b) An explanation to include:
- the sign of the charge on the nucleus is the same as that of an alpha particle;
- we know this because some are repelled. *(2 marks)*

c) A description to include four from:
- most of atom is empty space;
- as most particles are undeviated;
- mass / charge is concentrated / nucleus is small;
- very few are turned through large angles;
- rarity of back-scattering;
- force needed to turn through large angle is large *(4 marks)*

Section B will be marked out of a total of 32 marks
(so the whole exam is marked out of 122 marks).

# Index

# Index

# Index

Make sure you're not missing out on another superb CGP revision book that might just save your life...

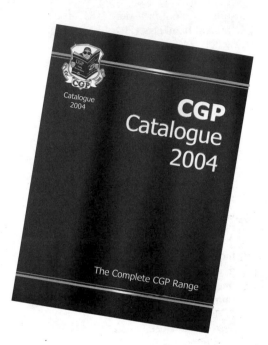

...order your **free** catalogue today.

## CGP customer service is second to none

We work very hard to despatch all orders the **same day** we receive them, and our success rate is currently 99.7%. We send all orders by **overnight courier** or **First Class** post.
If you ring us today you should get your catalogue or book tomorrow. Irresistible, surely?

PHS41